Proceedings of the International Conference on Computing, Mathematics and Statistics (iCMS 2015)

Abd-Razak Ahmad · Liew Kee Kor
Illiasaak Ahmad · Zanariah Idrus
Editors

Proceedings of the International Conference on Computing, Mathematics and Statistics (iCMS 2015)

Bridging Research Endeavors

Springer

Editors
Abd-Razak Ahmad
Universiti Teknologi MARA
Merbok, Kedah
Malaysia

Illiasaak Ahmad
Universiti Teknologi MARA
Merbok, Kedah
Malaysia

Liew Kee Kor
Universiti Teknologi MARA
Merbok, Kedah
Malaysia

Zanariah Idrus
Universiti Teknologi Mara
Merbok, Kedah
Malaysia

ISBN 978-981-10-2770-3 ISBN 978-981-10-2772-7 (eBook)
DOI 10.1007/978-981-10-2772-7

Library of Congress Control Number: 2016954590

Printed on acid-free paper

This Springer imprint is published by Springer Nature
The registered company is Springer Nature Singapore Pte Ltd.
The registered company address is: 152 Beach Road, #22-06/08 Gateway East, Singapore 189721, Singapore

Preface

The International Conference on Computing, Mathematics and Statistics (iCMS2015) is originally one of the flagship conferences under the previously known International Conference of Art, Science and Technology (iCAST), which had been successfully organized in 2010 and 2012. Building on the success of these two previous iCAST conferences, the Faculty of Computer and Mathematical Sciences, Universiti Teknologi MARA Kedah Campus decided to hold a more focused conference, the first iCMS in 2013 and the second in 2015. With a theme "Bridging Research Endeavour in Computer and Mathematical Sciences", the conference is a catalyst for researchers to present their ideas that bring the three different fields of computer science, mathematics and statistics together and bridge the gap by giving them the opportunity to gain insights into new areas and new perspectives to their respective fields. Given the rapidity with which computer science, mathematics and statistics are advancing in all areas of life, the conference gives an avenue for academics, researchers and students to share their diversified ideas, research findings and achievements. This meeting of great minds from various universities provides numerous opportunities for networking.

Merbok, Malaysia

Abd-Razak Ahmad
Liew Kee Kor
Illiasaak Ahmad
Zanariah Idrus

Organizing Committee

Abd-Razak Ahmad
Asmahani Nayan
Ida Normaya Mohd Nasir
Illiasaak Ahmad
Kamarul Ariffin Mansor
Liew Kee Kor
Noor Hafizah Zainal Aznam
Norin Rahayu Shamsuddin
Roshidah Shafeei
Rosidah Ahmad
Shahida Farhan Zakaria
Siti Nurbaya Ismail
Wan Siti Esah Che Hussain
Wan Zulkipli Wan Salleh
Zanariah Idrus

International Scientific Committee

Reviewers

Abd Razak Ahmad, Universiti Teknologi MARA (UiTM), Malaysia
Azman Yasin, Universiti Utara Malaysia (UUM), Malaysia
Chew Cheng Meng, Universiti Sains Malaysia (USM), Malaysia
Fadiya Bt Mohd Noor, Universiti Malaya (UM), Malaysia
Fahad Nazir Bhatti, Universiti Malaysia Perlis (UniMAP), Malaysia
Fam Pei Shan, Universiti Sains Malaysia (USM), Malaysia
Goh Kim Leng, Universiti Malaya (UM), Malaysia
Hajar Sulaiman, Universiti Sains Malaysia (USM), Malaysia
Illiasaak Ahmad, Universiti Teknologi MARA (UiTM), Malaysia
Jehan Zeb Shah, Optics Laboratories, Islamabad, Pakistan
Kinkar Chandra Das, Sungkyunkwan University, South Korea
Leony Tham Yew Seng, Universiti Malaysia Kelantan (UMK), Malaysia
Liew Kee Kor, Universiti Teknologi MARA (UiTM), Malaysia
Mahadzir Ismail, Universiti Teknologi MARA (UiTM), Malaysia
Min Wang, Clemson University, USA
Mohd Rijal Ilias, Universiti Teknologi MARA (UiTM), Malaysia
Noor Zahirah Mohd Sidek, Universiti Teknologi MARA (UiTM), Malaysia
Nor Idayu Mahat, Universiti Utara Malaysia (UUM), Malaysia
Norhashidah Awang, Universiti Sains Malaysia (USM), Malaysia
Norhayati Baharun, Universiti Teknologi MARA (UiTM), Malaysia
Norin Rahayu Shamsuddin, Universiti Teknologi MARA (UiTM), Malaysia
Roshidi Din, Universiti Utara Malaysia (UUM), Malaysia
Rozaini Roslan, Universiti Tun Hussein Onn Malaysia (UTHM), Malaysia
Shamsul Jamel Elias, Universiti Teknologi MARA (UiTM), Malaysia
Siti Ainor Binti Mohd Yatim, Universiti Sains Malaysia (USM), Malaysia
Sumarni Abu Bakar, Universiti Teknologi MARA (UiTM), Malaysia
Suraya Masrom, Universiti Teknologi MARA (UiTM), Malaysia
Teoh Sian Hoon, Universiti Teknologi MARA (UiTM), Malaysia
Waail Mahmood Lafta Al-Waely, Al-Mustafa University College, Iraq
Yahya Naji Saleh Obad, Universiti Malaysia Perlis (UniMAP), Malaysia
Yanti Aspha Ameira Mustapha, Universiti Teknologi MARA (UiTM), Malaysia

Yap Bee Wah, Universiti Teknologi MARA (UiTM), Malaysia
Zanariah Idrus, Universiti Teknologi MARA (UiTM), Malaysia
Zainura Idrus, Universiti Teknologi MARA (UiTM), Malaysia

Contents

Part IV Application and Education

Part I
Computing

Chapter 1
A Fuzzy Learning Algorithm for Harumanis Maturity Classification

Khairul Adilah bt Ahmad, Mahmod Othman, Ab. Razak Mansor and Mohd Nazari Abu Bakar

Abstract This paper presents an approach to classify the fruit maturity of fruit grading system when a higher level of accuracy and easy interpretability of the estimate model is desired. The proposed method automatically generates membership functions (MFs) and constructs the associated fuzzy rules (FRs). The proposed approach is applied to a case study of Harumanis mango fruit grading system. The task is to classify the fruit maturity and grade using agronomic image data set acquired by digital camera. The parameters of the MFs are adjusted by the learning algorithm for the training data. This MF is then used to generate the MFbox using hyperbox-type fuzzy partition of feature space to generate FRs from training instance to deal with the features data. FRs are extracted from the flexible MFs and MFbox. As a case study, the proposed method is applied to Harumanis data set with 108 instances (input–output pairs), two real-valued inputs, and one output. Analysis results show that the proposed maturity classifier yields an accuracy of 98 %. The developed maturity classifier can act as an instrument in determining the correct mango fruit maturity category.

Keywords Quality · Harumanis mango · Fuzzy learning algorithm · Image processing · Classification · Maturity

K.A. Ahmad (✉) · M. Othman · Ab. R. Mansor
Faculty of Computer and Mathematical Sciences, Universiti Teknologi MARA, Shah Alam, Malaysia
e-mail: adilah475@kedah.uitm.edu.my

M. Othman
e-mail: mahmod135085@perlis.uitm.edu.my

Ab.R. Mansor
e-mail: arazman@kedah.uitm.edu.my

M.N. Abu Bakar
Faculty of Applied Sciences, Universiti Teknologi MARA, Shah Alam, Malaysia
e-mail: mohdnazari@perlis.uitm.edu.my

A.-R. Ahmad et al. (eds.), *Proceedings of the International Conference on Computing, Mathematics and Statistics (iCMS 2015)*,
DOI 10.1007/978-981-10-2772-7_1

1.1 Introduction

Automatic fruit quality grading plays an important role to increase the value of produces in agriculture industry. The use of computer vision for the inspection of fruits has increased considerably in recent years. Computer vision is a technology for acquiring and analyzing an image of a real scene by computers. The fruit quality is determine based on the appearance of fruit which are color and geometric (shape, size, texture) attributes. For many crop products, color is the most important indicators to evaluate ripeness, maturity and/or quality for various fruits [1]. This is true for apples, strawberry, watermelon, and mango [2–5], but Harumanis mango color remains in green even in the matured or fully ripe or stage. Thus, this circumstance often leads to erroneous predictions to grade maturing and ripening Harumanis mango based on skin color. Therefore color is not a good feature for measuring the maturity, ripeness, or quality level for Harumanis. Hence, determining the quality of Harumanis mango is possible by analyzing the size and shape of the mango appearance [6].

Classification is vital for the evaluation of agricultural produce. Currently the Harumanis mangoes are classified by the expert graders according to grade Premium, A and B. Unfortunately, the grading processes are inspected manually and these expose the process to human mistakes. Therefore the manual grading inspection needs to improve with the advantages of high accuracy, consistency, uniformity in nondestructive testing. The agricultural data set usually is associated with imprecise and uncertain information with inherent pattern data that are difficult to describe. The fuzzy logic concept from [7] is able to deal with imprecise and uncertain information and imitate human expert mind and focus on pattern classification in decision-making [8]. Computer vision and fuzzy logic (FL) can be adapted as a good alternative for automating inspection and classification. But a problem appears when working with manual rules. The problem occurs with some tasks where the system has to be able to show results with good interpretability, specifically, the expert has to identify the reason of the given decision. Therefore, it is necessary to have an automatic system to construct interpretable rules since a human expert cannot be able to direct the knowledge about the problem in a rule form or impartial there are no available experts with an appropriate knowledge in the problem domain.

The objective of this study is to introduce an improved learning algorithm of fuzzy logic (FLA) classification to provide powerful tools in representing and processing human knowledge. In this study, a new fuzzy learning algorithm is proposed to automatically generate MF and construct FR that considers both the high accuracy of the model and easy interpretability. The results demonstrate the ability in distinguishing the three different classes of Harumanis with efficiency of 98 % compared with grades directly suggested by a human expert. The classification algorithm proposed in this paper proved a promising result and capable to be applied to improve and standardize the current fruit grading system. The rest of this paper is organized as follows: Sect. 1.2 presented the methods and materials and Sect. 1.3 explains the implementation of the method. Section 1.4 focuses on the analysis of the result. Finally, concluding remarks are discussed in Sect. 1.5.

1.2 Materials and Methods

The task of performing an automatic fruit grading involves the following steps: (1) image acquisition and image preprocessing; (3) image segmentation and feature extraction; and (5) leading to classification of objects of interest within images. In this work these steps are completed with the generation of the MFs and FRs. Therefore Step 5 (classification) is subdivided as follows: (1) Define data set; (2) Generation of the MFs; and (3) Construct FR. These steps are illustrated in Fig. 1.1. A brief description of each step is presented below, and a complementary description is presented in the implementation section.

1.2.1 Image Acquisition and Image Preprocessing

Samples of Harumanis images are acquired from the Perlis Agriculture Department's orchard in Bukit Bintang. A total of 108 images are taken in controlled environment using digital camera. All the images are saved in the Joint Photographic Expert Group (JPEG) format with resolution of 1624 $\times$ 2448 pixels. Then the image is converted into grayscale image. Conversion of image is crucial in image analysis to reduce the computational complexity as well as memory requirements. In grayscale image, background images tend to merge with the object being investigated. Therefore, image segmentation was performed to extract the object of interest from its background.

1.2.2 Image Segmentation and Feature Extraction

Segmentation was performed using an improved multilevel edge detection algorithm based on convolution. The aim of the algorithm is to extract crop contour. From the contour segmentation result the features are extracted. A set of features or attributes needs to be selected in order to contribute in the classification of the fruit

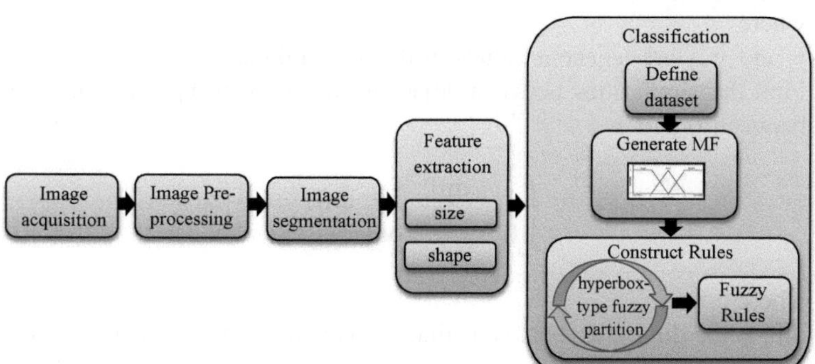

Fig. 1.1 Proposed structure of the FLA for Harumanis grading problems

grading. In this study, the geometric features are used to determine the size and shape of the Harumanis.

Size is the first parameter identified with quality attribute used by farmers [9]. The bigger size fruit is considered of better quality. The fruits are categorized as small (S), medium (M), and large (L). In mangoes, there is an interest in examining fruit shape in order to determine the maturity index [9, 10]. The fruits shapes are categorized as immature (IM), half mature (HM), and Full mature (FM).

1.2.3 A Fuzzy New Learning Method

In this step the FLA model is trained. The learning method, presented by the authors in [11], is used in this approach with additional changes to work with fruit grading problems. The method is detailed here in three sub-steps to construct MFs and generate FRs from training instances.

Step 1 Define data set: The aim of this step is to define a data set values for each attribute by selection of fuzzy inputs and outputs

Step 2 Generate MFs: Fuzzy set are represented by triangular MFs. The triangular MF for class j consists of three points: the central vertex point, b_j, and the two endpoints, a_j and c_j. The proposed learning algorithm for generating MFs proceeds as follows.

Input: V_i, values of are sorted in ascending order
Output: R_{Vi}: Regions fuzzy set with initial lower and upper parameters for values of V_i. Initially $R_{Vi} = \emptyset$.

Begin

1. Eliminate the duplicate raw data value to define a proper vector of interval values of fuzzy regions in the universe of discourse $V = [V_{\min}, V_{\max}]$
2. Determine the difference between adjacent values in the sorted data.
   $\text{diff}_i = y_{i+1} - y_i$, for $i = 1, 2, 3, \ldots n - 1$
   where
   y_i and y_{i+1} are adjacent values in the sorted data.
3. Find the similarities between adjacent values maps V_i into real numbers between 0 and 1.

$$s_i = \left\{ \begin{matrix} 1 - \frac{\text{diff}_i}{C*\sigma_s} & \text{for } \text{diff}_i \leq C*\sigma_s, \\ 0 & \text{otherwise}, \end{matrix} \right\}$$

where
diff_i is the difference between adjacent data, σ_s is the standard deviation of diff_i, and C is the control parameter is used to determine the shape of the MF.

4. The data is clustered according to similarities. The maximum similarities value is select as threshold value, s_t, divides adjacent values into classes. The number of classes determines the number of MFs. Determination of the number of classes can be summarized by a rule: If the similarity is greater than the threshold value, then the two adjacent values belong to the same class, otherwise the values are divided into different classes. That is, a new class is created. Expressed as a formula,

   $$\text{If } (s_i > s_t) \text{ then } \quad y_i, y_{i+1} \in C_i \quad \text{else} \quad y_i \in C_i, y_{i+1} \in C_{i+1}$$

   where
   C_i and C_{i+1} denote two distinct classes for the same input or output parameter.
5. Assign values to central vertex point, b_j, and the two endpoints, a_j and c_j of fuzzy set. The central vertex point is determined for each class and is calculated by the given formula:

   $$b_j = \frac{y_i * s_i + y_{i+1} * \frac{s_t + s_t}{2} + \cdots + y_{k-1} * \frac{s_t + s_t}{2} + y_k * s_{k-1}}{s_i + \frac{s_t}{2} + \cdots + \frac{s_t}{2} + s_{k-1}}$$

   where
   j represents the jth class, k represents the ending data index for this class, i.e., data y_i through y_k fall into class j, and s_i is the similarity between y_i and y_{i+1}. The endpoints of the MF, a_j and c_j, are obtained using the following equations to determine the left and right endpoints.

   $$a_j = b_j \frac{b_j - y_i}{1 - s_t}; \quad c_j = b_j + \frac{y_k - b_j}{1 - s_t}$$

   where
   a_j is the left endpoint and c_j is the right endpoint.
   End

Step 3 Construct Rules with the new MFs: The objective of this step is to construct a set of FRs. The new rule base is constructed with the new MF. In this section, we present a new algorithm called MFbox to generate FRs from training instance to deal with the features data. FRs can be determined by flexibleMFs and a hyperbox-type fuzzy partition of feature space. The major advantage of the fuzzy region approach is that only few overlapping fuzzy regions can cover all training patterns with high classification accuracy.

1.3 Implementation

In this section, the Harumanis mango fruit data were used to illustrate the new fuzzy learning algorithm.

1.3.1 Define Data Set for Each Attribute

The mangoes data contain 108 instances having two input attributes, which are *size* and *shape* and one output value with three grading categories which are Premium, Grade A, and Grade B as shown in Table 1.1.

Table 1.2 tabulates the linguistic variables of the fuzzy set for fruit grading system. The input value is fuzzified into two MF to obtain the confidence value of that particular value. The numeric value that is supplied into the system is converted to linguistic variables.

1.3.2 Generate MFs

The example shows the calculation of attribute shape

Input values M_i are sorted in ascending order

$$M_i = \{0.48, 0.49, 0.50, 0.51, \ldots 0.77\}.$$

1: Selected data.

$$Y = \{0.48, 0.51, 0.54, 0.60, 0.63, 0.66, 0.71, 0.74, 0.77\}$$

2: Difference sequence.

Table 1.1 Mango data

No	Shape	Size	Grading
M_1	0.77	533.79	Premium
M_2	0.51	351.79	B
…	…	…	…
M_{108}	0.64	482.1	A

Table 1.2 Fuzzy set definition

Fuzzy variables		Fuzzy set linguistic term
Input	Size	{Small, medium, large}
	Shape	{Immature, half mature, mature}
Output	Grading	{Premium, A, B}

$$0.51-0.48 = 0.03, 0.54-0.51 = 0.03, 0.60-0.54 = 0.06\ldots0.77-0.74 = 0.03$$

3: Find similarities
Let the control parameter C=8.
The standard deviation σ is first calculated as 0.011
MF value similarity is shown as follows:

$$s_1 = 1 - (0.03/(0.011 * 8)) = 0.66,$$

4: Clustered data (Table 1.3).
5: Find vertex.

(a) The first data cluster is used as an example = {0.48, 0.51, 0.54}. To calculate the point b_j

$$b_j = \frac{(0.48 * 0.66) + \left(0.51 * \frac{0.66+0.66}{2}\right) + (0.54 * 0.66)}{\left(0.66 + \frac{0.66+0.66}{2} + 0.66\right)} = 0.51$$

The endpoints a_j and c_j, are obtained:
(b) To calculate the point a_j

$$a_j = 0.51 - \frac{0.51-0.48}{1-0.66} = 0.42$$

(c) To calculate the point c_j

$$c_j = 0.51 + \frac{0.54-0.51}{1-0.66} = 0.60$$

Calculations for the group of size and shape for each fuzzy set were performed using the same approach. The results of the calculation for size and shape feature are shown in Table 1.4.

Two MFs for the classes of size and shape are shown in Fig. 1.2, respectively.

Table 1.3 Cluster data for each region class

Similarity, s_i		s_1	s_2	s_3	s_4		s_5	s_6	s_7
		0.66	0.66	*0.32*	0.66	0.66	*0.43*	0.66	0.66
Values, V_i	0.48	0.51	0.54	0.60	0.63	0.66	0.71	0.74	0.77
Region, R_i	R_1			R_2			R_3		

Table 1.4 MF value of size and shape instance

Size			Shape		
S	M	L	IM	HM	FM
0.42	0.54	0.65	228.54	308.54	418.54
0.51	0.63	0.74	330.00	410.00	520.00
0.60	0.72	0.83	431.46	511.4	621.46

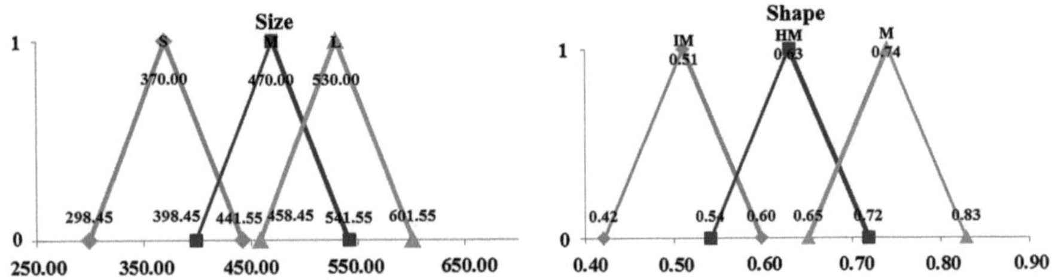

Fig. 1.2 MF for the size and shape features

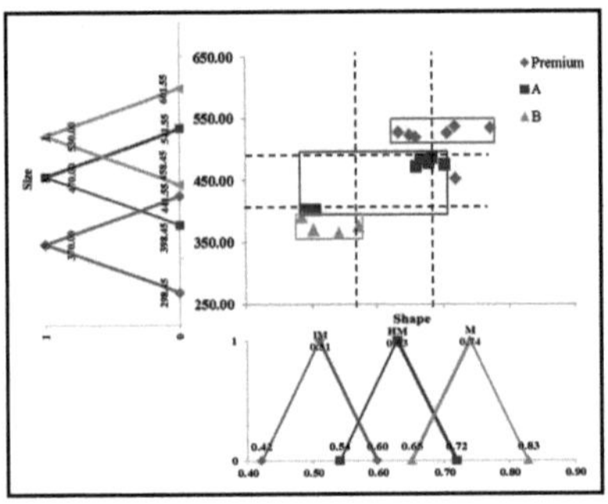

Fuzzy rules generated from the MFbox algorithm.

R_0: If *size* is *L* and *shape* is *FM* then Grade is Premium
R_1: If *size* is *L* and *shape* is *HM* then Grade is Premium
R_2: If *size* is *M* and *shape* is *FM* then Grade is A
R_3: If *size* is *M* and *shape* is *HM* then Grade is A
R_5: If *size* is *S* and *shape* is *HM* then Grade is B
R_6: If *size* is *S* and *shape* is *IM* then Grade is B

Fig. 1.3 MFbox and fuzzy rules

1.3.3 Construct Fuzzy Rules with the New MFs

Figure 1.3 illustrate MFbox and MFs of FRs extraction, with two input variables and one output intervals and have training data listed in Table 1.2.

1.4 Analysis of Results

Testing data are used to derive the grading decisions through the MFs and FRs. Figure 1.4 shows the rule viewers which consist of the system's input and output and the defuzzification result column. The first and second column is the inputs which are size and shape values while the last column is the grading column. The last row of the category column shows the defuzzification result where category of Harumanis mango fruit is obtained.

The result show Harumanis fruit fulfilled Rule R_2 where *size* is *M* and *shape* is *FM* then Grade is A. The value of the grading is calculated by using the centroid method. Then the classification of Harumanis fruit is determined based on the classification of Harumanis fruit grading given in Table 1.5. The value of category is 0.5. Therefore, the Harumanis used in this experiment is classified as grade A.

To demonstrate the effectiveness of the proposed fuzzy learning algorithm, 108 testing instances were classified and the results are shown in Table 1.6.

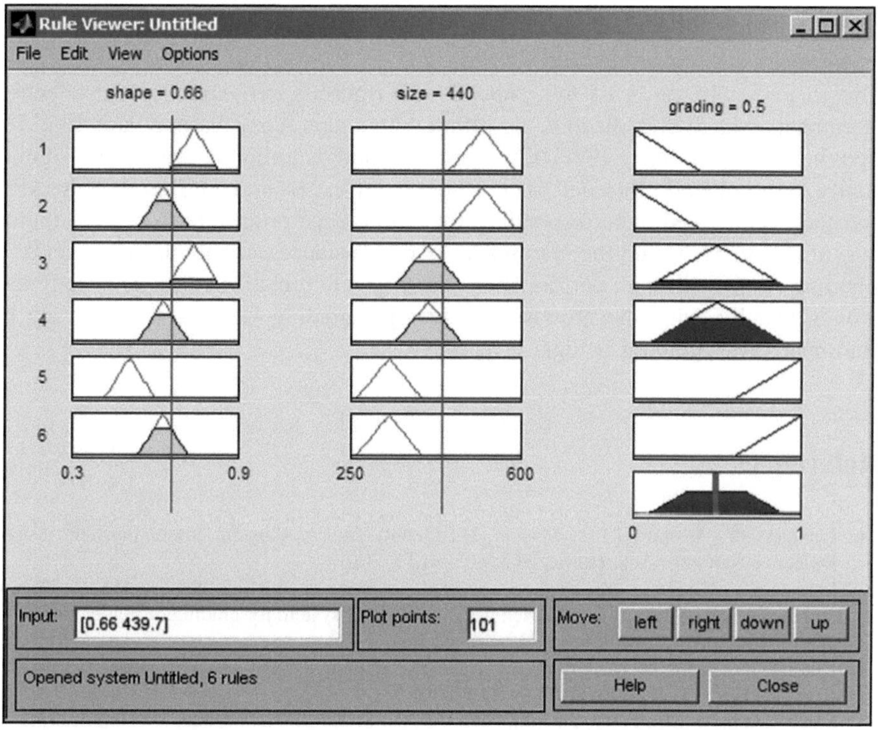

Fig. 1.4 Defuzzification result from the rule viewer

Table 1.5 Clasification of Harumanis fruit grading

Defuzzification output	Harumanis fruit grading
(Grade < 0.4)	Premium
(Grade > = 0.4) & (Grade <= 0.6)	A
(Grade > 0.6)	B

The resulting classification result accuracies obtained from FLA in comparison with the classification results from the expert are given in Table 1.6. FLA classified 98 % of Harumanis correctly. Misclassification errors observed were only one adjacent group. The result proved that MF and FR used in the proposed mango grading method are efficient and accurate.

Table 1.6 Comparison of FLA and human expert classification of Harumanis

	Fuzzy logic grading					
	Class	Premium	Grade A	Grade B	Total classification	%
Human expert	Premium	32	2	0	35	94.1
	Grade A	0	36	0	38	100
	Grade B	0	0	38	38	100
Image acquired		32	38	38	106/108	98.1
%		100	94.7	100	(Correctly classified)	

1.5 Conclusions

The major contribution of this paper is the rigorous derivation of the effective Harumanis mango classification algorithm with a new fuzzy learning algorithm for deriving MFs and fuzzy IF-THEN rules. The new learning algorithm can significantly reduce the effort needed to develop a fuzzy expert system. Based on the MFs and the FRs derived, a corresponding fuzzy inference procedure to process inputs was also applied. Using the Harumanis testing instances, the model gives an outstanding result with high performance and easy interpretation rules. The study has proven that by using the proposed fuzzy logic grading method, the accuracy of Harumanis fruit grading is high.

References

1. Leemans, V., Destain, M.F., Magein, H.: Quality fruit grading by colour machine vision: Defect recognition. Acta Hortic. **517**, 405–412 (2000)
2. Lorestani, A.N., Omid, M., Bagheri-Shooraki, S., Borghei, A.M., Tabatabaeefar, A.: Design and evaluation of a fuzzy logic based decision support system for grading of golden delicious apples. Int. J. Agric. Biol. **8**(4), 440–444 (2006)
3. Liming, X., Yanchao, Z.: Automated strawberry grading system based on image processing. Comput. Electron. Agric. **71**, S32–S39 (2010)
4. Shah Rizam, M.S.B., Farah Yasmin, A.R., Ahmad Ihsan, M.Y., Shazana, K.: Non-destructive watermelon ripeness determination using image processing and artificial neural network (ANN). World Acad. Sci. Eng. Technol. **50**, 538–542 (2009)
5. Razak, M.A., Mahmod, O., Mohd Nazari, A.B., Khairul Adilah, A., Rosli, R.T.: Fuzzy ripening mango index using RGB colour sensor model. J. Arts Sci. Commer. **5**(2), 1–9 (2014)
6. Slaughter, D.C.: Nondestructive Maturity Assessment Methods for Mango: A Review of Literature and Identification of Future Research Needs, pp. 1–18. National Mango Board, Orlando (2009)
7. Zadeh, L.A., Fellow, L.: Fuzzy logic = computing with words. IEEE Trans. Fuzzy Syst. **4**(2), 103–111 (1996)
8. Birle, S., Hussein, M.A., Becker, T.: Fuzzy logic control and soft sensing applications in food and beverage processes. Food Control **29**(1), 254–269 (2013)
9. Kader, A.: Mango quality attributes and grade standards: a review of available information and identification of future research needs. Technical report to the National Mango Board Davis, CA, USA. Kader Consulting Services (2008)
10. Moreda, G.P., Muñoz, M.A., Ruiz-Altisent, M., Perdigones, A.: Shape determination of horticultural produce using two-dimensional computer vision—a review. J. Food Eng. **108**(2), 245–261 (2012)
11. Hong, T., Lee, C.: Induction of fuzzy rules and membership functions from training examples. Fuzzy Sets Syst. **84**(95), 33–47 (1996)

Chapter 2
Dynamic Role Behavior for Managing Users Through A Networked Collaborative Monopoly Game Abstraction

Zainura Idrus, Siti Zaleha Zainal Abidin, Nasiroh Omar and Zanariah Idrus

Abstract e-Technology has enabled users from remote locations to collaborate together virtually and involve in e-activities to reach their common goals. The virtual environment is commonly known as Networked Collaborative Virtual Environment. It is a challenge to manage these remote located users since they are diverse in term of language, culture, and skill. Most of the managements are performed by external entity such as administrator. As a result, the external handling is hardly matched or coped. Thus, this paper investigates the possibility of having a method called role behavior to manage users both internally and externally. The initial finding shows that roles are dynamic and their dynamic behaviors are further analyzed through a collaborative monopoly game. As a result, a set of role behaviors has been extracted from the game. Each behavior is analyzed and mapped to a real-life application as an abstraction of user management. Finally, a mathematical model has been proposed to represent a comprehensive collection of roles which has to be used in managing dynamic environment.

Z. Idrus (✉)
Faculty of Computer and Mathematical Sciences, Universiti Teknologi MARA, Jasin, Melaka, Malaysia
e-mail: zainura501@melaka.uitm.edu.my

S.Z. Zainal Abidin · N. Omar
Faculty of Computer and Mathematical Sciences, Universiti Teknologi MARA, Shah Alam, Malaysia
e-mail: sitizaleha533@salam.uitm.edu.my

N. Omar
e-mail: nasiroh@tmsk.uitm.edu.my

Z. Idrus
Faculty of Computer and Mathematical Sciences, Universiti Teknologi MARA, Merbok, Kedah, Malaysia
e-mail: zanaidrus@kedah.uitm.edu.my

A.-R. Ahmad et al. (eds.), *Proceedings of the International Conference on Computing, Mathematics and Statistics (iCMS 2015)*,
DOI 10.1007/978-981-10-2772-7_2

Keywords e-Technology · Monopoly game · Role management · Role behavior · Real life abstraction · Dynamic role · Static role

2.1 Introduction

In networked collaborative virtual environments (NCVE), users work together from remote location. This is possible with the advance of e-Technology such as network communication (machine to machine), video, audio, and text communication (human to human). In such environment users are represented by digital elements such as avatars, texts, and graphic images. They might be from different organization or countries as such their ethnics, working culture, and skills are also different. In most cases, they do not have the opportunity to see each other in a face-to-face environment. Therefore, it is important to have an effective management system to manage these remote users through standard of principles, governing rules, and policies [1, 2].

Users in NCVE are dynamic, thus, role-based management has been identified as a suitable method to manage them. In such management, tasks for the collaborative works will first be identified and assigned to role. Then, access right to resources such as database and files are assigned to the role [3, 4]. For security purposes, only certain role has the permissions to access certain resources. The permissions are managed by role-based access control (RBAC) through policies [5]. The policies are a set of rules which specifies permission to resources. In other words, RBAC controls all the permission to resources as a means to protect resources [6]. RBAC also has to ensure that only qualified roles get access to permitted resources.

Moreover, it is essential for role to be managed and structured. The common organization structure for role is flat hierarchy, matrix, and peer-to-peer organization [7]. Only after the role organization structured has been form, qualified users can be searched and assigned to roles. Once assigned, users are accountable to the tasks and are granted with permission to the resources. Since the organization structure is independent of users, changes of users impose minimum impact to the roles and their structure. Hence, a vacant role can be replaced by simply searching a new qualified user.

From management perspective, role is defined as an abstract entity to establish communication, coordination, and cooperation among, tasks, users, and resources. It becomes the hub of management where its obligation is more than just an agent for assigning responsibility and permission to users. Essentially, a successful collaboration is highly determined by how well the role integrates the three main collaborative elements which are users, tasks, and resources. The elements have to be integrated in a right manner and at the right time. This is vital since users, tasks, and resources are dynamics where their states are unpredictable.

As a central of management, roles' behaviors must be identified to depict their full potentials. Thus, this paper has explored roles in NCVE to discover their behaviors. The exploration has identified a range of roles' behaviors which will be utilized by controlling agent in managing users. A collaborative monopoly game has been developed as a case study. The game is then simulated to extract numerous role abstractions which are then mapped to real-life applications.

The organization of this paper starts with Sect. 2.2 where it highlights some of the previous works on role management. Then, the paper moves on to the findings of role management which are in the form of formal notation. The findings are discussed in Sect. 2.3 followed by Sect. 2.3.4 which covers roles' behavior in the form of rules. Finally Sect. 2.4 concludes the paper.

2.2 Related Work

In 1992, Ferraiolo et al. [5] for the first time has introduced the concept of role as a mechanism to control access to resources. Since then, his concept continues to gain popularity and branded as RBAC. His concept also has grown to be the foundation to a range of access control models. With the main intention to protect resources, RBAC structured role hierarchically as such to inherit permissions. The structure is commonly referred to as "is a" relationship [6]. Even though role was first introduced as a medium between users and their access right to resources, role eventually evolves to become agents between users and their responsibilities. The responsibilities mandate users with the right to make decision and perform collaborative tasks. The RBAC role hierarchy concept has also easied up the distribution of tasks to users where a child row can inherit responsibility from a parent row [5, 6, 8–10].

In contrast, the concept of role hierarchy has a different meaning in the context of virtual organization. It is not meant for heritance but for the purpose to structure the flow of communication and command [7]. Roles are structured as such roles at the higher level hold higher authority as compared to role at lower level. The flow of command is from top to bottom where higher roles have more right over lower role. However, flow of communication takes place in two ways which are bottom up and top down.

Another entity in the role structure is the user who is to play the roles. For productivity purposes, the area of management consistently strive to ensure that the right users is assigned to play the right role and at the right time [11, 12]. Thus, various measures have been studied to reach the best approach. In addition, the research mostly focuses on optimization specifically on human resources structure [13]. However, the structure is designed for homogeneous setting where users' skill is similar. In heterogeneous environment where users' skill varies, optimization is not as important as searching for users with suitable qualification [10]. In addition, research has also been conducted to optimize number of users per role in order to achieve balance distribution of workload [14, 15].

Relationship among roles is vital in social networking. The roles are closely interconnected with each other where any changes will affect their relationship. Consistently, role graph model are used as a medium to manage the relationship [16, 17]. The main purpose of the model is to minimize the effect of the changes such as conflict relation, promotion, and inheritance.

The concept of role has been contributed in various research areas. In most cases, role is the medium for assigning property to users. In such cases, the true identity of users and even the users themselves are hidden. This is important in the event of information transferring for security purposes.

Role has also been employed as medium to represent users [15] where permissions and responsibilities are assigned to role prior to users. Thus, personal issue which can be an obstacle to equally distribute workforce among users can be avoided. The concept of role management is summarized in Fig. 2.1. In short, the concept of role management classifies roles into a few subgroup of:

- Manager—manages the flow of authorization and communication.
- Optimizer—manages human resources and their relationship.
- Protector—protects users' identities during information transfer.
- Task Allocator—manages workflow of tasks.
- Resource Allocator—manages access right to resources.

Most of the managements are performed by external entity such as administrator. The managements are static whose internal entities find it difficult to cope [17]. Thus, system malfunctions are not easily managed. However, the study on internal

Fig. 2.1 Role management

controlling is lacking attention. Thus, this paper intends to discover the dynamic behaviors of roles which can be utilized as internal as well as external controller. The finding is important in dynamic environment which requires flexible controlling.

2.3 Monopoly Game

The monopoly game was first created by Elizabeth Magie in 1903 who is from America. The game was called *The Landlord's Game* since it involves selling and buying properties. The main properties in the game are hotel, house, and land. By the year 1934, after releasing a few versions, a version similar to the version introduced by Parker Brothers was released. The version gains its popularity in twentieth and twenty-first centuries [18, 19]. Besides game, monopoly has also been utilized in many areas of researches such as strategic planning, economics, financing, mathematics, group behaviour, virtual community, and communication [20, 21].

2.3.1 Game Rules

In collaborative monopoly game, there are ten roles which are tourist, visitor, citizen, businessman, landlord, millionaire, tenant, prisoner, observer, and banker. The screenshot of the game is shown in Fig. 2.2.

The game is separated into two stages which are visiting and monopolizing stages

Visiting stage
All players start the game by playing the role of tourist in this visiting stage. They are authorized with basic rights which are throwing dices prior to making a move. No other authorizations are given.

The monopolizing stage
Once players have completed the visiting stage, they are moved to monopolizing stage. They are transited to citizen role and authorized to earn salary. They are also authorized to buy properties. Thus, to monopolize the game, they will strive to own as many properties as possible.

This is the stage where users consistently change their roles. For example, citizen is transmitted to landlord immediately after they buy a property. However, users have to leave the game when they are short of cash and incapable of paying other expenses.

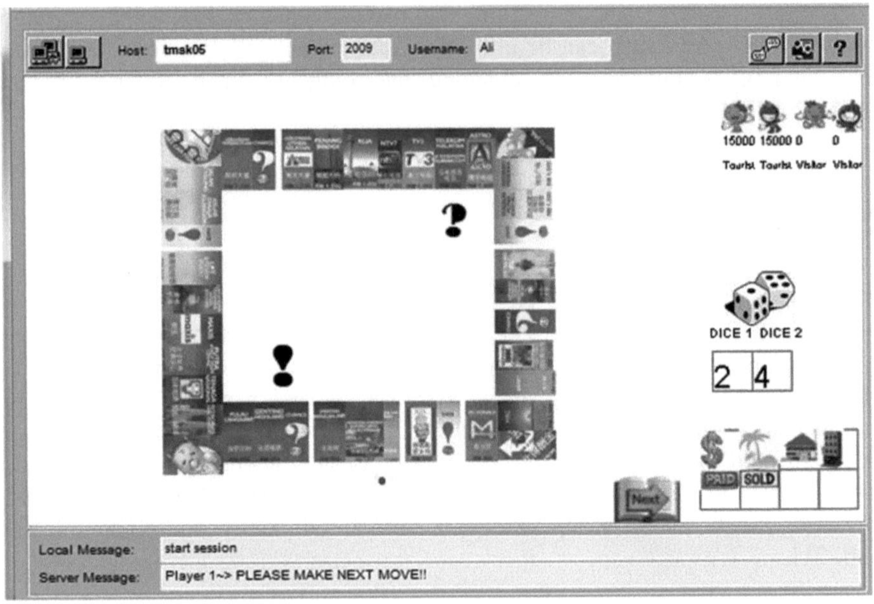

Fig. 2.2 Monopoly collaborative game

2.3.2 *Monopoly and the Analysis*

R denotes the set of roles in the game where

$$r \in R = \{\text{citizen, visitor, tourist, landlord, businessman, banker, millionaire, prisoner, tenant and observer}\} \quad (2.1)$$

R can be in either static or dynamic state. Let *S* be the set of two states which

$$s \in S = \{\text{static}, \text{dynamic}\} \quad (2.2)$$

When a role is in its static states, it blocks any current holder from releasing the role and blocks any outsider from entering the role. On the other hand, dynamic states allowed the role to freely accept or release players.

Time frames have been enforced to the state to control the role activities in that state. There are three time frames which are whole session, duration, and condition. Let TF represent the set of time frame, then

$$tf \in \text{TF} = \{\text{duration}, \text{whole session}, \text{condition}\} \quad (2.3)$$

With the time frames and states, there exist six possible combinations (static, whole session), (static, duration), (static, condition), (dynamic, whole session), (dynamic, duration), and (dynamic, condition). Therefore, the collaborative monopoly game, M is essentially a function which assigns R to their state and time frame which can be denoted as

$$S \times \mathrm{TR} \quad \text{where}\ M : R \rightarrow S \times \mathrm{TF} \tag{2.4}$$

The function, M, is said to *freeze* all users to their role if

$$\forall r \in R, S \times \mathrm{TF} = \{(\text{static, duration}), (\text{static, whole session}), (\text{static, condition})\} \tag{2.5}$$

Otherwise, if one or more roles are assigned to dynamic, the role is free to change its state (dynamic, static) if

$$\begin{aligned} &\exists r \in R\,\text{such that}\, S \times \mathrm{TF} \\ &\quad = \{(\text{dynamic, duration}), (\text{dynamic, whole session}), (\text{dynamic, condition})\} \end{aligned} \tag{2.6}$$

Since this research is interested in the changes of the roles' state over time, a parameter t is included into the function which extend a semi-bounded domain $T = [t_{\text{start}}, \infty)$, where t_{start} denotes the time at the beginning of the game. Hence, $M(t)$ is used to set the roles' state at time t and M having domain T x R.

During dynamic state and at any time of the game, a controller which can be a human or an automation machine can assign new players to the roles. Thus, the controller is a set of L = {external, internal}

Since there are only two types of actions that can be performed on role which are assigning state (static, dynamic) within time frame, the action are supported by the function change where

$$\text{change} : T \times R \rightarrow S \times \mathrm{TF} \times L \tag{2.7}$$

Let the graph of change be denoted by $C(t_a, t_b)$ and time t is bounded by $[t_a, t_b)$ where

$$C(t_a, t_b) = \{\langle t, r, l, \text{change}(t, r, l)\rangle | t_a \leq t < t_b, r \in R, l \in L\} \tag{2.8}$$

The function manages all the changes on roles' state throughout the time interval $[t_a, t_b)$. The changes directly affect players changing role. Therefore, a role *behavior management*, is actually a temporal function, ψ, which manages players changing roles that varies from time to time and it is defined as

$$M(t) = \psi(t, C(t_{\text{start}}, t), \text{ assign}(t, r, l)), \tag{2.9}$$

where function assign(t, r, l), make decision either to accept or reject any role-user assignment. However, the decisions made are based on C(t_{start}, t) outcome.

Table 2.1 Roles' static *behaviors* and applications

Roles' behaviors identified from monopoly	Type of behavior	Time frame
Banker cannot change player throughout the game	Static-sustainable	Whole session
There will only be one *banker* at all the time	Static-sustainable	Whole session
Visitor and *tourist* are *static-interval,* they are active only during the first phase	Static-interval	Duration
During first phase, *visitor, tourist, citizen, landlord, businessman, millionaire, tenant* and *prisoner* are static. No users are allowed to be signed into the roles for the duration of first phase	Static-interval	Duration
Banker is *static-sustainable,* but manager can force new user into the role	Static-conditional	Condition
Prisoners are static during phase L as long as no one get double dice three time consecutively	Static-conditional	Condition

2.3.3 Monopoly and Role Behaviors

Monopoly game abstractly captures various role behaviors that are common in collaborative applications. Tables 2.1 and 2.2 list the six possible behaviors of roles. *Role behavior* is a mechanism to be used in controlling users playing roles bounded by the time frame.

2.3.4 Role Behavior and Its Model

This section discusses each of the six *role behavior.* Table 2.3 depicts relationship between *behaviors* and their rules.

Roles with dynamic-interval, dynamic-sustainable, and dynamic-conditional are dynamic, however, their timeframe are different. Similarly, dynamic-conditional is also dynamic but based on a certain condition. In contrast, static-interval, static-sustainable, and static-conditional set role on static state based on timeframe while static-conditional is based on condition.

Next the behaviors will be described in further details.

(i) dynamic-sustainable role is set by $\exists r \in R, M(t_i)(r)$ such that

$$S \times \mathrm{TF} = \{(\text{dynamic, whole session})\} \quad \text{where } t_i \in T \tag{2.10}$$

Table 2.2 Roles' dynamic *behavior* and applications

Roles' behaviors identified from monopoly	Type of behavior	Time frame
Observers enter or leave the game unlimitedly	Dynamic-sustainable	Whole session
Players play *visitor* and *tourist* one time only	dynamic-interval	Duration
A player is substituted by his friend	Dynamic-interval	Duration
Roles are organized into hierarchy thus players are allowed to move up and down the rank	Dynamic-interval	Duration
Roles are distributed into phases and their activation varies with time	Dynamic-interval	Duration
Moving to phase 2. phase I's roles are set to inactive	Dynamic-interval	Duration
All roles except *banker, visitor* and *tourist* can be played unlimitedly	Dynamic-interval	Duration
Taking a role *prisoner,* a player will automatically drop temporarily all of his current roles	Dynamic-conditional	Condition
Prisoner cannot be played for more than three rounds consecutively	Dynamic-conditional	Condition

Table 2.3 Behavior rules

Behavior	Dynamic	Static	Whole collaboration	Duration	Cond
Dynamic-sustainable	√		√		
Dynamic-interval	√			√	
Dynamic-conditional	√				√
Static-sustainable		√	√		
Static-interval				√	
Static-conditional		√			√

(ii) function assign (t_i, r) will then be activated which allows, users to hold and release the role r during the whole session, t_j, where $t_i < t_j < \infty$ and $t_j \in T$.

Dynamic-interval allows users to enter and exit the role during the time frame.

(i) dynamic-interval role is set by $\exists r \in R, M(t_i)(r)$ such that

$$S \times \mathrm{TF} = \{(\text{dynamic}, \text{duration})\} \quad \text{where } t_i \in T \tag{2.11}$$

(ii) function assign(t_i, r) will then be set to active which will accept hold and release of role throughout the specified duration. Hence, the duration is denoted by t_j where

$$t_j = [t_a, t_b], t_a \leq t_b \quad \text{and} \quad t_j, t_a, t_a \in T$$

A dynamic-conditional role will accept users only when the conditions are met.

(i) dynamic-conditional role is set by $\exists r \in R, M(t_i)(r)$ such that

$$S \times \text{TF} = [(\text{dynamic}, \text{condition})\} \quad \text{where } t_i \in \text{T} \tag{2.12}$$

(ii) the condition will then activate function assign(t_i, r) which will hold and release the role as long as the conditions are met. Let $C = \{c_1, c_2, \ldots, c_k\}$ be a set of conditions. Therefore, assign*(t, r)* is active as long as c is TRUE where $c \in C$

Static-sustainable prevents users from entering and leaving the role.

(i) static-sustainable role is set by $\exists r \in R, M(t_i)(r)$ such that

$$S \times \text{TF} = \{(\text{static}, \text{whole session})\} \quad \text{where } t_i \in T \tag{2.13}$$

(ii) assign*(t_i, r)* will then be set to inactive which will block any hold and release of role throughout the whole session, t_j where $t_i \leq t_j < \infty$ and $t_j \in T$.

Static-interval is frequently used to stop users from entering the roles.

(i) static-interval role is set by $\exists r \in R, M(t_i)(r)$ such that

$$S \times \text{TF} = \{(\text{static}, \text{duration})\} \quad \text{where } t_i \in T \tag{2.14}$$

(ii) operation assign (t_i, r) will then be set to inactive which will block any hold and release of role for a time duration. Hence, the duration is denoted by t_j, where

$$t_j = [t_a, t_b], t_a \leq t_b \quad \text{and} \quad t_j, t_a, t_a \in T.$$

Finally, static-conditional permit users to change role only by authorized personal.

(i) static-conditional role is set by $\exists r \in R, M(t_i)(r)$ such that

$$S \times \mathrm{TF} = \{(\mathrm{static}, \mathrm{condition})\} \quad \text{where } t_i \in T \tag{2.15}$$

(ii) function assign*(t,, r)* will then be set to inactive and only be active if condition is set to true. Let C = $\{c_1, c_2, c_3, \ldots, c_k\}$ be a set of conditions. Therefore assign (t_i, *r*) is active as long as *c* is TRUE where $c \in C$.

In short, in temporal function ψ, there exists a method assign (*t, r*) which is authorized to monitor users taking new roles and the function is managed by

$$C(t_{\mathrm{start}}, t)$$

where $T = [t_{\mathrm{start}}, \infty)$, and t_{start}, is the starting of the game.

2.4 Conclusion

This paper has studied role management specifically in collaborative environment. Then six role *behaviors have been identified* from a collaborative monopoly game. The behaviors signify role's states which are *static* and *dynamic* and also the role's time frames. A clear *behavior will be* used to manage users through internal and external controller. This controlling entity also enforces flexibility in managing users.

Acknowledgment The authors would like to thank Universiti Teknologi MARA and Ministry of Higher Education Malaysia for the financial support under the national grant 600-RMI/RAGS 5/3 (20/2012).

References

1. Idrus, Z., Abidin, S.Z.Z., Hashim, R., Omar, N.: Social awareness: the power of digital elements in collaborative environment. WSEAS Trans. Comput. **9**(6), 644–653 (2010)
2. Abidin, S.Z.Z., Chen, M., Grant, P.: Designing interaction protocols using noughts and crosses type games. J. Netw. Comput. Appl. **30**, 586–613 (2007)
3. Zhu, H.: Role mechanisms in vollaborative systems. Int. J. Prod. Res. **44**(1), 181–193 (2006)
4. Redmiles, D., Al-Ani, A., Hildenbrand, B., Quirk, T., Sarma, S., Filho, A., de Souza, R., Trainer, C.: Continuous coordination: a new paradigm to support globally distributed software development projects. J. Wirtschafts Informatik **49**(1), 28–38 (2007)
5. Ferraiolo, D., Kuhn, R.: Role-based access control. In: Proceedings of the NIST–NSA National (USA) Computer Security Conference, pp. 554–563 (1992)
6. Sandhu, R.S., Cope, E.J., Feinstein, H.L., Youman, C.E.: Role- based access control models. IEEE Comput. **29**, 38–48 (1996)

7. Jones, G.R.: Organizational Theory, Design and Change, 4th edn. Prentice Hall, Upper Saddle River (2004)
8. Lupu, E., Sloman, M.: Reconciling role based management and role based access control. In: Proceedings of the Second ACM Workshop on Role-Based Access Control (RBAC '97), pp. 135–141 (1997)
9. Wang, H., Zhang, Y., Cao, J.: Effective collaboration with information sharing in virtual universities. IEEE Trans. Knowl. Data Eng. **21**(6), 840–853 (2009)
10. Idrus, Z., Abidin, S.Z.Z., Omar, N., Hashim, R.: Users management in interactive networked collaborative environment. Int. J. Comput. Commun. **5**(4), 254–262 (2011)
11. Lukasz, M., Mazur, S. Chen: A task-member assignment model for complex engineering projects. Int. J. Ind. Syst. Eng. **7**(1) (2011)
12. Wainera, J., Kumarb, A., Barthelmess, P.: DW-RBAC: a formal security model of delegation and revocation in workflow systems. Inf. Syst. **32**, 365–384 (2007)
13. Lin, L.H..: Electronic human resource management and organizational innovation: the roles of information technology and virtual organizational structure. Int. J. Hum. Resour. Manage. **22** (2) (2011)
14. Zhu, H.: Fundamental issues in the design of a role engine. In: International Symposium on Collaborative Technologies and Systems. pp. 399–407 (2008)
15. Noponen, S., Sihvonen, H., Salonen, J., Kurki, J.: Systematic literature review of virtual role. In: International Conference for Internet Technology and Secured Transactions (ICITST), pp. 738–743 (2011)
16. Wang, H., Osborn, S.L.: Static and dynamic delegation in the role graph model. Trans. Knowl. Data Eng. **23**(10), 1569–1582 (2011)
17. Fiala, J., Kratochvíl, J.: Locally constrained graph homomorphisms—structure, complexity, and applications. Comput. Sci. Rev. **2**(2), 97–111 (2008)
18. Dodson, E.J.: How Henry George's principles were corrupted into the game called Monopoly, http://www.henrygeorge.org/dodson_on_monopoly.htm
19. The history of Monopoly, http://www.worldofmonopoly.com/history/index.php
20. Korsgaard, M.A., Picot, A., Wigand, R.T., Welpe, I.M., Assmann, J.J.: Cooperation, coordination, and trust in virtual teams: insights from virtual games. In: Online Worlds: Convergence of the Real and the Virtual. Human–Computer Interaction Series, pp. 253–264 (2010)
21. Kalay, Y.E., Jeong, Y.: A collaborative design simulation game. Int. J. Archit. Comput., pp. 423–434 (2009)

Chapter 3
Experimental Model of Congestion Control Vehicular Ad Hoc Network Using OMNET++

Shamsul Jamel Elias, M.N. Mohd Warip, Shaifizat Mansor, Siti Rafidah Muhamat Dawam and Ab. Razak Mansor

Abstract Vehicular Ad hoc Networks (VANET) are important Vehicle-to-Infrastructure (V2I) communication systems where vehicles are communicated by broadcasting and delivering transmission scheme and traffic information to each other. Due to both high mobility and high dynamic network topology, congestion control needs to be executed in a distributed way. Optimizing the congestion control in terms of percentage for delay, packet delivery ratio (PDR), and throughput could minimize the traffic packet transmissions. These have not been studied thoroughly so far—but this attribute will be essential for VANET applications and network performance. This paper presents a novel technique for congestion control and packet forwarding through Service Control Channel (SCH) channels in VANET. The Taguchi method has been applied in getting the optimize set of parameter for congestion control in highway scenario. This concept reduces the unnecessary traffic information and reduces the possibility of congestion in network based on performance metrics for delay, PDR, and throughput. The proposed technique performance is measured with the normal VANET scenario in V2I communication in highway driving conditions and the simulation results show that the performance of proposed scheme provides the better performance and improves the network performance with efficient bandwidth utilization.

Keywords Non-safety applications · Service control channel · V2I · Bandwidth utilization · Taguchi method

---

S.J. Elias (✉) · S. Mansor · S.R. Muhamat Dawam · Ab.R. Mansor
Faculty of Computer and Mathematical Sciences, Universiti Teknologi MARA, Merbok, Kedah, Malaysia
e-mail: sjamel@gmail.com

Ab.R. Mansor
e-mail: arazman@kedah.uitm.edu.my

M.N. Mohd Warip
School of Computer and Communication Engineering, Universiti Malaysia Perlis, Pauh Putra Campus, Arau, Perlis, Malaysia
e-mail: nazri@unimap.edu.my

A.-R. Ahmad et al. (eds.), *Proceedings of the International Conference on Computing, Mathematics and Statistics (iCMS 2015)*,
DOI 10.1007/978-981-10-2772-7_3

3.1 Introduction

The Vehicular Ad Hoc Networks (VANET) is a progressively indispensable area in Mobile Ad Hoc Networks (MANETs). Recently, researchers have shown an increased interest in Vehicular communications. According to the definition provided by authors in [1], the VANETs comprise vehicle-to-vehicle (V2V) and vehicle-to-infrastructure (V2I) communications based on wireless local area network (WLAN) technologies. The distinctive set of application (e.g., collision warning and local traffic information for drivers), resources (licensed spectrum, rechargeable power source), and the environment (e.g., vehicular traffic flow patterns, privacy concerns) makes VANET a unique area of wireless communication.

The challenging part of VANETs is the high mobility where it can contributes to the high rate of topology changes and high variability of node density. It can be concluded that MANET routing protocols are difficult to implement, e.g., outdated neighbor information in routing table protocol [2].

Value-added applications could be categorized as on-demand services related to infotainment, multimedia, or non-safety applications where the author is going to focus on. Notification from point of interest such as hotel and lodging, list of available restaurants or e-map downloading may save time and thus reduce the fuel consumption.

Figure 3.1, illustrated the highway driving scenarios which explained the three lanes for each driving direction. These traffic congestion and road safety information can be published through service center to different RSU clusters for broadcasting. Every vehicle participates in sensing and updating latest road information.

Internet communication on vehicles, the data exchange between OBU (on board unit) in vehicles, and RSU (road side unit) assistance system, which is based on wireless networking technology. There are three wireless radio technology standards set for the vehicle communication [3]:

a. IEEE 802.11—WiFi,
b. IEEE 802.15.3—UltraWideband (high data rate),
c. IEEE 802.15.4—ZigBee (low data rate).

In order to facilitate the communication in VANETs, the IEEE 802.11p (WAVE —Vehicular Environments Wireless Access) and 802.11 protocols standard for

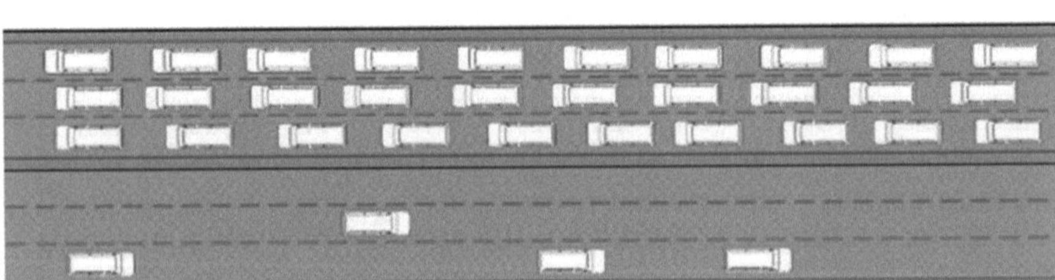

Fig. 3.1 Highway scenario showing the variety of requirements for both moving vehicles

Dedicated Short Range Communication (DSRC) were introduced. The DSRC was designed using a multichannel system. The standard 802.11p is intended for V2V communication [4]. The FCC divided the spectrum into seven channels, each with 10–20 MHz, in which six were identified as Service Channels (SCHs), and one as a Control Channel (CCH). The CCH channel is used for safety messages while non-safety services (WAVE-mode short messages) are expected to go through the other six SCH service channels available [5, 6].

3.2 WLAN Standards

Table 3.1 briefly summarizes the comparison of 802.11 standard and the enhancements related to data rate and the different standard supports in terms of functionality and operation of WLAN. The most popular are those defined by the 802.11b and 802.11g protocols, which are amendments to the original standard followed by 802.11n and 802.11p. 802.11p is a new multi-streaming modulation technique. The WLAN standard operates on the 2.4 and 5 GHz Industrial, Science and Medical (ISM) frequency bands.

From Table 3.1, we can conclude there are many standards related to WLAN accessibility in VANETs environments. There are various functionalities and operations among all the standards. These standards range from protocols that apply to RSU equipment and OBU communication protocols through security specification, routing, addressing services, and interoperability.

3.3 Congestion Control Algorithm Design Criteria

Congestion control algorithm can be adopted from many techniques to reduce congestion in VANETs. Some of them are based on messages, where the message is broadcasted to its neighbors to avoid congestion. Table 3.2 shows types of congestion control algorithms and their parameters which are mostly studied by many researchers based on broadcast warning messages.

Table 3.1 WLAN modes of operation

Standard	Ad hoc	Infrastructure	VANETs
802.11a/b/g/n/p	Yes	Yes	Yes
802.15.1/4/3	Yes	No	Yes
802.16 m/e/d	Yes	Yes	Yes
802.20	Yes	Yes	Yes

Table 3.2 Types of congestion control algorithms and parameters

Congestion algorithms	Parameters
Interplay between TVWS and DSRC: optimal strategy for QoS of safety message dissemination in VANET [12]	Interplay among TVWS and DSRC
DBFC [13]	Beacon frequency control
A congestion control scheme for fast and reliable dissemination of safety messages [14]	Priority assignment and transmission power and rate adjustment
Power-control-based broadcast scheme for emergency messages in VANETs [15]	Selects boundary nodes to relay data
Performance evaluation of Beacon congestion control algorithms for VANETs [16]	Transmit rate and/or transmit power control
Congestion control approach for event driven messages [17]	Prioritizing messages based
Efficient congestion control in VANET for safety messaging [18]	Transmission rate and transmission power
Congestion control to achieve optimal broadcast efficiency in VANETs [19]	Transmission power based
Utility-based congestion control broadcast messages [20]	Data rate based
Cooperative collision warning [21]	Cooperative collision warning
Measurement-based detection and queue freezing techniques [22]	Emergency messages in VANETs
MAC protocol makes use of a distributed beaconing scheme and a reservation based channel access (DRP) on SCH [23]	VMESH MAC protocol for enhancing the performance
A cooperative scheme for the reservation of service channels in 802.11p/WAVE-based [24]	A gossip-based reservation mechanism both proactive and reactive
Position information to select the data rate to be used for packet transmission during the SCH interval [25]	Data rate adaptation policies
Channel allocation scheme based on channel throughput analysis [26]	MAC and SCH allocation scheme to guarantee the QoS
Enhanced the throughput of SCHs and reduced the transmission delay of the service packets [27]	Adjustment of the length ratio between the CCH and SCHs
Transmission power and packet generation based on dynamic carrier sense threshold [28]	Proactive algorithm
The adoption, adaptation, and improvement of the multichannel architecture [29]	Multiple channels
Multiple service channels is based on vehicular environments using a single transceiver [6]	Dynamic service channels allocation (DSCA)
Enhanced multichannel MAC for VANETs [30]	VEMMAC protocol

3.4 Simulation and Testing Phase

Experiment efforts are carried out using OMNeT++ ver 4.6 simulator [7] running under UBUNTU 14.04.2 LTS. All medium access control (MAC) and routing protocols are based on INET framework [8, 9] and INET-MANET [10] of the OMNeT++. The combination of control factors and noise factors is presented in Table 3.3 as the experiments sets. Simulation time of each experiment sets are 250 s and 3 random seed generation is conducted [11]. This paper concerns at optimizing the control factors in VANET congestion controls in achieving minimum end-to-end delay, maximum PDR, and maximum throughput for highway test environment scenario. Simulation parameters for tested experiments are as stated in Table 3.3.

Simulation times are tested from 50 s up to 250 s. Five levels of RSU distance along the highway. As for MAC protocols we used 802.11p, while routing protocol AODV was selected. Levels of variations for control factors are stated in Table 3.4. Levels of variations of noise factors are displayed in Table 3.5.

To determine the effect that each factor has on the output, the signal-to-noise (SN) ratio needs to be calculated for each experiment conducted. The SN value represents the magnitude of the mean of a process compared to its variation. There are three types of SN ratio that could be calculated depending on different types of performance characteristics. As for the case of minimizing the performance of the system, the following SN ratio, which is called smaller-the-better, should be calculated using Eqs. 3.1, 3.2 and 3.3

$$SN_i = -10 \log \left(\sum_{u=1}^{N_i} \frac{y_u^2}{N_i} \right) \tag{3.1}$$

$$SN_i = -10 \log \left[\frac{1}{N_i} \sum_{u=1}^{N_i} \frac{1}{y_u^2} \right] \tag{3.2}$$

Table 3.3 Simulation parameters

Parameter	Values
Number of cars	60
Number of RSUs	5
Distance	1500 m
Direction	2 lanes each
Simulation times	250 s
Traffic type	UDP
Routing protocol	AODV
.bitrate	27 Mbps
.wlan	IEEE 802.11p
.message length	512 byte
Random number generator	3 [7]

Table 3.4 Level of variations for control factors

Parameters	Levels	
	Low	High
wifiPreambleMode	Long	Short
slotTime (μs)	5	25
rtsThresholdBytes	500 with rts	2346 without rts/cts
minSuccessTheshold	5	20
successCoef	2.0	8.0

Table 3.5 Level of variations for noise factors

	Levels				
Parameters	1	2	3	4	5
Simulation time (s)	50	100	150	200	250

$$SN_i = 10\log\frac{\bar{y}_i^2}{s_i^2} \tag{3.3}$$

where

y mean response of the experiment,
i experiment number,
u trial number,
N_i number of trials for experiment i

The third case is for nominal-the-best situation when a specified value is most desired. For maximizing the performance characteristics, the following SN ratio, which is called the larger-the-better should be applied as follows:

$$\bar{y}_i = \frac{1}{N_i}\sum_{u=1}^{N_i} y_{i,u} \tag{3.4}$$

$$s_i^2 = \frac{1}{N_i - 1}\sum_{u=1}^{N_i}\left(y_{i,u} - \bar{y}_i\right) \tag{3.5}$$

For this experiment, SN ratio larger-the-better is used for PDR and throughput evaluation. For optimal performance, the larger-the-better performance metric for both packet delivery ratio (PDR) and throughput sensitivity should be taken for obtaining optimal VANETs congestion control design for non-safety applications. Basically for non-safety applications demand of low level quality of service (QoS) is delay sensitive. The study is focused on non-safety applications in VANETs for highway driving scenario. Table 3.6 explained the performance metrics that is applied for this highway scenario experiments.

Figure 3.2 derived the whole experiments process that anticipates the Taguchi optimizations methodology for minimizing the delay, maximizing PDR and maximizing the throughput for highway scenario.

Table 3.6 Performance metrics

Name	Definition
Throughput	Total number of delivered data packets divided by the total duration of simulation time [31] $\text{Throughput} = \frac{\text{received packets in bits}}{\text{simulation time}}$
Packet delivery ratio (PDR)	According to [32], packet delivery ratio is the number of packet received at the destination over the packet generated by the source $\text{PDR} = \frac{\text{received packets}}{\text{sent packets}}$
End-to-end delay (Delay)	The time it takes for a data packets to reach destination from source [31]

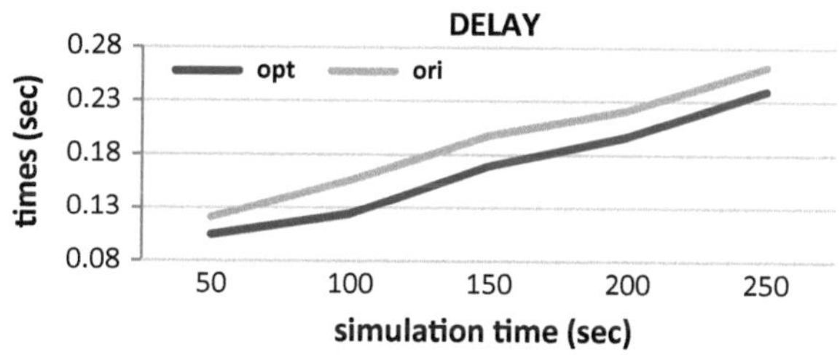

Fig. 3.2 Minimizing delay for congestion control before and after optimization based on simulation time

3.5 Results and Analysis

This section will present the results from the optimization design and simulation in OMNeT++. Based in the Fig. 3.2, the average improvement propagation for delay sensitive applications is 15.59 % after optimization. At the iteration point of 50 and 5100 s, as the network speed increases, there is the possibility of the connection generally suffering from latency from 150 to 250 s. Besides that, AODV multi-hop relaying and the point-to-multipoint data transmission of V2I can also cause additional delay as a packet may need to wait for prior missing packets during transmission.

However, this did not decrease the effective bandwidth utilization. The graphs also represents the S/N smaller is better which based Taguchi method which is to obtain optimum and smaller delay percentage with give the best fit parameter setting through the parameter screening step. From the analysis, the optimization of congestion control for vehicular network on minimizing end-to-end delay is clearly shown from the simulation time of 50 to 100 s. Simulation time shows that it has a small effect on congestion control performance as stated earlier, 15.59 % average in terms of optimization propagation.

The PDR and throughput performance characteristic are illustrated in Figs. 3.3 and 3.4. The two graphs above reflect the completeness and accuracy of the framework about the AODV protocol of VANETs over congestion control toward non-safety or multimedia applications after optimization. Based on Fig. 3.3, there is improvement after optimization around 26.4 % based on the average of PDR thus packet loss is also reduced. From 50 s onwards, the PDR is at downward trend that is lower for a particular time since the modified AODV protocol creates more latency of routing activity from source to the destination.

In principle, the characteristics of AODV protocol routing over congestion control on multimedia applications show feasibility impact on PDR of VANETs. Due to increase in simulation time at interval of 50 s up to 250 s, the PDR diminishes when there is considerable increase in data traffic activity, topology changes, and broken links to the next hops or transmissions. Also based in the Taguchi design, the mean S/N larger is better which is to obtain optimum and higher PDR as possible with the best fit parameter setting through the parameter screening step.

When simulation time is our main requirement then in high static and dynamic environment, after optimization the respective framework is showing better performance in terms of bandwidth efficiency in this respective framework based on the graph in Fig. 3.4. It is showing better data transmitted successfully to their final destination as compared to previous optimization step taken. The throughput improved also due to shorter hops as well as performing basic route recovery failure phase of AODV in VANET.

Figure 3.4 represents that the mean S/N larger is better in the Taguchi analysis which is to obtain optimum and higher capacity for throughput with the best fit parameter setting through the parameter screening step. Based on Fig. 3.4 too, there is improvement after optimization around 30.96 % on the average of throughput, packet loss is also decreased. From 50s onwards, the throughput is at upward trend that is high for a particular time since the modified AODV protocol creates less latency of routing activity from source to the destination. Therefore the number of packets could be transmitted increased as simulation time increase as well.

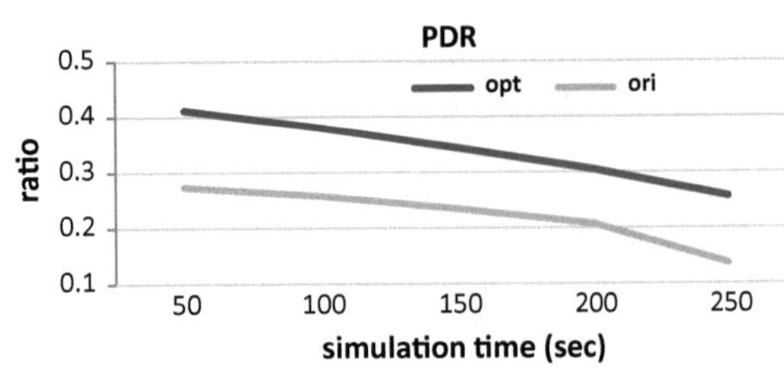

Fig. 3.3 Maximizing PDR for congestion control before and after optimization based on simulation time

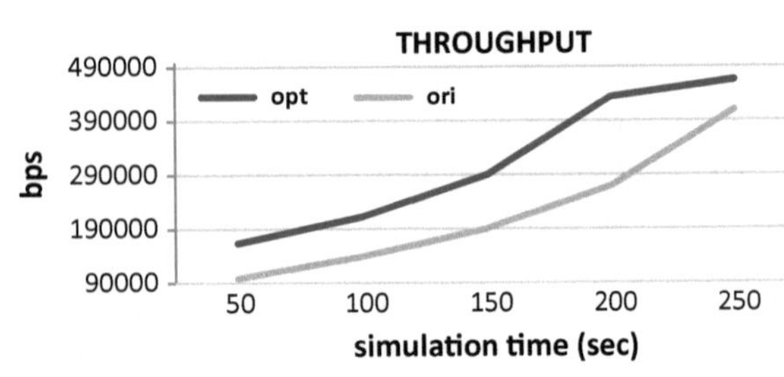

Fig. 3.4 Maximizing throughput for congestion control before and after optimization based on simulation time

3.6 Conclusion and Future Works

In this paper, we investigated technologies and methods realized by various researchers and proposed a framework for congestion control for SCH applications mainly for non-safety applications using Taguchi optimization method. The congestion control approach is one of the solid solutions to alleviate congestion in wireless communications channel. We have highlighted the algorithm for the non-safety messages mechanism to reduce the channel communications usage based on defined threshold.

This paper proposed Taguchi optimization method for optimizing the end-to-end delay applications sensitivity, PDR sensitivity, and throughput sensitivity. For future research direction, more parameters and scenarios can be tested for congestion control optimization. Simulation time is a tested factor in improving network throughput and PDR. Distance of RSU has the major factor effects to performance in terms of PDR. Our simulation results show that the AODV routing deployment has the positive effects in terms of throughput and PDR transmission over the network topology. Our simulation results show that the AODV routing deployment has the positive effects in terms of end-to-end delay transmission received. But looking on the bright side, the congestion control of the particular framework is improved drastically after optimization for the vehicular ad hoc network of user or application level.

As conclusion, non-safety or multimedia applications which are throughput and PDR sensitive have a significant impact on the simulation time which is the main control factor for this experiment. For future work, more parameters can be involved or tested for non-safety applications in optimizing the congestion control for VANETs.

References

1. Hartenstein, H., Laberteaux, K.: A Tutorial Survey on Vehicular Ad Hoc Networks (2008)
2. Blum, J.J., Eskandarian, A., Hoffman, L.J.: Challenges of Intervehicle Ad Hoc Networks (2004)
3. Welekar, A.R.: Comparative Study of IEEE 802.11, 802.15, 802.16, 802.20 Standards for Distributed VANET (2012)
4. IEEE Computer Society.: IEEE standard for: wireless LAN medium access control(MAC) and Physical Layer(PHY) Specifications: Amendment 6: Wireless Access in Vehicular Environments (2010)
5. Bouassida, M.S., Shawky, M.: On the congestion control within VANET. In: 1st IFIP Wireless Days, pp. 1–5, (2008)
6. Park, S., Chang, Y., Khan, F., Copeland, J.A.: Dynamic service-channels allocation (DSCA) in vehicular ad-hoc networks. In: 2013 IEEE 10th Consumer Communication Network Conference, pp. 351–357 (2013)
7. Varga, A., Hornig, R.: An overview of the OMNeT++ simulation environment. In: Proceedings of the 1st International Conference on Simulation Tools and Techniques for Communications, Networks and Systems & Workshops (2008)

8. Nagel, R., Eichler, S.: Efficient and realistic mobility and channel modeling for VANET scenarios using OMNeT++ and INET-framework. In: Proceedings of the 1st International Conference on Simulation Tools and Techniques for Communications, Networks and Systems & Workshops (2008)
9. Klein, D., Jarschel, M.: An OpenFlow extension for the OMNeT++INET Framework. In: Proceedings of the Sixth International Conference on Simulation Tools and Techniques (2013)
10. Kumar, N., Alvi, P.A., Singh, A., Swami, A.: A study of routing protocols for Ad-hoc network. In: Int. J. Appl. Innov. Eng. Manag. (2013)
11. Pawlikowski, K., Jeong, H.D.J., Lee, J.S.R.: On credibility of simulation studies of telecommunication networks. IEEE Commun. Mag. **40**(1), 132–139 (2002)
12. Lim, J.H., Kim, W., Naito, K., Gerla, M.: Interplay between TVWS and DSRC: optimal strategy for QoS of safety message dissemination in VANET. Int. Conf. Comput. Netw. Commun. ICNC **2013**, 1156–1161 (2013)
13. Humeng, L., Xuemei, Y., Li, A., Yuan. W.: Distributed Beacon frequency control algorithm for VANETs (DBFC). In: International Conference on Intelligent Systems Design and Engineering Application (2012)
14. Djahel, S., Ghamri-Doudane, Y.: A robust congestion control scheme for fast and reliable dissemination of safety messages in VANETs. IEEE Wireless Communication Network Conference WCNC, pp. 2264–2269 (2012)
15. Wei, L., Xiao, X., Chen, Y., Xu, M., Fan, H.: Power-control-based broadcast scheme for emergency messages in VANETs. In: 11th International Symposium Communication Information Technology Isc. no. Iscit, pp. 274–279 (2011)
16. Le, L., Baldessari, R., Salvador, P., Festag, A., Zhang, W.: Performance evaluation of beacon congestion control algorithms for VANETs. In: GLOBECOM—IEEE Global Telecommunication Conference (2011)
17. Darus, M.Y., Abu Bakar, K.: A review of congestion control algorithm for event-driven safety messages in vehicular networks. Int. J. Comput. Sci. **8**(2), 49–53 (2011)
18. Mughal, B.M., Wagan, A.A., Hasbullah, H.: Efficient congestion control in VANET for safety messaging. IEEE, pp. 654–659 (2010)
19. Ye, F., Yim, R., Zhang, J., Roy, S.: Congestion control to achieve optimal broadcast efficiency in VANETs. In: IEEE International Conference Communication, pp. 1–5 (2010)
20. Wischhof, L., Rohling, H.: Congestion control in vehicular ad hoc networks. IEEE Int. Conf. Veh. Electron. Safety **2005**, 58–63 (2005)
21. Sumra, I.A. Manan, J.A.B., Hasbullah, H., Iskandar, B.S.: Timing Attack in Vehicular Network 2 VANET Applications and Time, pp. 151–155 (2011)
22. Darus, M.Y., Bakar, K.A.: Congestion control framework for emergency messages in VANETs. Control **2**(3), 643–646 (2011)
23. Zang, Y., Stibor, L., Walke, B., Reumerman, H.-J., Barroso, A.: A novel MAC protocol for throughput sensitive applications in vehicular environments. In: 2007 IEEE 65th Vehicle. Technology Conference—VTC2007-Spring, pp. 2580–2584 (2007)
24. Campolo, C., Cortese, A., Molinaro, A.: CRaSCH : a cooperative scheme for service channel reservation in 802.11p/WAVE vehicular ad hoc networks (2009)
25. Cheng, N., Lu, N., Wang, P., Wang, X., Liu, F.: A QoS-provision multi-channel MAC in RSU-assisted vehicular networks. In: IEEE Vehicle Network Conference, pp. 193–197 (2011)
26. Amadeo, M., Campolo, C., Molinaro, A., Ruggeri, G.: A WAVE-compliant MAC protocol to support vehicle-to-infrastructure non-safety applications (2009)
27. Wang, Q., Leng, S., Fu, H., Zhang, Y., Member, S.: An IEEE 802.11p-based multichannel MAC scheme with channel coordination for vehicular ad hoc networks (2012)
28. Sattari, M.R.J., Noor, R.M., Keshavarz, H.: A taxonomy for congestion control algorithms in Vehicular Ad Hoc Networks. In: IEEE International Conference Communication Networks Satellite, pp. 44–49 (2012)
29. Campolo, C., Molinaro, A.: Multichannel communications in vehicular Ad Hoc networks: a survey. IEEE Commun. Mag. **51**(5), 158–169 (2013)

30. Hong, C.S.: An enhanced multi-channel MAC for vehicular ad hoc networks. In: IEEE Wireless Communication Network Conference, pp. 351–355 (2013)
31. Al-Maashri, A., Ould-Khaoua, M.: Performance analysis of MANET routing protocols in the presence of self-similar traffic. In: Proceedings 31st IEEE Conference Local Computing Networks (2006)
32. Jörg, D.O., Heissenbüttel, M.: Performance comparison of MANET routing protocols in different network sizes computer science project. Comput. Networks Distrib. Syst., pp. 1–31 (2003)

Chapter 4
Experimental Simulation of ZigBee's Multi-hops Wireless Sensor Network Using OPNET

Mohd Zaki Shahabuddin and Halabi Hasbullah

Abstract Wireless Sensor Network (WSN) is developed based on IEEE 802.15.4 consists of low-rate capabilities. The low-rate signifies the network is operated at low cost, using low power, and has nonexist physical topology infrastructure. Among the available technology is ZigBee, the most common standard of low-rate wireless sensor network. This work presents experimental simulations and analysis of ZigBee WSN using the OPNET network simulation tool. The work evaluates several configurations of ZigBee parameters particularly at the Physical, Media Access Control (MAC) and Application layers based on IEEE 802.15.4 standard. The objective of the study is to understand the OPNET modeler on how the connectivity is established between ZigBee devices. Further, the model of WSN scenarios are performed to understand the controlling parameter of how number of hops routing has the effects toward the end-to-end latency of ZigBee wireless sensor network.

Keywords WSN · IEEE 802.15.4 · ZigBee · OPNET · End-to-end latency

4.1 Introduction

The purpose of Wireless Sensor Network (WSN) is to overcome the inconveniences of cabling works due to logistic and costly implementation. WSN protocol standard is developed by IEEE 802.15.4 with intention to offer a fundamental of lower network layers (Physical and Media Access Control (MAC) layers) of low-rate wireless connectivity [1]. The term low-rate emphasizes on low cost, operated by low power, and has nonexist physical topology infrastructure. Among the available

M.Z. Shahabuddin (✉) · H. Hasbullah
Department of Computer and Information Sciences,
Universiti Teknologi PETRONAS, Perak, Malaysia
e-mail: mzakishah77@gmail.com

H. Hasbullah
e-mail: halabi@petronas.com.my

A.-R. Ahmad et al. (eds.), *Proceedings of the International Conference on Computing, Mathematics and Statistics (iCMS 2015)*,
DOI 10.1007/978-981-10-2772-7_4

wireless personal area network technologies are ZigBee and Bluetooth. The basic characteristic of ZigBee device is its ability to communicate up to 100 m of transmission range, with a transfer rate of up to 250 Kbps. For security, the real-time reservation of guaranteed time slots, through collision avoidance protocol CSMA/CA, coexists with symmetrical DES algorithm are applied for protected communication [2].

Regardless of low-rate characteristic, WSN is expected to operate for a relatively long time. This is the primary design goal of WSN, and it is achievable through many factors. Among the significant factors related to implementation of WSN topology control is the routing implementation that determines the way of data transmission, number of hops, and the size of data transferred [3].

The objective of this work is to perform ZigBee WSN simulation using OPNET network simulation tool for the purpose to understand the controlling behavior of WSN. Several scenarios are created using the OPNET to demonstrate ZigBee WSN with particular configuration to study the relation of network latency with number of multi-hops data transmission.

The work is organized with brief overview of ZigBee WSN specifications and its related network protocol. Then, procedures of OPNET simulation will be developed to replicate few WSN scenarios of multi-hops transmission versus end-to-end latency using ZigBee modeler. Finally, the result of the simulation will be analyzed and discussed, and the next actions of study will be furtherly considered.

4.2 Overview of ZigBee Technology

ZigBee standards are intended for wireless applications requiring low power consumption and tolerating low data rates. It is governed by ZigBee Alliance, a statutory body that governs the specifications at Application and Network Layers, whilst the IEEE 802.15.4 standard provides the specifications for MAC and Physical network layers [2].

Figure 4.1 shows the simplified ZigBee WSN protocol layers. There are segregated layers between ZigBee Alliance and IEEE 802.15.4 specification conformance.

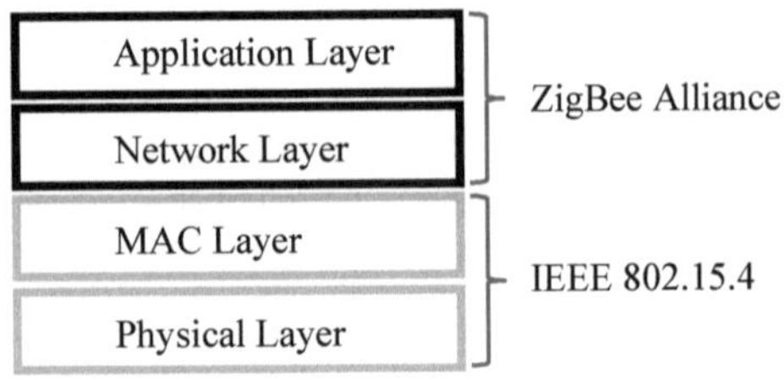

Fig. 4.1 ZigBee/IEEE 802.15.4 network protocol layers

4.2.1 ZigBee/IEEE 802.15.4 Network Protocol Layers

Application Layer The Application Layer is the top layer and it hosts the application objects that provide endpoint interface to the user. A single ZigBee node is able to cater up to 240 embedded applications [4] and they must conform to a standard application profile customized by the ZigBee Alliance to ensure the interoperability at ZigBee Application Layer.

Network Layer The Network Layer defines the function of ZigBee's network management and addressing scheme. This layer outlines tasks in network formation, starting on connectivity, managing devices in joining or leaving a network, and route and neighbor discoveries [2].

Media Access Control (MAC) Layer The MAC layer to provide association and dissociation of data packet frames arrangement. It determines the source and destination addressing of frames header through the form of Carrier Sense Multiple Access with Collision Avoidance (CSMA/CA) [1]. This layer defines four frame structures: Beacon, Data, Acknowledge, and MAC, where the synchronization of frames between sender and receiver is supported to provide reliable transmission between two devices [4].

Physical Layer The Physical Layer of the IEEE 802.15.4 is the protocol which controls the ZigBee devices to communicate using radio frequency transceiver. Located near to hardware, it handles all tasks involving the access to the hardware signal modulation, channel selection, link quality estimation, energy detection measurement, and clear channel assessment [4]. ZigBee devices support three frequency bands that have its specific data rate and channel numbers as depicted in the Table 4.1.

4.2.2 ZigBee Devices

WSN devices have homogenous features in its resources and capabilities. However, there are exceptions where different types of device capabilities are to be used in WSN deployment. The standard IEEE 802.15.4 defines two types of WSN devices:

1. *Full-function device* (*FFD*). It is equipped with proper power and memory resources to serve as coordinator of the network, cluster heads or as routers to relay data.

Table 4.1 IEEE 802.15.4/ZigBee basic specifications

Frequency band	Data rate (Kbps)	Channel numbers	Countries
868.3 MHz	20	0	European
902–928 MHz	40	1–10	USA
2.450 GHz	250	11–26	Worldwide

2. *Reduced-function devices* (*RFD*). It is meant to be a simple device, consume less power and memory resources. It can only communicate with FFDs to relay data. RFD is designed as an end-device with shorter transmission range and normally placed as the sensors.

Under ZigBee standard, there are three (3) types of devices defined at the Network Layer [2, 6] and these are:

1. *Coordinator*—The coordinator here is responsible in network initialization, proper channels selection, and controlling other devices connecting to the network. The coordinator is also in-charge on traffic routing algorithm of WSN. There can only be one coordinator (Access point) in ZigBee network.
2. *Router*—A router serves as intermediary node that connects a coordinator with other devices. The purpose of ZigBee router is to expand the WSN through a transmission relay between router nodes. It can have child nodes connected to it. It can also take place as end-devices where its routing capabilities would be useless in this case [1].
   Both ZigBee coordinator and router fall under the category of FFDs based on IEEE 802.15.4 standard.
3. *End-device*—ZigBee end-devices are the RFDs. They are not used for traffic relay, but it has adequate functions to talk to parent nodes, in this case the router or coordinator. It has limited power and capabilities other than to serve as application sensing devices. Hence, it has power saving features to extend its lifetime.
   In general, the ZigBee WSN supports network topologies of: Mesh, Star, Tree, and Hybrid [5]. The topology plays important function to indicate how the devices are logically connected.

4.3 ZigBee Simulation Using OPNET

4.3.1 ZigBee Model Properties

OPNET is a network simulation tool used in R&D of communication network and distributed systems. It provides reliable simulation of scalable communication model to present the behavior of applications and it components. Among the feature of OPNET is it supports IEEE 802.15.4 modeling, with built-in ZigBee library modeler. This makes the OPNET among a useful tool to understand WSN behavior based on IEEE 802.15.4 standard.

The ZigBee modeler supports contain the representation of device for ZigBee coordinators, routers, and end-devices. Figure 4.2 shows OPNET modeler representation of ZigBee devices.

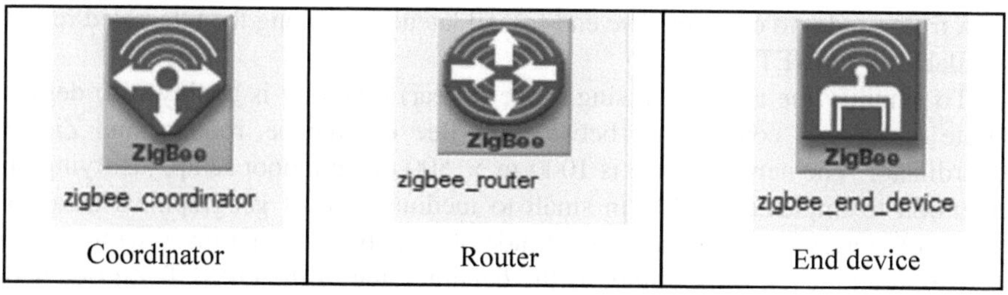

Fig. 4.2 OPNET presentation of ZigBee devices

4.3.2 *ZigBee Modeler Attribute Parameters*

The modeler provides an environment where the attributes of ZigBee can be configured, and used as inputs before the simulation is started. To demonstrate our simulation working model, the attributes of ZigBee at Physical, Application, and MAC Layers are set in Table 4.2.

All these parameters are set to control on how ZigBee devices to communicate as specified by IEEE 802.15.4, based on common usage of typical WSN applications [6, 7].

4.3.3 *OPNET's ZigBee Simulation*

The OPNET is used to simulate the ZigBee sensor network, with parameter to evaluate how the number of hops can affect the end-to-end latency of transmitted

Table 4.2 OPNET ZigBee modeler attribute parameters

Physical layer parameters	
Data rate (kbps)	250
Receiver sensitivity (dB)	−90
Transmission band (GHz)	2.45
Transmission power (W)	0.003
Application layer parameters	
Packet interval time/type (sec/constant)	1
Packet size/type (bits/constant)	1024
MAC layer parameters	
Acknowledge wait interval (s)	0.05
Maximum No. of retransmissions	5
Minimum value of the back-off exponent in the CSMA/CA (if the value is set to 0, collision avoidance is disabled)	3
Maximum number of back-offs the CSMA/CA algorithm will attempt before declaring a channel access failure	3
Channel sensing duration (s)	0.1

data from sender to receiver. The end-to-end latency is among the integrated reports available in OPNET.

To simplify the analysis, a single tree (linear) topology is deployed to demonstrate end-to-end connectivity between ZigBee end-device, routers, and ZigBee coordinator. The network scale is 1000 m × 500 m, an indoor setup, justifying our intention to implement WSN in small to medium size of geographical area (example of cluster of buildings). The altitude of all network scenarios is set to be 1 m above ground level. This altitude is the default value of the OPNET–ZigBee node model and can be varied according to the application being simulated. The simulation running time is 24 h for each scenario.

The following (Fig. 4.3) is the OPNET logical simulation design of ZigBee WSN.

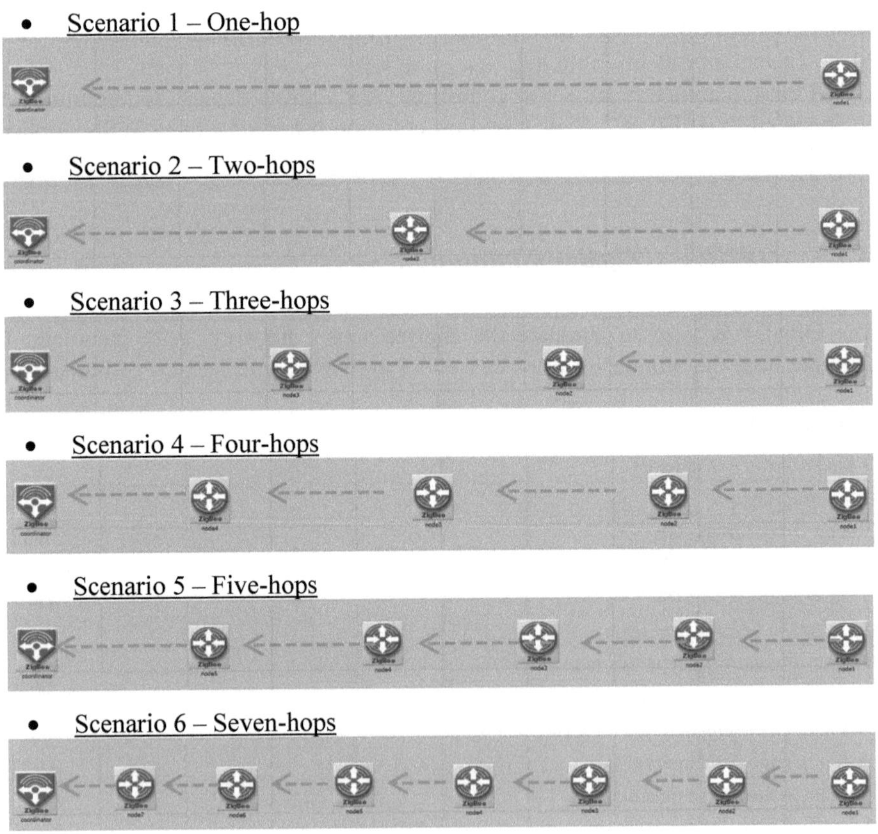

Fig. 4.3 Scenarios in multi-hops connectivity

4.4 Result and Analysis

The result of experimental simulation is compiled in the Fig. 4.4. The controlling parameter of the experiment would be the number of hops/relays for the transmitter node to convey the message towards the receiving node.

Out of six simulated scenarios, five scenarios (Scenario 2 to 6) show the existence of end-to-end delay against the simulation time. The Scenario 1 result is not available according to OPNET report of invalid configuration. This is due to OPNET marks that a single hop over 1000 m data transmission is not viable for ZigBee. ZigBee device has limited transmission range, and for standard ZigBee it should be in the range of 100 m but depending on manufacturers' specification and environmental factors the range can go up to 1000 m [8].

For Scenario 2 to Scenario 4 (two, three, and four-hops), the graph shows the end-to-end latency occurs almost at the same fold of time. But for Scenario 5 (five-hops) the latency drops drastically compared to previous scenarios. It shows that the number of hops in Scenario 5 is an optimal number to provide the least end-to-end latency in WSN in the area of 1000 m × 500 m.

Further, we try to increase the number of hops to seven as presented in Scenario 6. The result shows this network latency goes up to the same as in Scenario 2, 3, and 4. Moreover, Scenario 6 shows a shorter simulation time that might represent that this network has short lifespan. This is true as the nodes drain its energy fast perpetually to more number of hops data to travel. Hence, there should be a proper number of nodes to be deployed in a certain size of network, and the proper nodes deployment mechanism is important to ensure the network has long lifecycle.

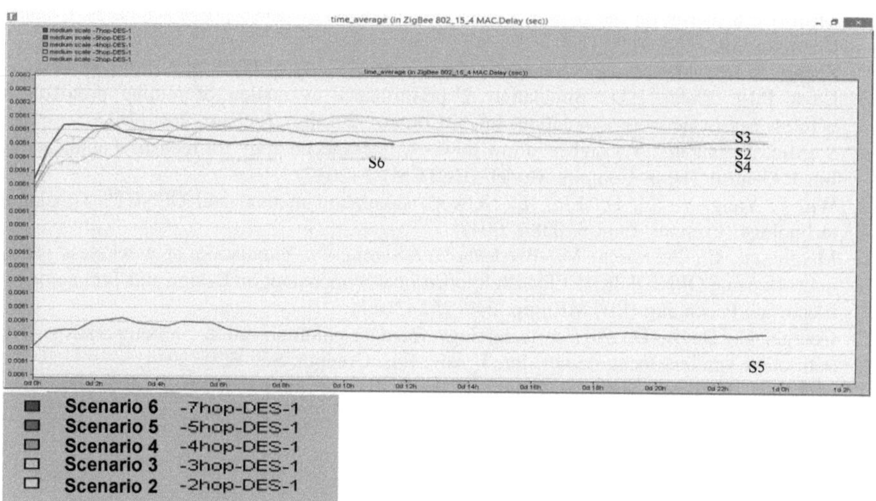

Fig. 4.4 End-to-end delay (in seconds) versus simulation time

4.5 Conclusion and Future Work

From this work we conclude that the OPNET provides comprehensive ZigBee modeling capabilities. It enables to configure technical parameters of ZigBee WSN according to IEEE 802.15.4 standard. This study on OPNET simulation of ZigBee WSN is intended for nonspecific application, but we are working on few simple scenarios to understand the relationship on multi-hops with transmission latency of ZigBee WSN. The result shows there is significance to acquire the optimal number of hops to improve end-to-end latency between two connected nodes.

In this work, the simplified setups of WSN topology and communication scenarios enable us to better understanding the OPNET and ZigBee itself. Moreover, the result outcome is produced in reasonable time compared to having complex scenarios. Hence, this study lacks in resembling to real-life application of IEEE 802.15.4 and ZigBee specifications. In next actions of study, we are going to thoroughly develop OPNET simulation of real-life WSN from a reliability point of view. These will include the MAC layer specification on routing mechanism as well the presentation of WSN in mesh topology corresponding to network robustness to achieve reliable WSN transmission.

Acknowledgment This research work is supported by the Ministry of Higher Education, Malaysia under MyRA Incentive Grant for Smart Integrated Low Carbon Infrastructure Model Program, and Universiti Teknologi PETRONAS.

References

1. Baronti, P., Pillai, P., Chook, V.W.C., Chess, S., Gotta, A., Fun Hu, Y.: Wireless sensor networks: a survey on the state of the art and the 802.15.4 and ZigBee standards. Comput. Commun. **30**, 1655–1695 (2007)
2. Zheng, J., Lee, M.J.: A Comprehensive Performance Study of IEEE 802.15.4 (2004)
3. Dave, P.M., Dalal, P.D.: Simulation & performance evaluation of routing protocols in wireless sensor network simulation. Int. J. Comput. Sci. Inf. Technol. **2**(3) (2013)
4. Somani, N.A., Patel, Y.: Zigbee: a low power wireless technology for industrial applications. Int. J. Control Theor. Comput. Model. (IJCTCM) **2** (2012)
5. Wu, J., Yang, Y., Li, H., Lin, X.: Network simulation method and OPNET's simulation technology. Comput. Eng. **5**, 038 (2004)
6. Marghescu, C., Pantazica, M., Brodeala, A., Svasta, P.: Simulation of a wireless sensor network using OPNET. In: IEEE 17th International Symposium on Design and Technology in Electronic Packaging (SIITME), pp. 249–252 (2011)
7. George, N., Davis, D.: ZigBee technology for data communication—a comparative study with other wireless technologies. Int. J. Adv. Res. Comput. Sci. **5**(7) (2014)
8. Mihajlov, B., Bogdanoski, M.: Overview and analysis of the performances of ZigBee-based wireless sensor networks. Int. J. Comput. Appl. **29**(12), 28–35 (2011)

Chapter 5
Online Recognition of Arabic Handwritten Words System Based on Alignments Matching Algorithm

Mustafa Ali Abuzaraida, Akram M. Zeki and Ahmed M. Zeki

Abstract Arabic language is considered as the primary language in most parts of the Arabic world. It is spoken as a first language by more than 280 million people, and more than 250 million as a secondary spoken language. In pattern recognition field, several studies were focused on Arabic language with textual or voice methods. In this paper, an online handwritten Arabic text recognition system using an alignment matching theory is presented. The proposed system deals with the handwritten words as one block instead of segmenting the words into characters or strokes. The system started with collecting the dataset of 120 common Quranic words. These words have been gone via some phases to be ready for use. These phases are: Preprocessing, features extraction, and recognition phase. In the first phase, the words went through some steps to be standardized. The second phase is about extracting the features of each word and to save them in the system database. In the third phase, the system uses matching technique to search for the testing word with the system database. The system was tested and the results reached up to 97 %, which were significantly accepted compared to the previous works in the same criteria.

Keywords Text recognition · Arabic text · Online recognition · Handwritten text

M.A. Abuzaraida (✉)
Computer Science Department, Faculty of Information Technology,
Misurata University, Misrata, Libya
e-mail: abuzaraida@it.misuratau.edu.ly

A.M. Zeki
Kulliyyah of Information and Communication Technology,
International Islamic University Malaysia, Kuala Lumpur, Malaysia
e-mail: akramzeki@iium.edu.my

A.M. Zeki
Department of Information Systems, College of Information Technology,
University of Bahrain, Sakhir, Kingdom of Bahrain
e-mail: amzeki@uob.edu.bh

A.-R. Ahmad et al. (eds.), *Proceedings of the International Conference on Computing, Mathematics and Statistics (iCMS 2015)*,
DOI 10.1007/978-981-10-2772-7_5

5.1 Introduction

The Arabic language is considered as the primary language in most parts of the Arabic world. It is spoken as a first language by more than 280 million people, and more than 250 million as a secondary spoken language. Therefore, the Arabic language is one of the most widely spoken languages in the world. In 2010, Arabic was ranked in the top five of the commonly spoken languages worldwide [1]. On the other hand, many other languages around the world are similar to the Arabic language [2]. These languages follow the Arabic language in the writing style and also in the way of speech. Many of these languages are the main language in Islamic countries like Persian in Iran; Jawi in Indonesia, Malaysia, and Brunei; Urdu in Pakistan; Pashto in Afghanistan; Bengali in Bangladesh; and others [3].

Looking at the Arabic text characteristics, there are differences between the Arabic texts and text from other languages with respect to the formatting and the way of writing. The written form of the Arabic language is summarized as follows: The 28 Arabic characters are written in different formatting. The character location in the word gives the character its formatting shape. In the Arabic text, there are four shapes of each character which are defined as the starting, middle, end, and isolated shape [4].

The Arabic word must be written cursively and the characters connect horizontally to give an understandable text [5, 6] as shown in Fig. 5.1.

The general objective of this research is to design an online handwritten Arabic text recognition system by using an alignment matching theory for recognizing handwritten Arabic words.

5.2 Architecture of the Proposed System

The proposed system followed the typical pattern recognition system architecture that contains four main phases which have been identified as text acquisition, preprocessing, feature extraction, and recognition phase as illustrated in Fig. 5.2 [8–10]. However, the segmentation step is not included in the system and the handwritten word will be processed as one block. This segmentation-free strategy can minimize the time process, help to overcome the segmenting overlapped characters' problem, and can enhance the rate of accuracy of recognition. Although having

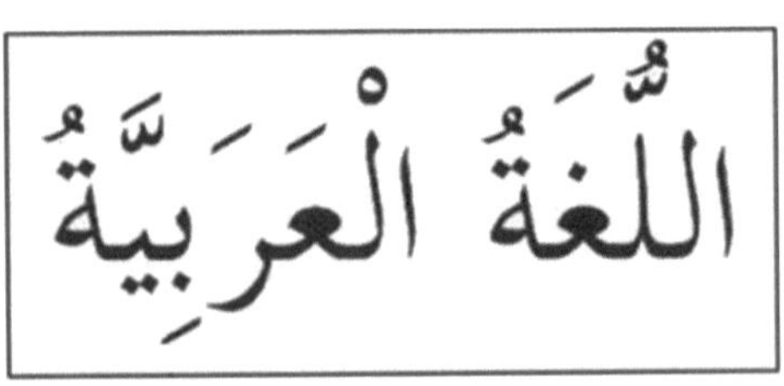

Fig. 5.1 The words (Arabic Language) written in Arabic

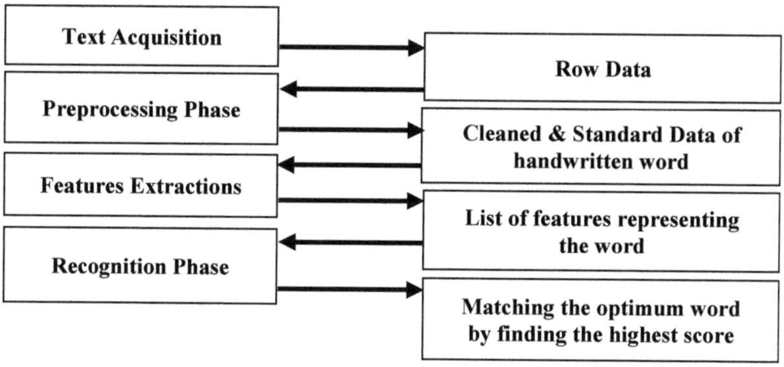

Fig. 5.2 Phases of the online Arabic handwriting recognition system [7]

stated that, each phase of the system has one or more objectives in order to reach the goal of the system and also to enhance the overall recognition accuracy rate.

5.2.1 Data Collection Stage

The data collection stage is the initial step of any pattern recognition system and aims to get raw data which will be used later by way of training and testing [11]. In this stage, the handwritten text is written by writing on an interface device that converts the handwritten text to time stamped coordinates of the stylus trajectory (x, y).

Here, for the purposes of collecting the training and testing databases [12], a 1.5 GHz core i3 Acer Tablet has been used to collect the dataset. This computer has a touch screen which can easily be used to acquire the Quranic handwritten words by a simple way of normal writing on the touch screen using a special stylus as illustrated in Fig. 5.3. The method of writing on the Tablet can minimize the noise and errors while recording on the Tablet's surface.

For collecting Quranic handwritten words, a platform was designed using a *Matlab* environment with a graphical user interface (GUI). Data collection from the

Fig. 5.3 Collecting data using the Acer tablet

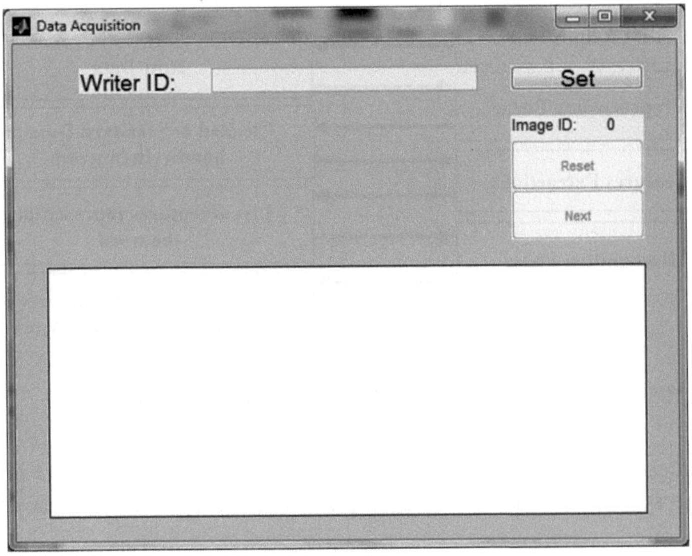

Fig. 5.4 Data collection platform

computer Tablets using this natural writing way can provide data which is identified as closely resembles, smoothed, and filtered. Figure 5.4 shows the data collection platform.

The next stage involved testing the system where the same procedure of training is performed. Global Alignment Algorithm (GAA) is used to match every handwritten word in the testing dataset with the whole training dataset. Accuracy rate and processing time is recorded. The most three highest accuracy words are presented. All these steps are explained in details in the following subsections.

5.2.2 *Preprocessing Phase*

The preprocessing phase is performed to minimize the noise which may occur in the handwritten text [13]. This phase includes several multiple steps and each step performs a specific function to filter the dataset. Besides that, it could improve the overall recognition rate, which is considered one of the essential phases of online handwriting recognition and most of the researchers have discussed its challenges in relation to the various texts from time to time [4, 14, 15].

Generally, the data collection for the online handwriting recognition system is made by storing the stylus movements on the writing surface. These movements are distributed at various positions on the writing area of the acquisition platform and then joined from the first position (x_1, y_1) to the last (x_n, y_n) to present the appearance of drawn text. Specifically, the stylus movements consist of three actions which comprise: pen-down, pen-move, and pen-up actions. The serial of points are collected when the writer presses, moves, and lifts the stylus up, consecutively. The pen-move function records the movements of the stylus on the writing Tablet from the writing starting point (x_1, y_1) until the last point (x_n, y_n) where n is the total number of points in the writing movements' list [16].

After recording the series of stylus movements, four essential steps were then performed in the preprocessing phase for this online handwritten Arabic text recognition system. These preprocessing steps are discussed in the following subsections:

Word Smoothing: In the proposed system, a smoothing technique was used to smooth the handwritten curves and this step is referred to as the Loess filter. This filter is based on conducting the local regression of the curves' points using a technique of a weighted linear least squares and a second degree polynomial model.

In this technique, each smoothed value is determined locally by neighboring data points defined within the writing curve. The process is weighted and a regression weight function is defined for each data point contained within the writing curve. The local regression smoothing algorithm is presented in the three steps indicated below for each data point [17].

Firstly, the regression weights for each data point in the writing curve by the tricube formula, calculated by using the equation below.

$$Wi = \left(1 - \left|\frac{x - xi}{\mathrm{d}(x)}\right|^3\right)^3 \tag{5.1}$$

where x is the predictor value associated with the response value to be smoothed, xi is the nearest neighbor of x as defined by the curve, and d(x) is the distance along the x-axis from x to the most distant predictor value within the curve. The weights have these characteristics. Accordingly, the data point to be smoothed has the largest weight and the most influence on the fit. Furthermore, data points outside the curve have zero weight and no influence on the fit.

Secondly, a weighted linear least squares regression is performed. Here, for the Loess method, the regression is based on a second degree polynomial.

Finally, the smoothed value is given by the weighted regression at the predictor value of interest.

Word Simplification: Douglas Peucker's algorithm [18] was adopted in this system to simplify the acquired handwritten word point sequence. Specifically, Douglas Peucker's algorithm is undertaken by considering an imaginary line between the first and the last point in a set of a curve points. The algorithm then checks which point is the furthest away from this line segment with the first and last points considered as end points. Although, if the point or all the other in-between points are closer than a given distance, it removes all these in-between points. However, if this outlier point is farther away from the imaginary line than a specific value known as a "tolerance", the curve is split into two parts. Here, Douglas Peucker's algorithm has been applied with a tolerance of 0.01 which is determined empirically.

Word Size Normalization: The size of the acquired handwritten word depends upon the way in which the writer moves the stylus on the designated writing area. The handwritten words are generally written in different sizes when the stylus is moved along the border of the writing area and this may cause some ambiguity in the next phases. Following on from that, size normalization is a necessary step that ought to be performed in order to recognize any type of text. This can be achieved by converting the acquired handwritten word with an assumed fixed-size format.

Centering of the Word: After resizing the acquired handwritten word, it is necessary for the current coordinates to be shifted to the centering axis ($X0$, $Y0$) to ensure that all points of the handwritten words are equal in formatting and to make certain that all the data is translated to the same spot relative to the origin. This step is undertaken using the following algorithm.

After passing the four steps of preprocessing phase, the points of the handwritten words are almost in a standard format. However, in this proposed system, a series of simple and less number of steps were performed to eliminate the complexity and to minimize any processing delay that may occur.

5.2.3 Features Extraction Phase

The proposed system takes the stylus trajectory directions as the main feature representing handwriting movements. Freeman's code is used to create the direction matrix for each handwritten word. Furthermore, Freeman's code [19] represents the directional movement of the stylus by a numeric code consisting of 8 digits. These directions are listed from 1–8 to represent the eight main writing directions as illustrated in Fig. 5.5.

5.2.4 Recognition Phase

In this study, a matching algorithm called GAA was used as the recognition engine to recognize the Arabic handwritten words. After conducting this phase, the system

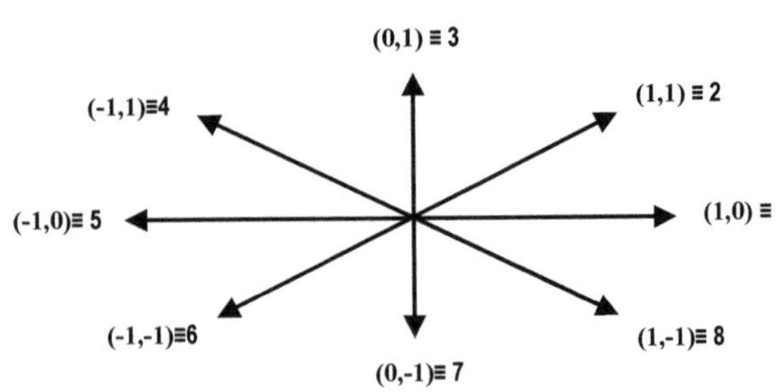

Fig. 5.5 Freeman's chain code [20]

can classify the proper word from the dataset of the system. In the following section, more details will be presented to explain the approach of the GAA method.

In fact, the most well-known and widely used methods for sequences alignments are the Local Alignment Algorithms and the Global Alignment Algorithms (GAAs). The GAA was developed by Saul B. Needleman and Christian D. Wunsch in 1970 [21]. Here, the alignment is carried out from the beginning until the end of the matched sequence to find out the best possible alignment [22].

GAA is basically a dynamic programming algorithm for sequence alignment. This dynamic programming can solve the original problem by dividing it into smaller independent subproblems. The algorithm explains global sequence alignment for aligning the nucleotide or protein sequences in general.

In general, dynamic programming is used to find the optimal alignment of two sequences. It finds the alignment in a quantitative way by giving score values for matches and mismatches. The alignment is accurately obtained by searching the highest scores in the matrix [23]. The procedure of GAA is explained in detail as following:

For matching two amino acid sequences, the algorithm is designed to find the highest score value of the sequences by building a two dimensional matrix. Basically, the algorithm procedure is defined with the three following steps in mind:

- Assuming an initialization score matrix with the possible scores;
- Filling the matrix with maximum scores; and
- For appropriate alignment, tracing back the previous maximum scores.

In the proposed system, the GAA uses the default values of 0, 1, and 1 for gap penalty, mismatching penalty, and matching score, respectively.

5.3 Presentation of the Results

For testing the system, 2400 handwritten Arabic words were fed into the system for recognition. These words were written by 40 writers who did not have any prior experience of writing by way of stylus on a digital surface. Each writer was asked to write 60 words of the same words of the dataset. Accordingly, each word was then written 20 times in total. The phases of the system were then applied to the testing dataset and then applied to the system's database for matching.

As a result of the GAA of matching every testing word with the system database, the system returns the word which gives the highest matching score and matches the sequence of the word examined. Furthermore, the matching algorithm is modified to give the first three highest scores of the first three words that match the word sequences analyzed.

5.4 Summary and Conclusion

The main goal of this research was to investigate the way of building an online Arabic handwriting recognition system using combination techniques for each phase of the proposed systems. The research also aimed to define how well the proposed system is able to resolve the Arabic handwriting recognition complexities.

In this study, the database contained 12,000 handwritten words. These handwritten words included more than 42,800 characters and 23,300 sub words written in different styles. A matching algorithm was used as a recognizer method using a global feature to describe the words. However, no segmentation step was included in the system.

The results of the experiment were statistically significant in comparison to the handwritten text recognition accuracy rates obtained from past works of online Arabic systems. Here, the results identified an accuracy of approximately 97 % in experiment I with an average of processing time about 3.034 s.

References

1. World 100 Largest Language in 2010, http://www.ne.se/spr%C3%A5k/v%C3%A4rldens-100-st%C3%B6rsta-spr%C3%A5k
2. Abuzaraida, M.A., Zeki, A.M., Zeki, A.M.: Difficulties and challenges of recognizing arabic text. In: Computer Applications: Theories and Applications. IIUM Press Malaysia, Kuala Lumpur (2011)
3. Versteegh, K., Eid, M., Elgibali, A., Woidich, M., Zaborski, A.: Encyclopedia of Arabic Language and Linguistics, vol. 1. Leiden Brill, Boston, USA (2006)
4. Al-A'ali, M., Ahmad, J.: Optical character recognition system for Arabic text using cursive multi-directional approach. J. Comput. Sci. **3**, 549 (2007)
5. Zeki, A.M., Zakaria, M.S.: Challenges in recognizing Arabic characters. In: The National Conference for Computer. Abu-al-Aziz king University, Saudi Arabia (2004)
6. Almohri, H., Gray, J.S., Alnajjar, H.: A real-time DSP-based optical character recognition system for isolated Arabic characters using the TI TMS320C6416T. In: Proceeding of the 2008 IAJC-IJME International Conference, pp. 25–35 (2008)
7. Abuzaraida, M.A., Zeki, A.M.: Segmentation techniques for online Arabic handwriting recognition: a survey. In: Proceeding of the International Conference on Information and Communication Technology for the Muslim World (ICT4M), pp. D37–D40. Jakarta, Indonesia (2010)
8. Hosny, I., Abdou, S., Fahmy, A.: Using advanced hidden Markov models for online Arabic handwriting recognition. In: Proceeding of the First Asian Conference on Pattern Recognition (ACPR), pp. 565–569 (2011)
9. Potrus, M.Y., Ngah, U.K., Sakim, H.A.M.: An effective segmentation method for single stroke online cursive Arabic words. In: Proceeding of the International Conference on Computer Applications and Industrial Electronics (ICCAIE, 2010), pp. 217–221 (2010)
10. Abuzaraida, M.A., Zeki, A.M., Zeki, A.M.: Recognition techniques for online Arabic handwriting recognition systems. In: Proceeding of the International Conference on Advanced Computer Science Applications and Technologies (ACSAT2012). Kuala Lumpur, Malaysia (2012)

11. Plamondon, R., Srihari, S.N.: Online and off-line handwriting recognition: a comprehensive survey. In: IEEE Trans. Pattern Anal. Mach. Intell. **22**, 63–84 (2000)
12. Abuzaraida, M.A., Zeki, A.M., Zeki, A.M.: Online database of Quranic handwritten words. J. Theor. Appl. Inf. Technol. **62** (2014)
13. Abuzaraida, M.A., Zeki, A.M., Zeki, A.M.: Online recognition system for handwritten Arabic mathematical symbols. In: Proceeding of the Second International Conference on Advanced Computer Science Applications and Technologies (ACSAT2013). Kuching, Malaysia (2013)
14. Razzak, M.I., Anwar, F., Husain, S.A., Belaid, A., Sher, M.: HMM and fuzzy logic: a hybrid approach for online Urdu script-based languages' character recognition. Knowledge-Based Syst. **23**, 914–923 (2010)
15. Harifi, A., Aghagolzadeh, A.: A new pattern for handwritten Persian/Arabic digit recognition. World Acad Sci Eng Technol **3**, 249–252 (2005)
16. Abuzaraida, M.A., Zeki, A.M., Zeki, A.M.: Problems of writing on digital surfaces in online handwriting recognition systems. In: Proceeding of the 5th International Conference on Information and Communication Technology for the Muslim World (ICT4M). Rabat, Morocco, pp. 1–5 (2013)
17. Loader, C.: Local Regression and Likelihood, vol. 47. Springer, New York (1999)
18. David, D., Thomas, P.: Algorithms for the reduction of the number of points required to represent a digitized line or its caricature. Cartographica Int. J. Geogr. Inf. Geovisualization **10**, 112–122 (1973)
19. Herbert, F.: Computer processing of line-drawing images. ACM Comput. Surv. **6**, 57–97 (1974)
20. Abuzaraida, M.A., Zeki, A.M., Zeki, A.M.: Feature extraction techniques of online handwriting Arabic text recognition. In: Proceeding of the 5th International Conference on Information and Communication Technology for the Muslim World (ICT4M), pp. 1–7. Rabat, Morocco (2013)
21. Needleman, S.B., Wunsch, C.D.: A general method applicable to the search for similarities in the amino acid sequence of two proteins. J. Mol. Biol. **48**, 443–453 (1970)
22. Durbin, R., Wddy, S., Korgh, A., Mitchison, G.: Biological Sequence Analysis: Probabilistic Models of Proteins and Nucleic Acids. Cambridge University Press, Cambridge (1998)
23. Jones, N.C., Pevzner, P.A.: An introduction to bioinformatics algorithms, illustrated ed. Massachusetts Institute of Technology Press, Cambridge, MA/London (2004)

Chapter 6
Parallelization of Simultaneous Algebraic Reconstruction Techniques for Medical Imaging Using GPU

M.A. Agmalaro, M. Ilyas, A.D. Garnadi and S. Nurdiati

Abstract This paper presents SCILAB implementation of Simultaneous Algebraic Reconstruction Technique (SART) which is one of a class algorithm iterative method for discretization of inverse problems in medical imaging. These so-called row-action methods rely on semi-convergence for achieving the necessary regularization of the problem. However, because iterative algebraic methods are computationally expensive, we improved utilization of computing resources and parallelize this method using CUDA platform implemented in NVIDIA Graphics Processing Units (GPUs) and SCILAB toolbox. Expansion of GPUs made it possible to explore use of SART. In this paper, we provide GPU-based SART implementation modified for a few simplified test problems in medical imaging with parallel beams and fan beams tomography techniques to test the solver.

Keywords Systems of linear equations · Simultaneous algebraic reconstruction methods · Parallelization · Medical imaging

6.1 Introduction

As a group of algorithms for reconstructing 2D and 3D images in certain imaging techniques, iterative reconstruction method have been used for decades to compute solutions of inverse problems such as in medical imaging, geophysics, material science, and many other disciplines [1]. For computerized tomography (CT) imaging, one of iterative reconstructions algorithm, simultaneous algebraic reconstruction techniques (SART) has had a major impact because can generates a good reconstruction in just one iteration where the projection data is limited [1–3].

M.A. Agmalaro (✉) · S. Nurdiati
Department of Computer Science, Bogor Agricultural University, Bogor, Indonesia
e-mail: agmalaro@ipb.ac.id

M. Ilyas · A.D. Garnadi · S. Nurdiati
Department of Mathematics, Bogor Agricultural University, Bogor, Indonesia

A.-R. Ahmad et al. (eds.), *Proceedings of the International Conference on Computing, Mathematics and Statistics (iCMS 2015)*,
DOI 10.1007/978-981-10-2772-7_6

This method uses all the rows simultaneously in one iteration. This paper presents SART to solve large linear systems of the form

$$Ax \cong b,\ A \in \mathbb{R}^{m \times n} \tag{6.1}$$

used in tomography and many other inverse problems. We assume that the elements of matrix A are nonnegative and contain no zero rows or columns. There are no restrictions on the dimensions.

We can write the system of linear equations with M equations with N unknown variables

$$\mathbf{a}_i^T \mathbf{x} = b_i, \quad i = 1, 2, \ldots, M \tag{6.2}$$

with $\mathbf{a}_i$ and $\mathbf{x}$ are column vectors. Each of M equation will be called hyperplane in N-dimension Euclidean space, $\mathbb{R}^N$.

6.2 Parallel Computing

The idea of parallel computing is to share a calculation to some processing unit and calculate it in each processing unit independently [4]. The restriction of this technique is the fully sequential method, for example a calculation that is based on previous calculation. In this case, it is based on programmer judgment when and where parallel computing will be used.

Parallel computing used in this paper is based on NVIDIA GPU. NVIDIA is a well-known graphic cards manufacturer and with its most recent innovation, graphic processing unit (GPU) which can be accessed by CUDA (Compute Unified Device Architecture), programmer can easily use GPU to calculate some mathematical problems. Programming language used in CUDA is similar to C/C++ with some specific instruction to access the task to processing unit.

Parallel computing with CUDA could be used with sciGPGPU package, a toolbox in SCILAB. With this package, programmer can access CUDA commands along with SCILAB command, which makes calculation easier and more flexible [5].

To evaluate how parallel algorithm is performed against sequential algorithm, we need to calculate speedup measurement. It is calculated by dividing time elapsed for sequential algorithm with parallel algorithm. The result can be perceived as how many calculations performed in parallel algorithm when sequential algorithm done single calculation.

$$S = \frac{T_{\mathrm{S}}}{T_{\mathrm{P}}} \tag{6.3}$$

where S is speedup measurement, T_S is time elapsed for sequential algorithm, and T_P is time elapsed for parallel algorithm.

6.3 Simultaneous Algebraic Reconstruction Techniques

In this section, we provide an introduction to the SART [2, 3]. SART is one of the methods used in commercial machine. This iterative algorithm presented by Anders Andersen and Avinash Kak in 1984 in their paper "*Simultaneous Algebraic Reconstruction Technique (SART): A Superior Implementation of ART,*" which is claimed more superior to standard algebraic reconstruction technique (ART).

From Sect. 6.1, we have the system of linear equations $Ax = b$. The objective of the reconstruction technique is to find a solution x of this system (Fig. 6.1). The core of SART method is reconstruction technique with row-action method that treats the equation use all the rows of a simultaneously in one at a time during the iterations. Here, we have the assumption that all elements in A are nonnegative. Even if this assumption is not satisfied, we still obtain the same convergence analysis without any difficulty.

SART is a method used to find one solution, which depends on the initial guess defined as $A_{i,+}$ which is the summation of all elements in the ith row, and $A_{+,j}$ is the summation of all elements in the jth column, shown in Fig. 6.2.

$$\begin{aligned} A_{i,+} &= \sum_{j=1}^{N} A_{i,j} \quad \text{for } i = 1, \ldots, M, \\ A_{+,j} &= \sum_{i=1}^{M} A_{i,j} \quad \text{for } i = 1, \ldots, N. \end{aligned} \tag{6.4}$$

Let V be the diagonal matrix with diagonal elements $A_{+,j}$, and W be the diagonal matrix with diagonal elements $A_{i,+}$. The SART method proposed in [2, 3] is

$$x^{k+1} = x^k + \omega V^{-1} A^T W^{-1} (b - Ax^k) \tag{6.5}$$

for $k = 0, 1, \ldots$, where ω is a relaxation parameter in (0, 2), and the starting point is x^0. By this equation, SART method is nearly full sequential algorithm. Therefore, the parallelization could be done only in one iteration, especially in row-action calculation.

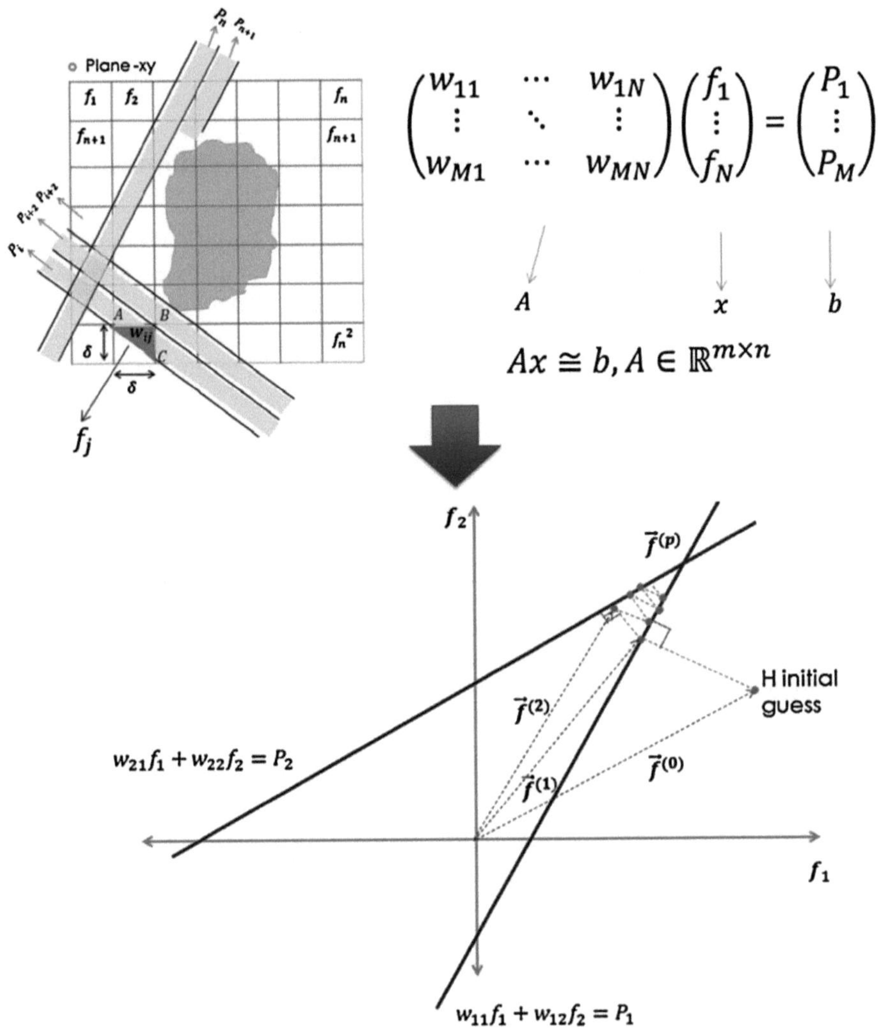

Fig. 6.1 Simple illustration of reconstruction technique

Fig. 6.2 SART method

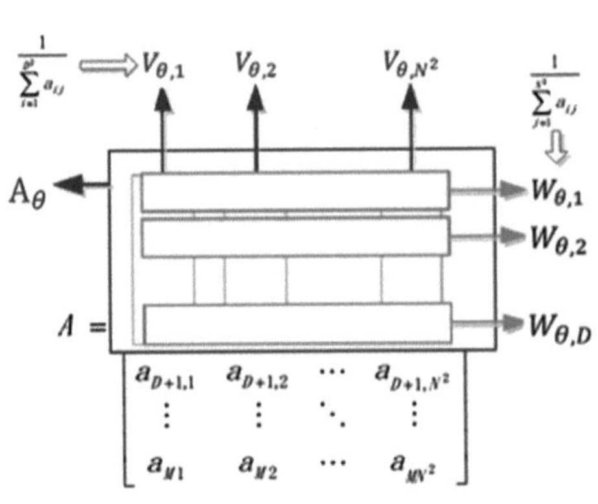

6.4 Computerized Tomography (CT) Imaging

Parallel beams tomography arise from medical sciences and used in medical imaging, X-ray CT scan geometries. We use parallel rays in some degrees to calculate the position and the interior of an object (Fig. 6.3).

These techniques also known as first generation, translate-rotate pencil beam geometry [6, 7].

Fan beams tomography also arise from medical sciences and used in medical imaging, X-ray CT scan geometries. In this technique, we use point beam with fan-like forms rays through an object. These techniques also known as second generation, translate-rotate fan beam geometry (Fig. 6.4).

The object used in the experiment is modified Shepp Logan phantom with the discretization [6, 7] shown in Fig. 6.5. Table 6.1 gives characteristics of matrix used in experiment.

Both parallel beams and fan beams tomography coefficient matrix is based on *AIR Tools – A MATLAB package of algebraic iterative reconstruction methods* [1]. By multiply the desired coefficient matrix with modified Shepp Logan phantom, we obtain the RHS vector.

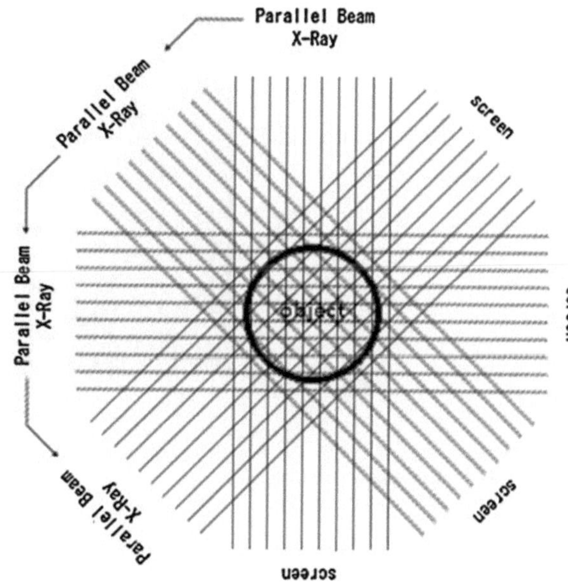

Fig. 6.3 Illustration of parallel beams tomography

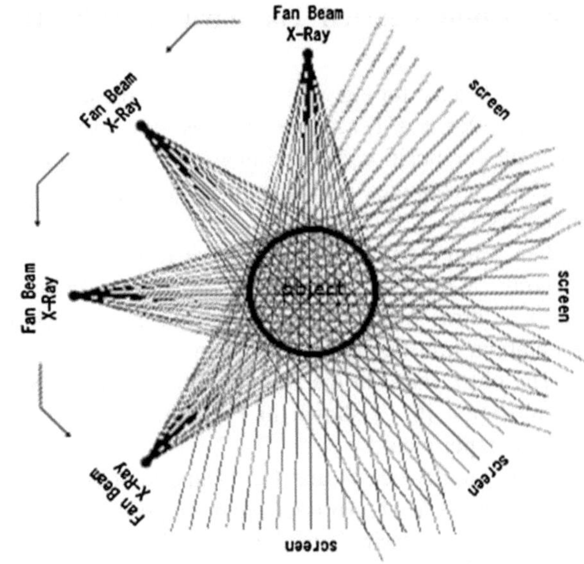

Fig. 6.4 Illustration of fan beams tomography

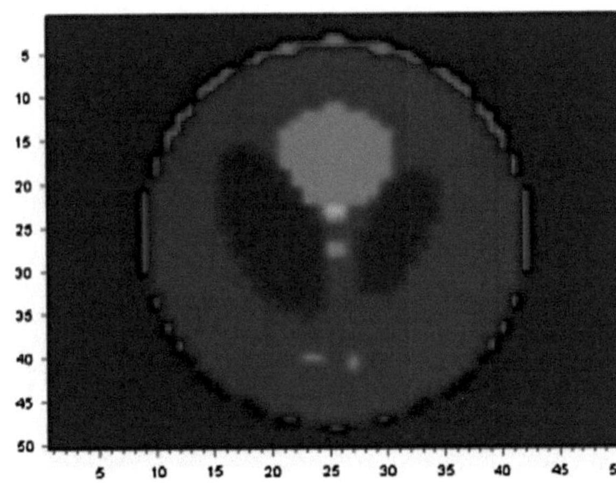

Fig. 6.5 Illustration of modified Shepp Logan phantom

Table 6.1 Matrix characteristics

Characteristics	Parallel beams tomography	Fan beams tomography
Size of coefficient matrix	2700 × 2500	2700 × 2500
Number of nonzero elements of coefficient matrix	114,484 (1.69 %)	132,604 (1.96 %)
Maximum value of coefficient matrix	1.1412	1.4142
Minimum value of coefficient matrix	0.0002069	0.0001421
Maximum value of RHS vector	12.9894	13.8333
Minimum value of RHS vector	0.0004451	0.0004804

6.5 Computational Result

All the computational experiment in this paper is conducted in computer laboratory, Department of Mathematics, Faculty of Mathematics and Natural Sciences, Bogor Agricultural University. Computer specifications used are: Processor Intel Core i5-3317U 1.7 GHz, RAM 4 GB, VGA NVIDIA GeForce GT 635 M. Experiment done in Windows 7 with sciGPGPU 1.3, SCILAB 5.4.2 and CUDA 4.0.

SART produces an iterative solution with given coefficient matrix and right hand side vector. We reproduce approximate solution with increased SART iteration numbers by using original coefficient matrix and right hand side vector. Both tomography techniques produce different results below. Additionally, we test the problems by giving some perturbation to the right hand side vector in the form of random error vector. With an additional vector, we also reproduce approximate solution with increased iteration numbers.

$$\mathbf{b}_{\text{new}} = \mathbf{b} + \eta \frac{||\mathbf{b}||}{||\boldsymbol{e}||}\boldsymbol{e} \tag{6.6}$$

where $\boldsymbol{e}$ is a random error vector and η is some small number (in this experiment we use $\eta = 0.05$).

6.5.1 Parallel Beams Tomography Result

Below (Fig. 6.6a, b) are approximate solution to the parallel beams tomography problems with increased SART iterations, with original and perturbed right hand side vector.

From Table 6.2, with exponentially increased SART iterations, value of $||Ax^* - b||$ become smaller which means better approximation to the exact solution. This is also matched with approximation data resized to matrix form and represented by scaled image shown in Fig. 6.3. On the other hand, perturbed right hand side also give same result, shown by smaller value of $||Ax^* - b||$, even with slower rate of convergence.

Time elapsed by both SART approximation is not quite different, also increased proportionally with more SART iterations. We could also see that time elapsed for both SART iteration using CPU and GPU is not significantly different. The maximum speedup measurements time we get is 1.4065, which we get at 1000 of SART iteration with perturbed right hand side. This probably because that SART iteration is nearly full sequential algorithm so that more iteration needed to observe the speedup time.

(a)

Iteration	Original RHS vector	Perturbed RHS vector
10 SART iterations	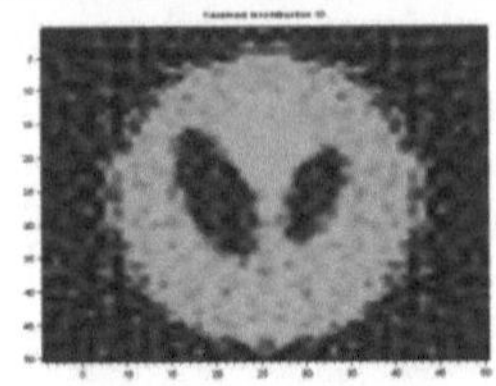	
100 SART iterations	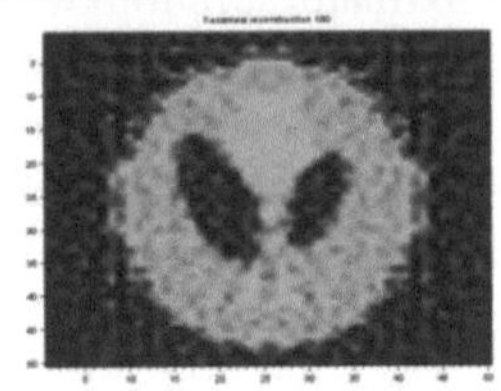	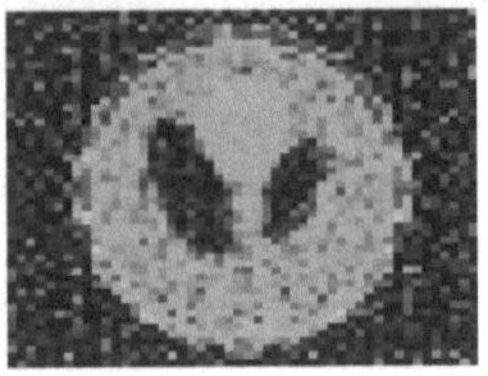

(b)

Iteration	Original RHS vector	Perturbed RHS vector
1000 SART iterations	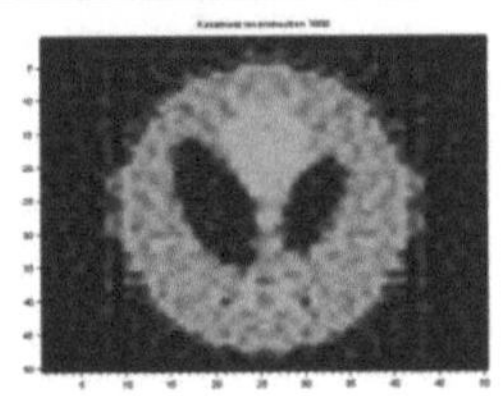	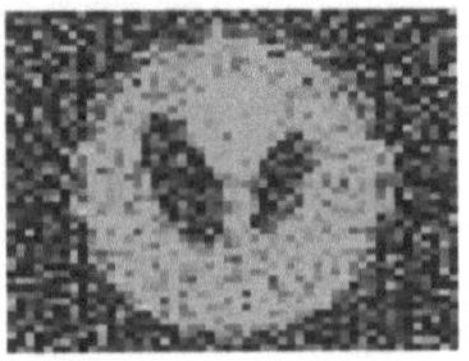

Fig. 6.6 **a**, **b** Approximate solution figure with exponentially increased SART iterations

Table 6.2 Approximate solution norm and time elapsed for CPU and GPU process

Iterations	Original RHS vector	Perturbed RHS vector
10 SART iterations	$\|\|Ax^{*} - b\|\| = 39.356$	$\|\|Ax^{*} - b_{new}\|\| = 41.391$
	CPU time = 1.95 s	CPU time = 1.95 s
	GPU time = 1.44 s	GPU time = 1.145 s
100 SART iterations	$\|\|Ax^{*} - b\|\| = 5.639$	$\|\|Ax^{*} - b_{new}\|\| = 11.146$
	CPU time = 14.523 s	CPU time = 14.258 s
	GPU time = 10.218 s	GPU time = 10.171 s
1000 SART iterations	$\|\|Ax^{*} - b\|\| = 1.040$	$\|\|Ax^{*} - b_{new}\|\| = 8.767$
	CPU time = 137.81 s	CPU time = 137.24 s
	GPU time = 97.67 s	GPU time = 97.57 s

6.5.2 Fan Beams Tomography Result

Figure 6.7 shows approximate solution of the fan beams tomography problems with increased SART iterations, with original and perturbed right hand side vector.

From Table 6.3, with exponentially increased SART iterations, value of $\|Ax^{*} - b\|$ become smaller which means better approximation to the exact solution. This is also matched with approximation data resized to matrix form and represented by scaled image shown in Fig. 6.3. On the other hand, perturbed right hand side also give the same result, shown by smaller value of $\|Ax^{*} - b\|$, even with slower rate of convergence.

Time elapsed by both SART approximation is not quite different, also increased proportionally with more SART iterations. We could also see that time elapsed for both SART iteration using CPU and GPU is not significantly different. The maximum speedup measurements time we get is 1.4049, which we get at 1000 of SART iteration with perturbed right hand side. This is probably because that SART iteration is nearly full sequential algorithm so that more iteration is needed to observe the speedup time.

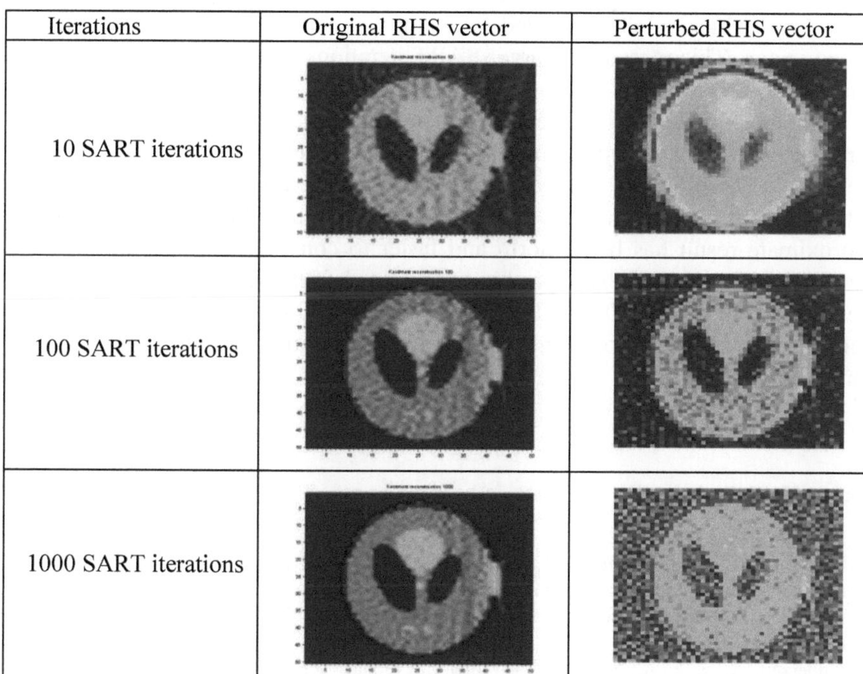

Fig. 6.7 Approximate solution figure with exponentially increased SART iterations

Table 6.3 Approximate solution norm and time elapsed for CPU and GPU process

Iterations	Original RHS vector	Perturbed RHS vector
10 SART iterations	$\|\|Ax^* - b\|\| = 43.576$	$\|\|Ax^* - b_{new}\|\| = 45.786$
	CPU time = 2.53 s	CPU time = 1.996 s
	GPU time = 1.47 s	GPU time = 1.44 s
100 SART iterations	$\|\|Ax^* - b\|\| = 6.784$	$\|\|Ax^* - b_{new}\|\| = 11.739$
	CPU time = 14.242 s	CPU time = 14.242 s
	GPU time = 10.2181 s	GPU time = 10.3 s
1000 SART iterations	$\|\|Ax^* - b\|\| = 1.293$	$\|\|Ax^* - b_{new}\|\| = 8.204$
	CPU time = 138.04 s	CPU time = 137.12 s
	GPU time = 98.77 s	GPU time = 97.6 s

6.6 Conclusion

Both parallel beams tomography and fan beams tomography are widely used in CT scan geometries. This experiment shows that the parallelization of SART to solve system of linear equations is not too good on the speed measurement, but approximate solutions is closer to exact solution shown by smaller value of norm $||Ax^* - b||$, which is also matched with approximation data that represented by scaled image. This means that larger SART iterations give better approximation to exact solution.

With an additional error vector in hand right side vector, which we called perturbed systems, the approximate solution is harder to converge, again, shown by the value of norm $||Ax^* - b||$ even though time elapsed for solving the equation is not significantly different. From the result shown, perturbed fan beams tomography approximate result has better norm and faster to converge to exact solution than parallel beams tomography. Speedup time gained from GPU calculation enhancement only 1.4065 in 1000 SART iteration (perturbed) in parallel beam tomography. This makes SART iteration, which is likely full sequential algorithm, is not working very well on GPU.

References

1. Hansen, P.C., Saxild-Hansen, M.: AIR Tools—A MATLAB package of algebraic iterative reconstruction methods. J. Comput. Appl. Math. (2012)
2. Andersen, A.: Algebraic reconstruction in CT from limited views. IEEE Trans. Med. Imaging **8**, 50–55 (1989)
3. Andersen, A., Kak, A.: Simultaneous algebraic reconstruction technique (SART): a superior implementation of the ART algorithm. Ultrason. Imaging **6**, 81–94 (1984)
4. Grama, A., Gupta, A., Karypis, G., Kumar, V.: Introduction to Parallel Computing. Addison Wesley, London (2003)
5. sciGPGPU QuickStart, http://forge.scilab.org/index.php/p/sciCuda

6. Andersen, A., Kak, A.: Simultaneous algebraic reconstruction technique (SART): a superior implementation of ART. In: Ultrasonic Imaging (1984)
7. Kak, A.C., Slaney, M.: Principles of Computerized Tomographic Imaging. SIAM, Philadelphia (2001)

Chapter 7
Redundancy Elimination in the Future Internet

Ikram Ud Din, Suhaidi Hassan and Adib Habbal

Abstract Information-Centric Networking (ICN), also called Content-Centric Networking, is the most ascendant architecture for the Future Internet. In ICN, network nodes cache contents for a time before forwarding them to the end users. One of the applicable proposals to in-network caching, called Networking Named Content (NNC), suggests that caching contents at every node along the path can improve caching gain in terms of hit rate and content retrieval delay. However, storing content items at every node along the path can increase the chances of redundancy. Some strategies investigated that caching contents at only one intermediate node along the path can reduce the chances of redundancy. In addition, some strategies suggested that storing only popular contents can do better in performance, especially in reduced content redundancy. In this paper, we present a scheme, named Content Placement Mechanism (CPM), based on popularity of the contents that is suitable for redundancy elimination.

Keywords Information-centric networking · Redundancy · Caching · Content placement

7.1 Introduction and Related Work

Unlike the current Internet architecture that was designed for source to destination communication, customers are only interested in actual content rather than its location. For this reason, novel architectures of Information-Centric Networking

I.U. Din (✉) · S. Hassan · A. Habbal
InterNetWorks Research Laboratory, School of Computing,
Universiti Utara Malaysia, Sintok, Malaysia
e-mail: ikram@internetworks.my

S. Hassan
e-mail: suhaidi@uum.edu.my

A. Habbal
e-mail: adib@uum.edu.my

A.-R. Ahmad et al. (eds.), *Proceedings of the International Conference on Computing, Mathematics and Statistics (iCMS 2015)*,
DOI 10.1007/978-981-10-2772-7_7

(ICN) such as Publish-Subscribe Internet Technology (PURSUIT) [1], Publish-Subscribe Internet Routing Paradigm (PSIRP) [2], Network of Information (NetInf) [3], and Content-Centric Networking (CCN) [4] have been defined, which give high priority to effective content dissemination at large scale. Among all these novel architectures, the research community has been considerably attracted by CCN [5].

CCN is an architecture that efficiently distributes popular contents to a possible number of clients [2, 6]. A fundamental characteristic of CCN is cache management with caching strategies, which determines what to cache and where. Content storing schemes play a vital role in CCN's overall performance. Caching is one of the critical issues in the introduction of ICN, i.e., the major issue in ICN is that what content should be recommended for caching and where it should be placed. The key goal of caching contents is to reduce the utilization of bandwidth and minimize the server overhead as well as retrieval delay of contents.

Research interest about ICN gained a vast popularity in the research community after 2009. Caching, which is one of the most vital features of ICN, attained enormous attraction by the researchers, and therefore many caching strategies have been designed so far, the most famous of which are Leave Copy Everywhere (LCE) [7], Leave Copy Down (LCD) [7], Centrality-based caching [8], and Most Popular Content (MPC) [9]. In LCE, the content is stored in all nodes along the path all the way from the node where hit occurred to the subscriber. In LCD, the content is cached only at the node below the one where the content is hit. In Centrality-based caching, the node which has the maximum outgoing interfaces is considered for caching the content; while MPC stores only popular contents, especially when their popularity reaches 5, i.e., if a content is requested five times. For more understanding about content placement, the readers are referred to [10–16]. In this study we will compare our proposed scheme with LCE and MPC because LCE is the default CCN strategy and MPC outperforms the existing ones.

7.2 Proposed Scheme

As in Fig. 7.1, if a request is arrived, it is first checked in the *Pending Interest Table* (PIT) and then forwarded to *Popularity Table* (PT). PT contains the requested content ID, name, and its popularity value. If the popularity value of a content reaches its minimum value (which is 2), i.e., if $C_P \geq 2$, it is forwarded to *Comparison Table* (CT). In contrast to PT, CT has only popular content items; therefore, this table is shared with all neighbor routers. In CT, the counter V_P will check the value of content, if $V_P > sV_P$ and content = mV_P, then it is forwarded to CS. Here, CS means that router which has maximum outgoing interfaces, in this case router R2 in Fig. 7.2.

Here a question arises that if the popularity value of content should never be less than 2 in CT then what is the need of checking content popularity here? This check

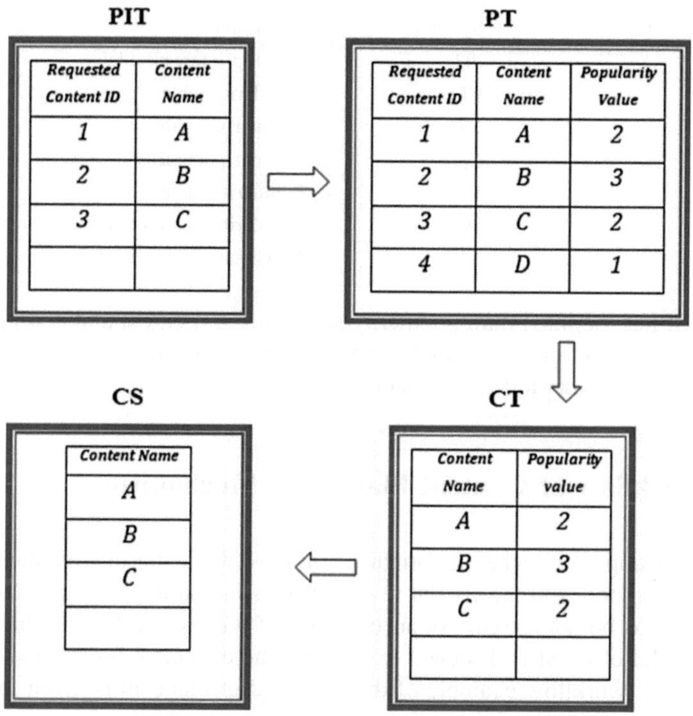

Fig. 7.1 CPM workflow example

Fig. 7.2 A network topology

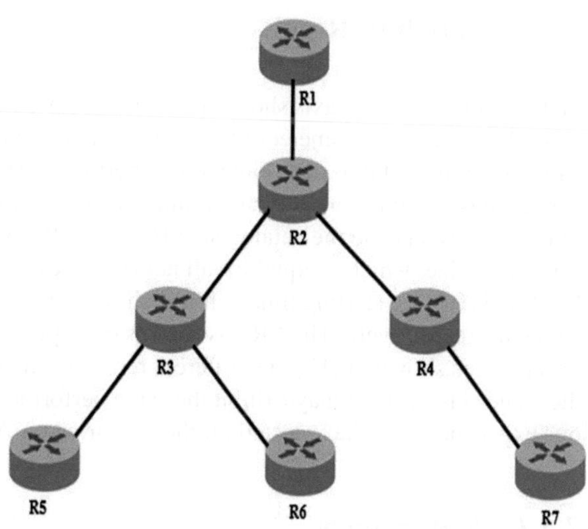

is needed because if, for example, there are three content items in CT and all have different popularity value (definitely greater than 2), then to decide which content item should be forwarded to content store if cache is full. In this case, that content

Table 7.1 Notation

Symbol	Definition
PIT	Pending interest table
PT	Popularity table
CT	Comparison table
C_P	Content popularity
V_P	Popularity value
sV_P	Popularity value of stored contents
mV_P	Maximum popularity value

item which has the maximum popularity value in CT (i.e., content = mV_P) and the value of content is greater than the value of all contents in content store (i.e., $V_P > sV_P$) will be forwarded for caching (Table 7.1).

7.3 Workflow of Content Placement Mechanism

In order to improve cache performance in ICN while basing on the proposed performance model and the critical review of existing strategies, redundancy is the major point to be addressed. As an outcome of addressing ICN caching, a conceptual model of CPM is designed to describe the expected desired and improved situation. The workflow example of the proposed mechanism is given in Fig. 7.1, where its explanation is given in Sect. 7.3.

7.4 Simulation Results

In the given figures, X-axis shows the cache size, while Y-axis represents the ratio of redundancy. The α parameter of the Zipf probability distribution varies between 0.70 and 2.0. For the catalog size, some researchers consider its value 10^4 elements in their simulations, but it is very difficult to imagine the Internet with such limited size [10]. Therefore, we choose the catalog size 10^6 as in [10], which is also a common value. The cache size, which is equal for all nodes, varies between 100 and 1000 elements. On X-axis 0 is the starting time, while 1 shows 100 elements and 10 represents 1000 elements, respectively. The LRU is used as the replacement policy in the comparison because it is shown in [17] that different replacement policies may be combined into the same class as they may exhibit the same performance. For validation purpose, in our experiments we have tested all the scenarios on Abilene topology.

7.4.1 Redundancy

Redundancy is an important metric in measuring the performance of ICN. Figure 7.3 shows the average percentage redundancy of the cached contents in all network nodes. In low popularity scenario (α = 0.70) the percentage of redundant contents was almost 100 % in LCE on all cache sizes (i.e., 100 to 1000). In MPC, this percentage was reduced from 92 % (with cache size 100) to 76 % (with cache size 1000); while in CPM the percentage was a little lower than MPC, i.e., 90 % with cache size 100–72 % with cache size 1000 elements. As the value of α parameter is increased from 0.70 to 2.0 in Fig. 7.4, the percentage of redundant contents is decreased dramatically in CPM. The figure shows that with cache size 100–1000, the percentage value in LCE is almost constant (100 %) for all cache sizes; in MPC, it is reduced from 42 to 18 %; while in CPM, the average percentage was recorded with reduction from 32 to 5 %. Thus, it confirms that using CPM with high popularity and higher cache sizes, the redundancy of the cached contents can be drastically reduced as compared to MPC and LCE.

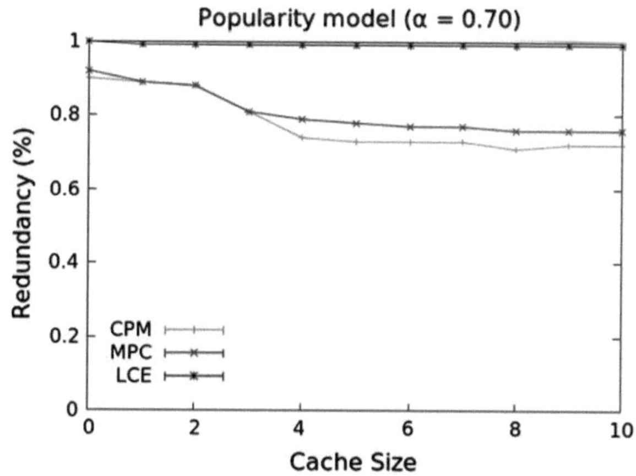

Fig. 7.3 Redundancy with α = 0.70

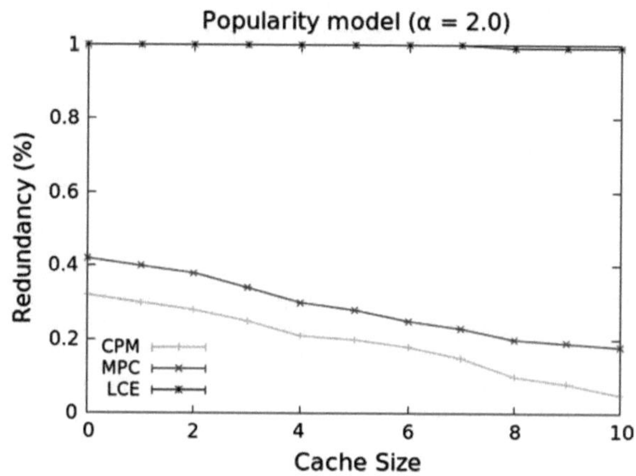

Fig. 7.4 Redundancy with α = 2.0

7.5 Conclusion

In this paper, we presented CPM: a popularity-based content placement scheme for the Future Internet. CPM has been designed by keeping in mind the shortcomings of the existing caching strategies which made them inappropriate for redundancy elimination. We compared our work with the CCN default strategy called LCE and MPC, which dominantly outperforms the existing caching strategies. Our results show that CPM is the best candidate for the elimination of redundancy compared to LCE and MPC.

Acknowledgment This work is sponsored by the InterNetWorks Research Laboratory, School of Computing, Universiti Utara Malaysia.

References

1. FP7 PURSUIT project, http://www.fp7-pursuit.eu/PursuitWeb
2. Jokela, P., Zahemszky, A., Esteve Rothenberg, C., Arianfar, S., Nikander, P.: Lipsin: linespeed publish/subscribe inter-networking. In: ACM SIGCOMM Computer Communication Review, vol. 39, no. 4, pp. 195–206. ACM (2009)
3. Dannewitz, C.: NetInf: an information-centric design for the future internet. In: 3rd GI/ITG KuVS Workshop on the Future Internet (2009)
4. Jacobson, V., Smetters, D.K., Thornton, J.D., Plass, M.F., Briggs, N.H., Braynard, R.L.: Networking named content. In: ACM CoNEXT, pp. 1–12 (2009)
5. Jacobson, V., Smetters, D.K., Thornton, J.D., Plass, M.F., Briggs, N.H., Braynard, R.L.: Networking named content. In: Proceedings of the 5th International Conference on Emerging Networking Experiments and Technologies, pp. 1–12. ACM (2009)
6. CCNx protocol, http://www.ccnx.org
7. Laoutaris, N., Che, H., Stavrakakis, I.: The lcd interconnection of lru caches and its analysis. Perform. Eval. **63**(7), 609–634 (2006)
8. Chai, W.K., He, D., Psaras, I., Pavlou, G.: Cache less for more in information-centric networks (extended version). Comput. Commun. **36**(7), 758–770 (2013)
9. Bernardini, C., Silverston, T., Festor, O.: Mpc: popularity-based caching strategy for content centric networks. In: IEEE International Conference on Communications (ICC), pp. 3619–3623. IEEE (2013)
10. Bernardini, C.: Stratégies de Cache basées sur la popularité pour Content Centric Networking. PhD Thesis, University of Lorraine, France (2015)
11. Wang, J.: A survey of web caching schemes for the Internet. ACM SIGCOMM Comput. Commun. Rev. **29**(5), 36–46 (1999)
12. Wessels, D., Claffy, K.: Internet Cache Protocol (ICP), Version 2, RFC 2186, September 1997, http://tools.ietf.org/html/rfc2186
13. Wessels, D., Claffy, K.: "Application of Internet Cache Protocol (ICP), Version 2", RFC 2187, September 1997, http://tools.ietf.org/html/rfc2187
14. Michel, S., Nguyen, K., Rosenstein, A., Zhang, L., Floyd, S., Jacobson, V.: Adaptive web caching: towards a new global caching architecture. Comput. Netw. ISDN Syst. **30**(22), 2169–2177 (1998)
15. Nygren, E., Sitaraman, R.K., Sun, J.: The akamai network: a platform for high-performance internet applications. ACM SIGOPS Operating Syst. Rev. **44**(3), 2–19 (2010)

16. Krenc, T., Hohlfeld, O., Feldmann, A.: An internet census taken by an illegal botnet: a qualitative assessment of published measurements. ACM SIGCOMM Comput. Commun. Rev. **44**(3), 103–111 (2014)
17. Rosensweig, E.J., Menasche, D.S., Kurose, J.: On the steady-state of cache networks. In: INFOCOM, 2013 Proceedings IEEE, pp. 863–871. IEEE (2013)

Chapter 8
Solving the Randomly Generated University Examination Timetabling Problem Through Domain Transformation Approach (DTA)

Siti Khatijah Nor Abdul Rahim, Andrzej Bargiela and Rong Qu

Abstract Amongst the wide-ranging areas of the timetabling problems, educational timetabling was reported as one of the most studied and researched areas in the timetabling literature. In this paper, our focus is the university examination timetabling. Despite many approaches proposed in the timetabling literature, it has been observed that there is no single heuristic that is able to solve a broad spectrum of scheduling problems because of the incorporation of problem-specific features in the heuristics. This observation calls for more extensive research and study into how to generate good quality schedules consistently. In order to solve the university examination timetabling problem systematically and efficiently, in our previous work, we have proposed an approach that we called a Domain Transformation Approach (DTA) which is underpinned by the insights from Granular Computing concept. We have tested DTA on some benchmark examination timetabling datasets, and the results obtained were very encouraging. Motivated by the previous encouraging results obtained, in this paper we will be analyzing the proposed method in different aspects. The objectives of this study include (1) To test the generality/applicability/universality of the proposed method (2) To compare and analyze the quality of the schedules generated by utilizing Hill Climbing (HC) optimization versus Genetic Algorithm (GA) optimization on a randomly

S.K.N. Abdul Rahim (✉)
Universiti Teknologi Mara, Tapah, Perak, Malaysia
e-mail: sitik781@perak.uitm.edu.my; khyx8skn@nottingham.edu.my

S.K.N. Abdul Rahim
University of Nottingham (Malaysia Campus), Selangor, Malaysia

A. Bargiela · R. Qu
University of Nottingham (UK Campus), Nottingham, UK
e-mail: abb@cs.nott.ac.uk

R. Qu
e-mail: rxq@cs.nott.ac.uk

A.-R. Ahmad et al. (eds.), *Proceedings of the International Conference on Computing, Mathematics and Statistics (iCMS 2015)*,
DOI 10.1007/978-981-10-2772-7_8

generated benchmark. Based on the results obtained in this study, it was shown that our proposed DTA method has produced very encouraging results on randomly generated problems. Having said this, it was also shown that our proposed DTA method is very universal and applicable to different sets of examination timetabling problems.

Keywords Examination scheduling · Domain transformation approach · Granular computing · Randomly generated problem

8.1 Introduction

Timetabling can be defined as a process of generating timetables or schedules that contain information about some events and the times at which they are planned to take place. Timetabling is normally considered a tiresome and laborious task. In some organizations, the personnel responsible in preparing the timetables usually do it manually and mostly using a trial-and-error approach.

Amongst the wide-ranging areas of the timetabling problems, educational timetabling was reported as one of the most studied and researched areas in the timetabling literature. Example of educational timetabling includes school timetabling (course/teacher timetabling), course timetabling, examination timetabling, etc.

In this paper, our focus is the university examination timetabling. Many universities in the world nowadays offer modular courses (across faculties) to their students, resulting in very strong interdependencies between students and exams data (many–many relationships in the data files). This has contributed to the complexities of the problem, and there were many claims in the literature that university timetabling problem as an NP complete problem [1, 2]. Besides satisfying hard constraints which is necessary to make the timetables feasible, the general objective of examination timetabling that is widely used in the literature is to reduce the cumulative inconvenience implied by the temporal proximity of consecutive exams taken by individual students. Even with the extensive research done in this area, there is still room for improvements for the state of the art methods especially in making sure the methods always able to reproduce good quality timetables consistently.

In the university examination timetabling research, the quality of the timetable is normally referred to as a 'cost' which is measured using objective function proposed by Carter [3] as below:

$$\frac{1}{T}\sum_{i=1}^{N}\sum_{j=i+1}^{N} s_{ij} w_{|pj-pi|} \tag{8.1}$$

where N is the number of exams, S_{ij} is the number of students enrolled in both exam i and j, pj is the time slot where exam j is scheduled, pi is the time slot where exam i is scheduled and T is the total number of students. According to this cost function, a student taking two exams that are $|pj - pi|$ slots apart, where $|pj - pi| = \{1, 2, 3, 4, 5\}$, leads to a cost of 16, 8, 4, 2, and 1, respectively.

8.2 Review of the Proposed Approach

In order to solve the university examination timetabling problem systematically and efficiently, we have proposed an approach that we called a *Domain Transformation Approach* (*DTA*) which was described in detail in our previous publications [4, 5]. DTA is an approach which is underpinned by the insights from Granular Computing concept [6–9, 10]. In the proposed DTA, the examination scheduling problem is transformed into smaller sub problems, therefore is easier to be solved systematically and efficiently.

Our previous works [4, 5, 11–13] have tested DTA on some benchmark examination timetabling datasets, and the results obtained were very encouraging. We have managed to produce good and feasible quality timetables on all datasets in the experiments. Besides, our proposed DTA which avoided the utilization of random-search has resulted in the generation of consistent timetables (able to be reproducible) and shown a very deterministic optimization pattern.

8.3 Objectives of the Study

Motivated by the encouraging results obtained as stated in the previous section, we will be analyzing the proposed method in different aspects. In this section, we will list the objectives of doing the study in this paper. Each objective is elaborated in brief as below.

(a) To test the generality/applicability/universality of the proposed method by testing it on a randomly generated benchmark datasets.
Note: In the previous works, the proposed DTA was tested on benchmark datasets that were extensively tested by other researchers. However, it is agreed with [14], that when certain benchmark datasets are relied upon to evaluate an algorithm, the resulting algorithm could be inclined towards the criteria of the benchmark datasets. Therefore, by testing on a randomly generated datasets, the universality of the method could be analyzed.

(b) To compare and analyze the quality of the schedules generated by utilizing Hill Climbing (HC) optimization versus Genetic Algorithm (GA) optimization on a randomly generated benchmark datasets (with the hypothesis that HC will maintain to outperform GA based on previous outcome).

Note: Minimization of the schedule cost by slot swapping is a Hill Climbing (HC) procedure which was implemented in the proposed DTA. Good quality feasible examination schedules were generated successfully using this method. In order to analyze whether a global search procedure could improve the quality of the schedules generated, we have also implemented Genetic Algorithm search procedure and incorporated it into the proposed DTA framework. It was observed that for Toronto benchmark datasets, HC has outperformed GA on all instances [12, 5]. Hence, we will see whether the good performance of HC is consistent on the newly tested datasets.

8.4 Domain Transformation Approach (DTA)

In our proposed DTA, after retrieval of datasets, data verification and standardization is done to alleviate the initial problem of dataset and format variety and convert it into a standard format that will be used as an input to the preprocessing stage. The preprocessing of data and constraints from the original problem space will provides important information granules which consecutively provide valuable information for scheduling. The new aggregated data (which was explained in greater detail in [4, 12, 13]) will reduce the subsequent cross-checking and cross-referencing in the original data thus expediting scheduling stage.

In the scheduling stage, which utilizes an allocation method using a Graph Colouring Heuristic coupled with a backtracking procedure (a modified version of Carter et al. (1996)'s backtracking approach [3]), is adopted as a basic scheduling process [4, 12, 13]. It is expected to produce only feasible solutions with a total number of slots that will satisfy the minimum requirements given in the problem. The last stage in the proposed algorithm is the optimization stage. This stage involves three procedures: minimization of the overall slots conflict, minimization of schedule cost by slot swapping and minimization of schedule cost by reassigning exams.

Our previous proposed examination timetabling method has generated very encouraging results (i.e., good quality timetables) on all examination timetabling benchmark datasets tested [4, 5, 11–13]. However, in this paper our aim is to further analyze the proposed method as stated in the objectives section and for this purpose we will be focusing on the procedure to minimize the schedule cost by slots swapping. This is because it was observed that out of the three optimization procedures, the slot swapping has recorded the most significant reduction of the cost.

8.4.1 Minimization of the Schedule Cost by Slot Swapping

Optimizing the feasible examination schedule obtained by the scheduling stage by doing slots swapping is explicitly focused on minimization of the cost function.

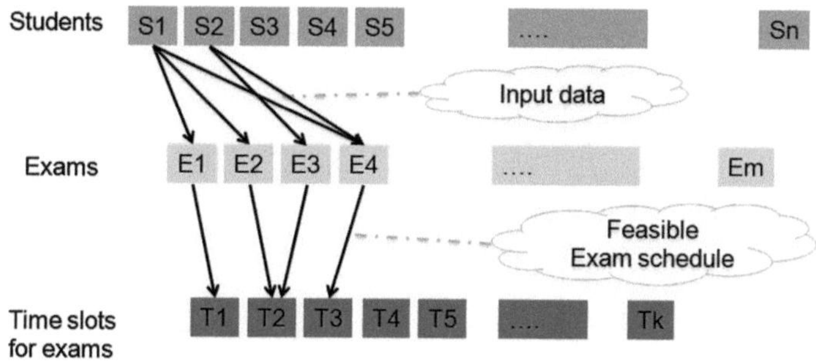

Fig. 8.1 An example of a feasible examination schedule

This process is also known as permutations of exams slots. Below, with the aid of some diagrams, we will demonstrate and briefly explain how slot swapping can reduce the schedule cost and hence increase the quality of the schedule.

As can be seen in Fig. 8.1, there are a few students: S1, S2, S3, S4, S5 … Sn and a few exams: E1, E2, E3, E4 … Em together with a few time slots: T1, T2, T3, T4 … Tk. In this example, student S1 has registered for exams E1, E2 and E4; and student S2 has registered for E3 and E4. Therefore, exams E1, E2 and E4 are the set of conflicting exams for student S1 and because of this, they cannot be assigned to the same time slots. The diagram below shows that these three exams are not assigned to the same time slot (they are assigned to time slots T1, T2 and T3 respectively) and thus this is considered a feasible examination schedule.

According to the above example, although student S2 has a feasible examination schedule, the timetable does not satisfy the soft constraint in terms of putting a gap between one exam and the next exam that student will have to sit. This is not necessary but satisfying this would improve the schedule quality by benefiting the student, since it allows the student to have more revision time between exams. Thus, if exam E3 is now reassigned to time slot T1, the quality of the schedule can be improved, as illustrated in Fig. 8.2.

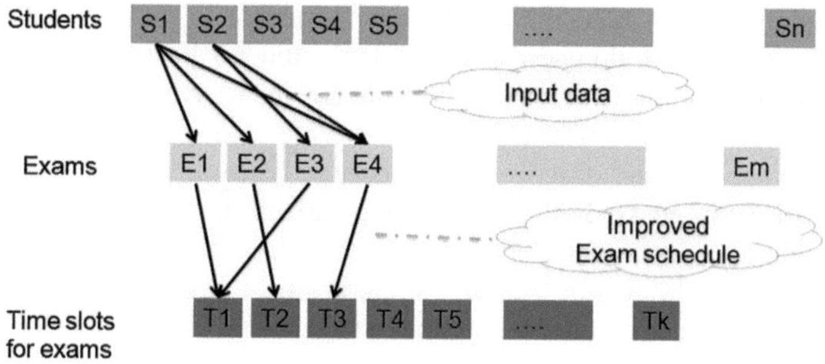

Fig. 8.2 An example of an improved examination schedule

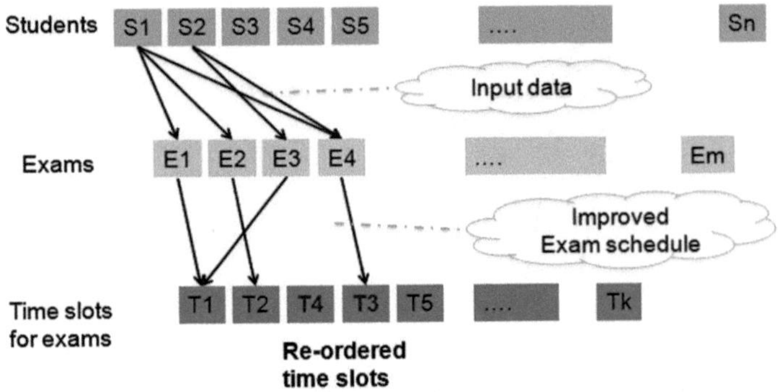

Fig. 8.3 An improved examination schedule after optimization (permutations of exam slots)

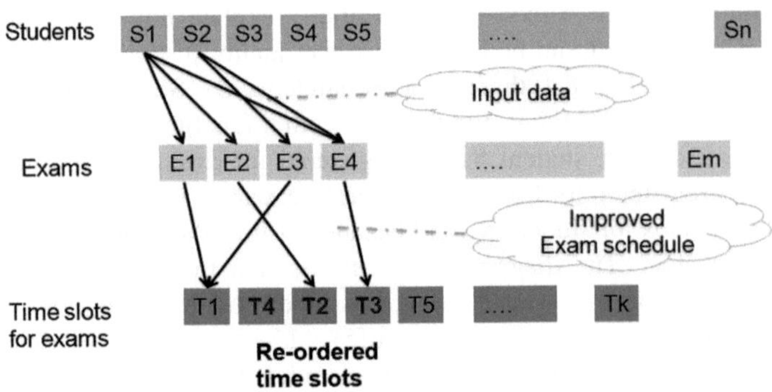

Fig. 8.4 Re-ordered time slots via permutations of slots with greater effect

Figure 8.3 shows how the permutation of exam slots has changed the original ordering of the slots in Fig. 8.2, and consequently an improved schedule has been generated. By this permutation, a time slot has been added between time slot T2 and T3, and thus giving extra time for the students to do their revision.

However, adding an extra time slot between T1 and T2 will have a greater effect as illustrated in Fig. 8.4 than adding it between time slot T2 and T3 as illustrated in the previous diagram. In this new example, both students have more time between exams as compared to the previous example, hence the newly generated timetable is considered a better quality timetable.

Following are the brief descriptions of the two types of permutations of slots.

Permutations of Slots: Hill Climbing Optimization. Shuffling the exam slots has the potential to reduce the cost of the schedule. This can be done by doing permutations of exam slots in the spread matrix [4, 5, 11–13]. Spread matrix will provide the information on how many students taking an exam from slot '*i*' and '*j*'. Permutations process involves the shuffling of slots or columns as block shifting and swapping. Each slot will be swapped with another slot in the provisional swapping stage, where by the Carter cost (1) is evaluated. The swap will be remembered and the exam proximity matrix will be updated if the swap resulted in a cost reduction. This kind of optimization procedure is called a greedy Hill Climbing (HC) because if a swap operation manages to improve the cost function, the swap is straight away accepted and the exam slots were rearranged accordingly. A few repetitions of block shift and swapping are done, besides restarting the optimization from several initial orderings of exam slots, in order to ensure that the greedy optimization does not lead to local optima.

Permutations of Slots: Genetic Algorithm Optimization. Realizing that the above proposed Hill Climbing which is a kind of a local search procedure may not direct the search to global optima, we have also implemented Genetic Algorithm (GA) to be integrated (substitute HC) into the proposed DTA. Having said that, this indirectly indicates that DTA is a flexible framework where different kinds of search can be used in one of the stage while maintaining other procedures (before and after the search) in timetable generation. Though GA is considered an 'old-fashioned' optimization procedure, but it has been confirmed that hybridizations of GA with some local search have led to some success in this area [15].

8.5 Experimentations, Results and Conclusion

We have tested the proposed DTA on the randomly generated university examination timetabling problem which was obtained from the University of Nottingham website (http://www.cs.nott.ac.uk/~rxq/data.htm). The following table (Table 8.1) listed the characteristics of each problem.

The datasets were used in our solution as an input and we have obtained the following results as summarized in Table 8.2. We have observed that HC has outperformed GA in all problems as predicted according to the results obtained in the previous study [5, 12]. Even though HC is a local search procedure compared to GA, the repetitions of block shift and swapping, and restarting the optimization from several initial orderings of exam slots, have managed to ensure that the greedy optimization does not end in local optima.

Table 8.1 Characteristics of the randomly generated problems (small problems). **a** Name of dataset. **b** No of exams. **c** No of students. **d** No of enrollments. **e** Conflict density. **f** Required no of slots

(a)	(b)	(c)	(d)	(e) (%)	(f)
SP5	80	66	194	7	15
SP10	100	100	359	11	15
SP15	80	81	314	17	15
SP20	80	83	344	19	15
SP25	80	119	503	26	15
SP30	80	126	577	32	15
SP35	100	145	811	36	19
SP40	81	168	798	42	19
SP45	80	180	901	47	19

Table 8.2 Results for randomly generated University examination timetabling problems using hill climbing (HC) and genetic algorithm (GA) in the optimization stage of DTA

Problem	HC	GA
SP5	3.4242	4.1212
SP10	10.8900	12.0000
SP15	16.0617	16.8025
SP20	18.9277	20.0723
SP25	23.5042	25.7899
SP30	32.4762	33.4365
SP35	45.2345	47.6552
SP40	27.2083	28.5714
SP45	29.9778	33.1889

Based on the results obtained in this study, it was also shown that our proposed DTA method has produced very encouraging results on randomly generated problems. Having said this, it was also shown that our proposed DTA method is very universal and applicable to different sets of examination timetabling problems. Besides generating good quality timetables, the DTA always produced consistent performance and demonstrate deterministic optimization pattern on all problems.

References

1. Cooper, T.B., Kingston, J.H.: The Complexity of Timetable Construction Problems. Springer, Heidelberg (1996)
2. Even, S., Itai, A., Shamir, A.: On the complexity of time table and multi-commodity flow problems. In: 16th Annual IEEE Symposium on Foundations of Computer Science (1975)
3. Carter, M.W., Laporte, G., Lee, S.Y.: Examination time-tabling: algorithmic strategies and applications. J. Oper. Res. Soc. 373–383 (1996)
4. Rahim, S.K.N.A., Bargiela, A., Qu, R.: Domain transformation approach to deterministic optimization of examination timetables. Artif. Intell. Res. **2**(1), 122 (2012)
5. Rahim, S.K.N.A., Bargiela, A., Qu, R.: A study on the effectiveness of genetic algorithm and identifying the best parameters range for slots swapping in the examination scheduling. In: iSMSC (2013)
6. Bargiela, A., Pedrycz, W.: Granular Computing: An Introduction. Springer Science & Business Media, Berlin (2003)
7. Bargiela, A., Pedrycz, W., Hirota, K.: Logic-based granular prototyping. In: Proceedings of 26th Annual International Computer Software and Applications Conference P IEEE (2002)
8. Bargiela, A., Pedrycz, W., Hirota, K.: Granular prototyping in fuzzy clustering. IEEE Trans. Fuzzy Syst. 697–709 (2004)
9. Bargiela, A., Pedrycz, W.: Toward a theory of granular computing for human-centered information processing. IEEE Trans. Fuzzy Syst. 320–330 (2008)
10. Pedrycz, W., Smith, M.H., Bargiela, A.: A granular signature of data. In: 19th IEEE International Conference of the North American. Fuzzy Information Processing Society (2000)
11. Rahim, S.K.N.A., Bargiela, A., Qu, R.: Granular modelling of exam to slot allocation. In: ECMS (2009)

12. Rahim, S.K.N.A., Bargiela, A., Qu, R.: Hill climbing versus genetic algorithm optimization in solving the examination timetabling problem. In: 2nd International Conference on Operations Research and Enterprise Systems, ICORES 2013 (2013)
13. Rahim, S.K.N.A., Bargiela, A., Qu, R.: Analysis of backtracking in university examination scheduling. In: ECMS (2013)
14. Lewis, R.: A survey of metaheuristic-based techniques for university timetabling problems. OR Spectrum **30**(1), 167–190 (2008)
15. Qu, R., et al.: A survey of search methodologies and automated system development for examination timetabling. J. Sched. 55–89 (2009)

Chapter 9
Synchronize Speed Control for Multiple DC Motors Using Linear Quadratic Regulator

M.S. Saealal, A.T. Ramachandran, M.F. Abas and N. Md. Saad

Abstract Direct Current (DC) motors have been extensively used in many industrial applications. Therefore, the control of the speed of a DC motor is an important issue and has been studied since the early decades in the last century. Proportional-Integrated-Derivative (PID) controller is one of the methods to control synchronize multiple motor. But the settling time is slow for this controller and its affect the performance too. This paper presents Linear Quadratic Regulator (LQR) controller which is another method of control synchronize multiple motor to reduce the settling time of the multiple motor using MATLAB software. The settling time result of the LQR controller is compared with PID controller.

Keywords Linear quadratic regulator · LQR · Proportional-Integrated-Derivative · DC motor · Synchronize multiple motor

9.1 Introduction

DC motor can be control by directly by hardware or software as well. The soft-ware that used to control the motor needs a computers which are big enough and not all are affordable, thus hardware controls more encouraged. In hardware if there is programmable device available then it is chosen because it can be modelled according to the prerequisites of the user. This type of system can be controlled using PID, Fuzzy, LQR and others. Based on previous project, PID controller is

M.S. Saealal (✉)
Faculty of Engineering Technology, Universiti Teknikal Malaysia Melaka,
76100 Durian Tunggal, Melaka, Malaysia
e-mail: salihin@utem.edu.my

A.T. Ramachandran · M.F. Abas · N.Md.Saad
Faculty of Electrical and Electronics, Universiti Malaysia Pahang,
26600 Pekan, Pahang, Malaysia

A.-R. Ahmad et al. (eds.), *Proceedings of the International Conference on Computing, Mathematics and Statistics (iCMS 2015)*,
DOI 10.1007/978-981-10-2772-7_9

used for synchronization purpose which results in a slow settling time. This is because PID takes some time to stabilizing the system.

In this project Linear Quadratic Regulator (LQR) controller is introduced in order to control four DC motor speeds as required in low settling time compare to PID. MATLAB/SIMULINK software is used to plan and fine-tune the LQR controller and be computer-generated to mathematical model of the DC motor. Linear Quadratic Regulator (LQR) is a controller where it uses methods which control the motor. LQR is theory of best control concerned with operating a dynamic system at low cost.

The advantages of using LQR controller is an easy way to develop and increase the precision of the state variables by estimating the state. Best characteristic of the LQR controller as per compare to pole position is that instead of having to specify where n eigenvalues must be placed a set of performance weighting are specified that would have more intuitive request. The result will be a control which is guaranteed to be stable.

9.2 DC Motors

Figure 9.1 presents an equivalent electrical circuit of a DC motor. It can be illustrated by a voltage supply (V_a) over the coil of an armature. The electrical equivalent of an armature coil could described by resistance (R) in series form with an inductance (L) in series form with an induced voltage (V_c) that oppose the voltage source. The rotation of an electrical coil is then generating the induced voltage between the permanent magnets which are fixed flux lines. This voltage is usually concerned as the electromotive force, Emf (Table 9.1).

The constant R, K, J, B are calculated using the datasheet given by the manufacturer while L is measured by using the inductance meter.

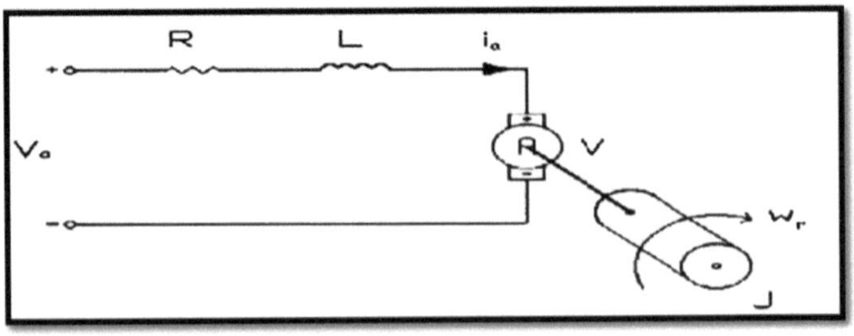

Fig. 9.1 Condition when wheel zigzag gets over block

Table 9.1 Constant for motor

Parameters	Values	Parameters
Electric resistance, *R*	5.36 Ω	Electric resistance, *R*
Electric inductance, *L*	0.00112 H	Electric inductance, *L*
Electromotive force constant, *K*	0.01467 Nm A^{-1}	Electromotive force constant, *K*
Moment of inertia of the rotor, *J*	0.00602 kg m^2	Moment of inertia of the rotor, *J*
Damping ratio of the inertia of the rotor, *B*	0.0022005 mNms rad^{-1}	Damping ratio of the inertia of the rotor, *B*

9.2.1 *Electrical Characteristic*

From the Fig. 9.1, the following equations are based on Kirchhoff's Law and Newton's Law.

This block diagram represents the DC Motor Model for the project. The electrical part of this system will be consider as first order system which consists of the element motor resistance, *R* and motor inductance *L*. According to the Fig. 9.2 the electrical model then multiplied with torque constant, *K* then it's added up with torque disturbance where the torque disturbance is negligible. Then the summing is multiplied with or as an input for mechanical part. Here the mechanical part consists of load inertia, *J* and friction, BM. The output is then multiplied with k_{gear} ratio of gear as shown in Fig. 9.3 to produce angular speed output from the motor.

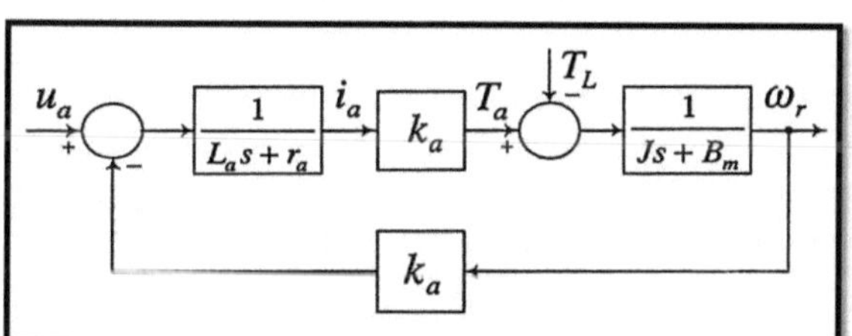

Fig. 9.2 Model of the permanent-magnet DC motor

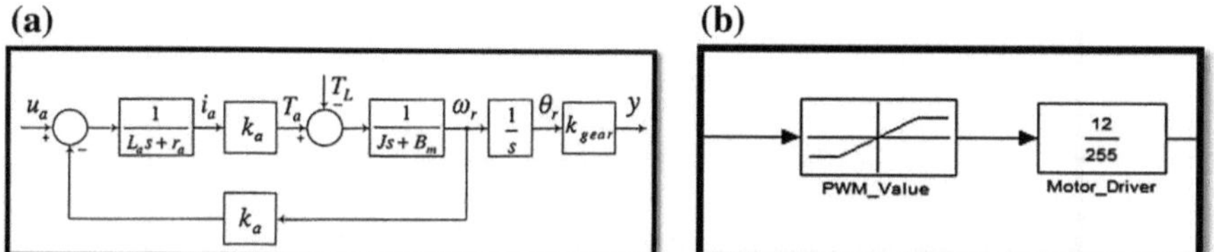

Fig. 9.3 **a** Model of the permanent-magnet DC motor with velocity output and **b** DC Motor driver model

9.3 Proportional-Integrated-Derivative (PID)

PID Control is most common and familiar type of automatic control used in industry. Although it has a relatively simple algorithm/structure, there are various suitable variations depends how it is used in industry [1]. A PID controller is a general control loop feedback mechanism mostly used in industry of control systems [2]. A PID controller will adjust the error between the output value and the desired input or set point by calculating and give the corrected output which will adjust the process accordingly. A PID controller has the general form [3].

$$u(\mathrm{t}) = K_{\mathrm{P}}e(t) + K_i \int_0^t e(\tau)\mathrm{d}\tau + K_d\frac{\mathrm{d}e}{\mathrm{d}t} \tag{9.1}$$

where K_p is proportional gain, K_i is the integral gain, and K_d is the derivative gain.

The PID controller with the calculation (algorithm) has three separate parameters; the Proportional, the Integral and Derivative values. Proportional values determine the reaction to the current error, the Integral determines the response based on the sum of latest errors. Whereas, Derivative determines the response to evaluate at which the error change. The weighted total of these actions are used to regulate the process via a control element such as the spot of a control valve, the power supply of a heating element or DC motor speediness and position.

9.4 Linear Quadratic Regulator (LQR)

LQR is a design method which is well known in modern best control theory and widely used for various applications. Below the hypothesis of all state variables are available for feedback, LQR controller design technique begin with a defined set of states which are to be controlled, as shown in Fig. 9.4.

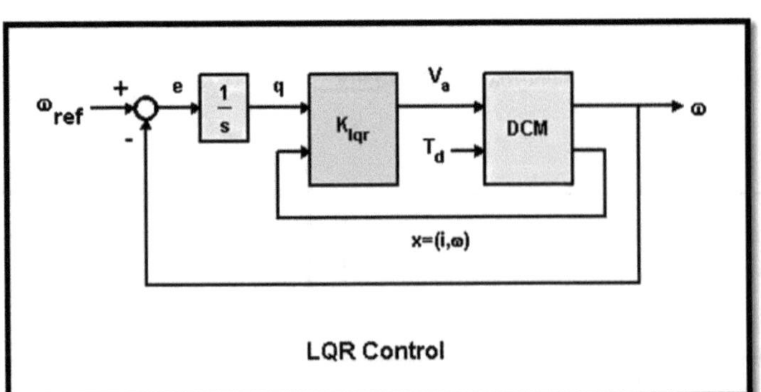

Fig. 9.4 LQR block diagram

The study of optimal control is monitored when operating a dynamic system at low cost. In this case where the dynamic systems are well explained by a set of linear differential equations and the cost is explained by a quadratic functional is called as LQ problem. The pioneer result in the theory is that the solution is given by the linear-quadratic regulator (LQR), a feedback controller with the equations are given below.

Generally, the system model can be described in state space equation as below:

$$X = Ax + Bu \tag{9.2}$$

The LQR controller is given by

$$U = -Kx \tag{9.3}$$

where K is the constant feedback gain obtained from the solution of the discrete algebraic Riccati equation. The gain matrix K which solve the LQR problem is

$$K = R - 1BTP^{*} \tag{9.4}$$

where $P*$ is unique, positive semi definite solution to the Riccati equation;

$$ATP + PA - PBR - 1BTP + Q = 0 \tag{9.5}$$

9.5 Result and Discussion

9.5.1 Data Analysis

In order to verify the validity of the linear quadratic regulator controller, several simulation tests are carried out using MATLAB/SIMULINK software package. The performance of the LQR controller has been investigated and compared with the conventional PID controller. Simulation tests are based on the facts that whether the LQR controller is better and more robust than the traditional PID controller or not. For the comparison, simulation tests of the speed response were performed on both controllers with synchronization between all four DC motors, with the same step response input, with or without a disturbance at one of the motor. The amplitude at y axis is set to position of 10 as the step response input for both controllers.

PID

For PID analysis 4 different models are simulated to find vary the step response. It is a closed-loop control system. The controller will maintain the speed at de-sired speed by using PID controller. The simulation begun by conducted for various values of P, I and D starting from value 0 until 15 in PID controller increment by 1 for each time to reach the set point. The proportional controller (K_p), integral controller (K_i), and derivative controller (K_d) is tuned to obtain fast responses and good stability for the control system. Then the suitable K_p, K_i, and K_d value has been chose which resulting a under damped response for this PID controller. The analysis is then continued with four different motor where each of it connected to a different scope and also one of the motor is add with step disturbance. In this paper, the best value for parameter K_p, K_i and K_d chose is 10, 10, and 0 respectively.

It is monitored that the motor stabilizes in the vertically upright position quickly and smoothly after some of minor overshoot for the case without disturbance input. Besides that, it also stabilizes in the vertically upright position with slight overshoot for the case of disturbance input. The amplitude at y axis reaches smoothly as per the desired position of 10 quickly in approximately 0.11 s.

LQR

Two parameters are defines for use LQR method. First the state-cost weighted matrix (Q) and second the control weighted matrix (R). For easy access purpose control weighted matrix set as 1 (R = 1) and the state-cost matrix (Q) equal to $pC'C$. By adding the vector C from the output equation it will considers those states in the output in deciding the cost. The weighting factor (p) will be different in order to upgrade the step response. In this case, R is a scalar since have a single input system.

In order to find the control matrix (K) it can be done by employing the MATLAB command of LQR. Firstly, the weighting factor (p) should be equal 1. Followed by the commands by adding it to the m-file and run it in the MATLAB command window. This will produce the following output (Fig. 9.5).

Finally the gain matrix K is obtained. Now the LQR controller is designed in Simulink by using the K value. The elements of the weighting matrices Q and R are chosen as: Q = [147.3572 0; 0 0] and R = 1, respectively.

Comparison

Figure 9.6 presents the result for PID controller of four motors synchronization. The time response specifications of the PID controller and LQR technique gained from the simulation of the separately excited DC motor speed control is shown in Fig. 9.6. From the simulation, LQR method has the fast settling time of 39.67 s while the PID controller has the slow settling time of 43.67 s. From the simulation results, linear quadratic regulator technique gives low settling time compared to traditional PID controller as summarized in Table 9.2. Even though the rise time of PID is much better, LQR provide a better characteristic that needed by a robust controller.

Fig. 9.5 K matrix code

```
p = 1;
Q = p*C'*C
R = 1;
[K] = lqr(A,B,Q,R)
```

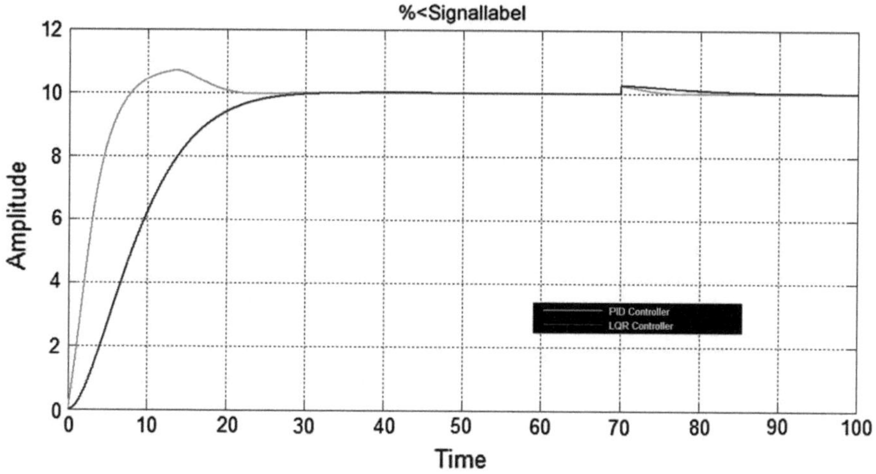

Fig. 9.6 The step response comparison of the separately excited DC motor between proposed PID and LQ

Table 9.2 Metrics of PID and LQR

Time response specification	PID	LQR
Rise time (Tr)	0.1143	5.9002
Settling time (Ts)	43.6700	39.6700
Settling min	68	95
Settling max	122	119
Overshoot (%)	20.7921	17.8218

9.6 Conclusion

The LQR based synchronize speed controller for multiple motor application was investigated to acquire low settling time compare to PID based synchronize speed controller. To control the speed of DC motor PID and LQR controller are chosen. The performance of the two different controllers is tested with simulations. Some of the simulation results are presented for comparison. Based on the comparative simulation results, it can be concluded that the linear quadratic regulator controller have a low settling time compared to conventional PID controller.

Although the objective is achieved without error some problem are faced. When the desired input sets as 20 constant for LQR after so many tuning couldn't get as the input step response. After setting 20 for both PID and LQR the range for the y-axis will be 0.7 only.

Besides that, even the LQR settling time low compared to PID but the value of settling time 41 s which is quite longer.

Acknowledgment This work is supported by the Kementerian Pelajaran Malaysia and Universiti Teknikal Malaysia Melaka (UTeM) through a financial support under project no. RAGS/1/2014/TK03/FTK/B00084. This financial support is gratefully acknowledged.

References

1. Zhang, Z., Chau, K.T., Wang, Z.: Chaotic speed synchronization control of multiple induction motors using stator flux regulation. IEEE Trans. Magn. **48**(11), 448–4490 (2012)
2. Zhao, D.Z., Li, C.W., Ren, J.: Speed synchronisation of multiple induction motors with adjacent cross-coupling control. IET Control Theory Appl. **4**(1), 119–128 (2010)
3. Marro, G., Prattichizzo, D., Zattoni, E.: Geometric insight into discrete-time cheap and singular linear quadratic Riccati (LQR) problems. IEEE Trans. Autom. Control **47**(1), 102–107 (2002)

Chapter 10
The Effect of Noise Elimination and Stemming in Sentiment Analysis for Malay Documents

Shereena M. Arif and Mazlina Mustapha

Abstract The growth of technology has changed the way of communicating opinions on services and products. In consumerism, the real challenge is to understand the latest trends and summarize the state or general opinions about products due to the diversity and size of social media data such as Twitter, Facebook and online forum. This paper discusses sentiments analysis in Malay documents from three perspectives. First, several alternatives of text representation were investigated. Second, the effects of the pre-processing strategies such as normalization and stemming with two type of Malay stemmer algorithm were highlighted. And lastly, the performance of Naïve Bayes (NB), Support Vector Machine (SVM) and K-nearest neighbour (kNN) classifiers in classifying positive and negative reviews, were compared. The results show that our selection of pre-processing strategies on the reviews slightly increases the performance of the classifiers.

Keywords Sentiment analysis · Malay documents · Normalization · Stemming

10.1 Introduction

The availability of Internet access, especially via mobile phone and micro-blogs has become one of the most important social platform. People give negative or positive reviews on products, and the reviews are significant for customers, companies and

S.M. Arif (✉)
Faculty of Information Science and Technology, Universiti Kebangsaan Malaysia, Bangi, Malaysia
e-mail: shereena.arif@ukm.edu.my

M. Mustapha
Politeknik Sultan Mizan Zainal Abidin, Terengganu, Malaysia
e-mail: mazlina.kch@gmail.com

A.-R. Ahmad et al. (eds.), *Proceedings of the International Conference on Computing, Mathematics and Statistics (iCMS 2015)*,
DOI 10.1007/978-981-10-2772-7_10

governments to make the right decision. However, the text content of micro-blogs is usually ambiguous and rich of acronym, slang and fashion word.

Malay Language (ML) is a part of Austronesian family language. It is used in several countries in South East Asia such as Malaysia, Indonesia and Brunei as the official language, and also used in Singapore and Thailand as the communication medium. There are approximately 300 million people using ML in the world to date [1]. Thus, this work is important as the number of people who used ML and actively engaged in social media keep increasing.

The main motivation of our study focuses on pre-processing phase in Malay's document sentiment classification. Many previous studies analyzed the users' opinions based on English language. There has been a very limited amount of research that focuses on SA in the Malay language. The main goal of this work is to determine the importance of some pre-processing tasks used for SA documents in ML.

10.2 Background

Sentiment Analysis (SA) is a study that analyzes the opinions, sentiments, judgment, behaviour and emotions of someone in writing [2]. It involves classifying the polarity of text in document or sentence that express the opinion in positive, negative or neutral form [3]. Generally, SA is the process of classification, but in reality, it is not as easy due to the language nature issues, such as choice of words, no annotation in a text and the complexity of the language itself [4].

10.2.1 Document-Level Sentiment Classification

Sentiment classification has three levels. Document-level sentiment classification is a task of labelling a document either positive or negative sentiment. This is different from other sentiment classification level (i.e. sentence level and aspect level), where it assumes that the opinionated document expresses opinions on a single target and the opinions belong to a single person, as discussed by Hajmohammadi [5]. The task of document-level sentiment classification is to predict whether the reviewer wrote a positive or negative review, based on analysis of the review text.

There are three approaches for sentiment classification of texts: (a) using a machine learning based text classifier—such as Naïve Bayes (NB), Sequential Minimal Optimization, Support Vector Machine (SVM) and K-Nearest Neighbour (kNN); (b) using the unsupervised semantic orientation scheme of extracting relevant *n*-grams of the text and then labelling them either as positive or negative and consequentially the document; and (c) using the SentiWordNet based on publicly available library that provides positive, negative and neutral scores for words [6].

10.2.2 Previous Work in Normalization of Noisy Text

Samsudin [7] used 1200 Malay reviews from several Malaysian online blogs and forums. They found that the data is unstructured and noisy. Thus, they proposed the normalization techniques to handle these issues [8].

Next, they introduced Malay Mixed Text Normalization Approach (MyTNA) to normalize noisy texts in ML used for social medias. The results show that accuracy values and document processing time using NB, kNN and SVM increased with feature selection based on Immune Network System (FS-INS) [9]. Saloot [10] proposed an approach to normalize Malay Twitter messages based on the features of colloquial and standard Malay where it outperformed the state-of-art system.

10.2.3 Previous Work Using Stemmer in Pre-processing Task

Stemming process will reduce a derivative word to its root word by removing all the prefixes, suffixes and infixes. It also reduces the indexing file size and speeds up the performance for retrieval [11–13].

Duwairi and El-Orfali [14] studied the effect of pre-processing strategies on SA for Arabic text. Two datasets of Politics and Movie review dataset has been used. They found that stemming combined with stop words removal improved the performance of the classification for both datasets. Santos and Ladeira [15] developed a similar work using Portuguese reviews extracted from the Google Play. However, they found that the proposed pre-processing methods consist of normalization, spell checking and stemming have not resulted in a significant improvement. Yussupova and Bogdanova [16] found that lemmatization has positive influence for Sentiment classification of text in Russian, but not for text in English.

Few studies in regard of stemmer usage for Malay SA documents have been conducted. In [12], they performed Sentiment Mining for Malay Newspaper (SAMNews) using Artificial Immune System (AIS). They found that stemming using Reverse Porter Algorithm (RPA) yields high accuracy of 88.5 %. Another study [11] also used AIS for extracting sentiment from Malay newspaper articles. They focused on stemming Malay text using RPA, Back–Forward Algorithm (BFA) and Artificial Immune Network, and found that there is no significant difference between RPA and BFA method in stemming.

Our initial studies showed that the previous study using Malay documents focus to only one of the pre-processing phase (i.e. normalization or stemming process). Thus, the limited resources of the ML in SA inspired us to find the effect of the pre-processing task with focus on normalization and stemming process.

10.3 Methodology

This section discusses the methodology used in our Malay SA system. First, the incomplete, ambiguous and noisy data were eliminated through pre-processing phase. The pre-processing phase consists of tokenization, stop words removal, normalization and stemming process. Next, a supervised machine learning classifier was used to classify the sentiments into positive and negative classes. Finally, the result of classification will be evaluated and discussed further. Our Malay corpus consists of four types of data where each level have extra pre-processing step. These will be later known as DTS, DNTS, DNTS/OSA and DNTS/FSA dataset.

10.3.1 Data Collection

Data were collected from several online forum and blogs of Malaysian Websites, mainly from http://www.cari.com.my and http://www.mesra.net, as there are no dedicated Malay movie reviews dataset available on the internet [8]. This corpus has 2000 reviews that were taken from 1000 positive reviews and 1000 negative reviews. Table 10.1 below shows samples of Malay review statements before pre-processing task.

10.3.2 Pre-processing

Pre-processing is a process to prepare the data for the classification process. We have used four types of pre-processing process such as tokenization, stop word removal, normalization and stemming process. The summary for each dataset are as following:

Table 10.1 An example of sentiments before pre-processing task

Example of a positive review	Example of a negative review
aaaaaaa betul gak tu anwarnuwa… tidak terfikir plak itu salah satu interpretasi seorang penonton bila tengok poster dan title. bagus komen tu. Bila saya bgtau feedback awak semalam, diorang terima tapi masih tak faham apa masalah dengan poster tu. hopefully, this time they get a clear picture	Sbb @missfake kan suka pakai high heel Sidahnya nampak buruk sgt sheola berlari2 bila nak kejor sam sbb fake nmpk sam jalan dgn pompuan n anak Mohon fake berlati2 depan cermin sambil pakai high heel…ko akan tgk betapa buruknya gaya ko berlari… mcm itik pulang petang

(a) Data with tokenization and stop word removal only (DTS);
(b) Data with tokenization, stop word removal and normalization only (DNTS);
(c) Data with tokenization, stop word removal, normalization and stemming using Othman Stemmer Algorithm (DNTS/OSA); and
(d) Data with tokenization, stop word removal, normalization and stemming using Fatimah Stemmer Algorithm (DNTS/FSA)

10.3.2.1 Normalization

Normalization is a process to reduce the ambiguity and noise in the sentiments. Several trends on how the Malaysian shortens the Malay terms in social media writings had been identified [17, 18]. The normalization process for our studies is based on the work of Samsudin [8]. Where appropriate, sentences were translated manually using official website of Malay Literary Reference Center available online at [19]. In detail, the normalization steps were carried out as described below:

i. English sentences were removed from the comments containing both English and Malay sentences in a paragraph.
ii. Comment that included simple English word and Malay in one sentence have been translated to Malay language.
iii. Comments that use English language only have been discarded.
iv. Comments that have a link address beginning with "*http://*" or "*www*" have been removed.
v. Symbols such as smiley animated icons, alias (@), a hash (#) and question mark (?) have been removed from the comment sentences.
vi. Sentences that combine more than one word and contain punctuation have been separated.
vii. Comments that contain dialects will be translated into formal ML.
viii. Words ending with numbers that bring meaning to the words will be spelt correctly. Numbers will also be separated properly from the word when it does not contribute to the meaning of the word.
ix. The words that were written in one character will be translated to formal ML.
x. The words that were simplified by using several characters in front of the correct spelling will be translated to formal ML.
xi. The words that used consonant only will be translated to formal ML.
xii. The words using the letters in front of and behind of the word will be translated to formal ML.
xiii. The words using the final syllable will be translated to formal ML.
xiv. The words using abbreviations will be translated into formal ML by reference in Pustaka [18].
xv. Letters repeated in one word will be limited to three syllables of repetition.
xvi. The ambiguous words will be translated into formal ML.

10.3.2.2 Tokenization

Each comment will be broken down into a series of words based on the detection of empty spaces between two words by removing all symbols and punctuations such as commas, exclamation marks, colons and others.

10.3.2.3 Stop Word Removal

This step includes removing unimportant words that have high frequency. This step is required to reduce the number of features by deleting terms that do not carry any information such as *dan* (i.e. and), *atau* (i.e. or), *kerana* (i.e. because) and others.

10.3.2.4 Stemming

Stemming process will reduce derivative word to its root words by removing all the prefixes, suffixes and infixes. For example, words "*makan*", "*pemakanan*" and "*makanan*" will have the same stem "*makan*". We have chosen two types of stemming algorithms, which are Othman [20] and Ahmad [21]. The Malay root-word list was adopted from a list used by official website of Malay Literary Reference Center available online at [19].

10.3.3 Experiments for Classifications Performance

The traditional classification techniques such as NB, SVM and kNN [5–7, 22–24] have been applied to the SA in the document level of the classification. These classifiers were chosen as the algorithm showed good performance for the SA and data mining in the past [25, 26].

10.3.4 Evaluations

All the algorithm for this work were evaluated through *k*-fold cross validation. We evaluate performance of sentiment classification based on the precision, recall and F1-measure.

$$\text{Precision} = \text{TP}/(\text{TP} + \text{FP}) \tag{10.1}$$

$$\text{Recall} = \text{TP}/(\text{TP} + \text{FN}) \tag{10.2}$$

The F1-measure is a combination of precision and recall represented by the function;

$$\text{F1}-\text{Measures} = 2 \times \text{precision} \times \text{recall}/(\text{recall} + \text{precision}) \qquad (10.3)$$

10.4 Analysis of the Results

This section will discuss the analysis of experimental results for each of the selected techniques. The performances of four Malay datasets were compared. This is done using precision, recall and F1-measure using three machine learning techniques (i.e. NB, SVM and kNN) to classify positive and negative reviews. Figure 10.1 shows the precision for NB, SVM and kNN for DTS, DNTS, DNTS/OSA and DNTS/FSA dataset. The lowest precision can be seen when kNN is used for DTS dataset classification (62.88 %), and the highest precision is SVM using DNTS/FSA dataset (74.11 %). Overall, the results show that SVM always perform the best, while kNN is the least.

Figure 10.2 shows the recall for DTS, DNTS, DNTS/OSA and DNTS/FSA dataset when using NB, SVM and kNN as classifiers. Interestingly, the best and worst recall is when using DNTS/FSA dataset, with SVM (77.70 %) and kNN (64.80 %) show the highest and lowest recall, respectively.

Figure 10.3 shows comparison on F1-measure for DTS and DNTS datasets. All the classifiers show slight increment when applied to the DNTS dataset. This result indicates that normalization step is important in order to improve the performance of opinion mining and agree with previous study reported before [7]. SVM classifier shows the best performance with highest increment of 3.27 % while kNN classifiers have the lowest increment of 0.78 %.

Figure 10.4 shows the F1-measure comparison for DNTS/OSA and DNTS/FSA datasets. Highest F1-measure can be seen for SVM classifier used on DNTS/FSA

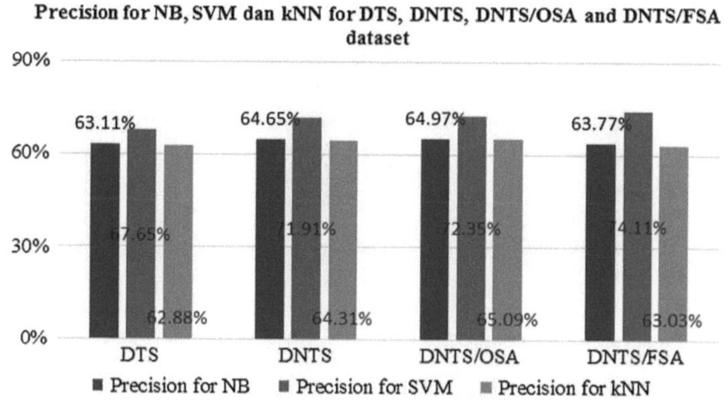

Fig. 10.1 Precision for NB, SVM and kNN for DTS, DNTS, DNTS/OSA and DNTS/FSA dataset

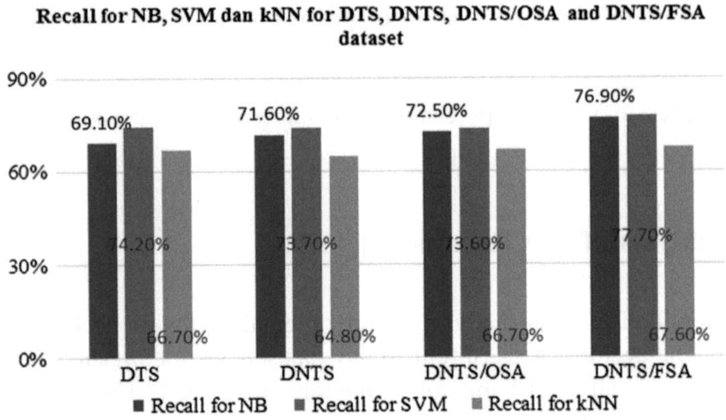

Fig. 10.2 Recall for NB, SVM and kNN for DTS, DNTS, DNTS/OSA and DNTS/FSA dataset

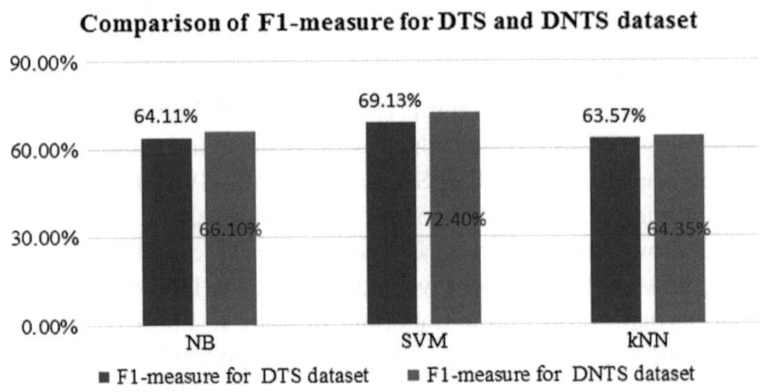

Fig. 10.3 Comparison of F1-measure for DTS and DNTS dataset

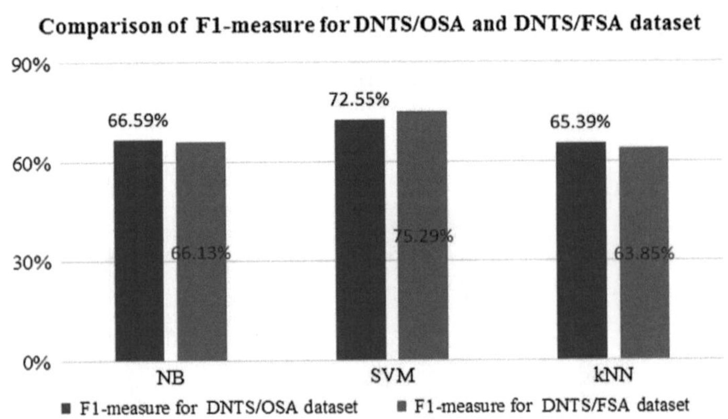

Fig. 10.4 Comparison for F1-measure DNTS/OSA and DNTS/FSA dataset

dataset (75.29 %), while kNN classifiers have the lowest F1-measure (63.85 %). It is also apparent that SVM works better with Fatimah algorithm, while Othman algorithm works well with NB and kNN classifiers.

Overall, all datasets obtain small improvements after applying different types of pre-processing, as also proved by Haddi [27]. The proposed pre-processing tasks do not show a significant improvement for classification as the reviews have short sentences and thus produced smaller size of documents. This caused the classifier having inadequate hint to produce high classification results for the dataset [15].

10.5 Conclusion and Future Work

This paper presents the effectiveness of pre-processing task for Malay corpus. We attempt to determine which pre-processing tasks contribute to sentiment classification on Malay documents. This study also investigates the effect of pre-processing task to improve the classification performance for the three machine learning classifiers on Malay SA. Finally, the results demonstrate that SVM classifier for DNTS/FSA dataset achieves the best performance for the Malay documents classification. This indicates that pre-processing tasks, particularly normalization, tokenization, stop word removal and stemming are important in order to improve performance of SA for this dataset.

Our future work will focus on developing a Malay sentiment lexicon and Malay large sentiment corpus and implementation of some optimization algorithm for Malay sentiment classification. Further work can also be carried out to implement SA work using Malay documents in various domains and sectors such as products, services and business. Construction of auto-translation system also needed to transform dialects into formal Malay language. This kind of system can support the process of normalization during pre-processing phase as it takes a long time to understand, identify and determine each word and sentence from human reviews.

Acknowledgment This study is supported by the UKM Grant ICONIC-2013-007 and FRGS-2-2013-ICT02-UKM-02-2.

References

1. Office of Prime Minister Putrajaya Malaysia Website. http://www.pmo.gov.my/home.php?menu=speech&page=1676&news_id=721&speech_cat=2
2. Liu, B.: Sentiment analysis and subjectivity. In: Handbook of Natural Language Processing (2010)
3. Pang, B., Lee, L.: Opinion mining and sentiment analysis. Found. Trends Inf. Retrieval **2** (1/2), 1–114 (2008)
4. Liu, B.: Sentiment analysis: a multi-faceted problem. IEEE Intell. Syst. (2010)

5. Hajmohammadi, M.S., Ibrahim, R., Ali Othman, Z.: Opinion mining and sentiment analysis: a survey. Int. J. Comput. Technol. **2**(3) (2012)
6. Singh, V.K., Piryani, R., Uddin, A.: Sentiment analysis of movie reviews a new feature-based heuristic for aspect-level sentiment classification. IEEE (2013)
7. Samsudin, N., Puteh, M., Hamdan, A.R.: Best or xbest: mining the Malaysian online review. In: Conference of Data Mining and Optimization (DMO), pp. 28–29, Selangor, Malaysia, (2011)
8. Samsudin, N., Puteh, M., Hamdan, A.R., Ahmad Nazri, M.Z.: Normalization of common noisy terms in Malaysian online media. In: Knowledge Management International Conference (KMICe), Johor Bahru, Malaysia (2012)
9. Samsudin, N., Puteh, M., Hamdan, A.R., Ahmad Nazari, M.Z.: Mining opinion in online messages. Int. J. Adv. Comput. Sci. Appl. **4**(8) (2013)
10. Saloot, M.A., Idris, N., Mahmud, R.: An Architecture for Malay Tweet Normalization. Information Processing and Management, vol. 50, pp. 621–633. Elsevier Ltd (2014)
11. Isa N., Puteh M., & Raja Kamarudin M.H.R.: Sentiment Classification of Malay Newspaper Using Immune Network (SCIN). In: Proceedings of the World Congress on Engineering 2013 Vol III. WCE (2013)
12. Puteh, M., Isa, N., Puteh, S., Redzuan, N.R.: Sentiment mining of Malay newspaper (SAMNews) using artificial immune system. In: Proceedings of the World Congress on Engineering 2013 Vol III, WCE (2013)
13. Vallbé, J., Martí, M.A., Blaz Fortuna, A., Dunja Mladenic J., Casanovas P.: Stemming and lemmatization: improving knowledge management through language processing techniques. In: Trends in Legal Knowledge, the Semantic Web and the Regulation of Electronic Social Systems (2007)
14. Duwairi, R., El-Orfali, M.A.: Study of the effects of pre-processing strategies on sentiment analysis for Arabic text. J. Inf. Sci. 1–14 (2013)
15. Santos, F.L.D., Ladeira, M.: The role of text pre-processing in opinion mining on a social media language dataset. In: Brazilian Conference on Intelligent Systems IEEE (2014)
16. Yussupova, N., Bogdanova, D.: Applying of sentiment analysis for texts in russian based on machine learning approach. In: The second International Conference on Advances in information Mining and Management, pp. 8–14. Venice, Italy (2012)
17. Hussin, S.: Blog. https://supyanhussin.wordpress.com/2009/07/11/bahasa-sms/
18. Panduan Singkatan Khidmat Pesanan Ringkas. Dewan Bahasa dan Pustaka. http://www.dbp.gov.my/khidmatsms.pdf
19. Official Website of Malay Literary Reference Center. http://prpm.dbp.gov.my/
20. Othman, A.: Pengakar Perkataan Melayu untuk Sistem Capaian Dokumen. Unpublished master's thesis, Universiti Kebangsaan Malaysia, Bangi, Malaysia (1993)
21. Ahmad, F., Yusoff, M., Sembok, T.M.T.: Experiments with a stemming algorithm for Malay words. J. Am. Soc. Inf. Sci. **47**(12), 909–918 (1996). USA
22. Omar, N., Albared, M., Al-Shabi, A.Q., Almoslmi, T.: Ensemble of classification algorithms for subjectivity and sentiment analysis of Arabic customers' review. Int. J. Adv. Comput. Technol. (IJACT) **5**(14) (2013)
23. Bharati, P., Kalaivaani, P.C.D.: Incremental learning on sentiment analysis using weakly supervised learning technique. Int. J. Eng. Sci. Innovative Technol. (IJESIT) **3**(2) (2014)
24. Anjaria, M., Reddy Guddeti, R.M.: Influence factor based opinion mining of twitter data using supervised learning. IEEE (2014)
25. Wu, X., Kumar, V., Quinlan, J.R., Ghosh, J., Yang, Q., Motoda, H., Mclachlan, G.J., Ng, A., Liu, B., Philip, S.Y.: Top 10 algorithms in data mining. Knowl. Inf. Syst. **14**, 1–37 (2008)
26. Vinodhini, G., Chandrasekaran, R.M.: Sentiment analysis and opinion mining: a survey. IJARCSSE **2**(6) (2012)
27. Haddi, E., Liua, X., Shib, Y.: The role of text pre-processing in sentiment analysis. In: Procedia Computer Science, vol. 17, pp. 26–32, ELSEVIER (2013)

Part II
Statistics

Chapter 11
A Hybrid *K*-Means Algorithm Combining Preprocessing-Wise and Centroid Based-Criteria for High Dimension Datasets

Dauda Usman and Ismail Bin Mohamad

Abstract Data clustering is an unsupervised classification method aimed at creating groups of objects, or clusters that are distinct. Among the clustering techniques, *K*-means is the most widely used technique. Two issues are prominent in creating a *K*-means clustering algorithm—the optimal number of clusters and the center of the clusters. In most cases, the number of clusters is predetermined by the researcher, thus leaving out the challenge where to put the cluster centers so that scattered points can be grouped properly. However, if it is not chosen correctly it will increase the computational complexity especially for high dimensional data set. To obtain an optimum solution for *K*-means cluster analysis, the data needs to be preprocessed. This is achieved by either data standardization or using principal component analysis on a scale data to reduce the dimensionality of the data. Based on the outcomes of the preprocessing carried out on the data, a hybrid *K*-means clustering method of center initialization is developed for producing optimum quality clusters which makes the algorithm more efficient. The result showed that *K*-means with preprocessed data performed better, judging from the sum of square error. Further experiment on the hybrid *K*-means algorithm was conducted simulated datasets and it was observed that, the sum of the total clustering errors reduced significantly whereas inter distances between clusters are preserved to be as large as possible for better clusters identification.

Keywords Basic *K*-means · Cluster center · Hybrid *K*-means · Principal component analysis · Sum of square error · *Z*-score

D. Usman (✉)
Department of Mathematics and Computer Science, Faculty of Natural and Applied Sciences, Umaru Musa Yar'adua University, Katsina, Nigeria
e-mail: dauusman@gmail.com

I.B. Mohamad
Department of Mathematical Sciences, Faculty of Science, Universiti Teknologi Malaysia, Johor Bahru, Malaysia
e-mail: ismailm@utm.my

A.-R. Ahmad et al. (eds.), *Proceedings of the International Conference on Computing, Mathematics and Statistics (iCMS 2015)*,
DOI 10.1007/978-981-10-2772-7_11

11.1 Introduction

Data exploration is a suitable way of extracting patterns, comprising learning implicitly kept in large datasets and concentrates on issues relating to their usefulness, scalability, feasibility, and effectiveness. This can be seen as a vital step in the process of knowledge discovery. Data are normally preprocessed through data transformation, data integration, data selections, and data cleaning, and prepared for the mining tasks. Data exploration might be performed on different sorts of databases and data storehouses, yet the kind of patterns to be discovered are specified by various data exploration functionalities, for example, correlation analysis, classification analysis, cluster analysis, and so on.

Cluster analysis is one of the major data exploration techniques widely used for numerous practical applications in emerging areas. Cluster exploration consist of a process in discovering groups of objects such that the objects in a group will be similar (or related) to one another and dissimilar with (or unrelated) to the objects in other groups. A good clustering technique will generate a high quality clusters and high intra-cluster similarity with low inter-cluster similarity. The quality of clustering results relies on the similarity measure used by the technique and its implementation and also by its ability to discover some or all of the hidden patterns.

K-means clustering algorithm is one of the most popular methods for clustering multivariate observations [1]. It is a system ordinarily used to directly segment sets of data into k groups. *K*-means algorithm generates a fast and efficient solution. The basic *K*-means algorithm works with the objective to minimize the mean square distance from each data point to its nearest center. However, there are two important issues in creating a *K*-means clustering algorithm: the optimal number of clusters and the center of the cluster. In many cases, the number of clusters is given, thus the important issue is where to put the cluster center so that scattered points can be grouped appropriately. Center of the cluster can be obtained by first assigning any random point and then optimizing the mean distance to the center. The process is repeated until all the mean square distances are optimized.

The drawback of the basic *K*-means algorithm is that it is sensitive to the selection of the initial partition and may converge to a local minimum of the criterion function value if the initial centroids are not properly chosen. A local minimum is the least value that is located within a set of points, which may or may not be a global minimum and it is not the lowest value in the entire set. Its computational complexity is also very high, especially for large data set. In addition, the number of distance calculations increases exponentially with the increase of the dimensionality of the data. An ad hoc solution to these problems is by choosing a set of different initial partition and the initial partition that gives the smallest sum of squares error is taken as the solution, but this ad hoc solution does not guarantee the solution will give the smallest sum of square error (SSE) because it is just a mere guessing approach.

When a random initialization of centroids is used, different runs of *K*-means typically produce different total SSEs, therefore choosing the proper initial centroids is the key step of the basic *K*-means procedure [2]. The result of the *K*-means algorithm is highly dependent upon its initial selection of cluster centers and before clustering, it must be previously known and fixed [1].

Thus, to obtain an optimum solution for *K*-means clustering, the data need to be preprocessed before the *K*-means cluster analysis [3]. This preprocessing process consists of data standardization method to scale the data set to fall within a specified range of values so that any attribute with larger value will not dominate the attribute with a smaller value. Moreover, for a very high dimensional data set, PCA can be used initially to reduce this dimension [4]. Furthermore, as *K*-means is highly dependent on its initial center position [5], an alternative way of center initialization method for *K*-means cluster analysis is also required to make the algorithm more effective and efficient. As fallout from the above drawback, this research focused on developing a *K*-means clustering algorithm by data preprocessing techniques and center point initialization for high dimensional data set.

Many attempts were made by researchers to improve the effectiveness and efficiency of the *K*-means algorithm. Su and Dy [6] motivated the use of a deterministic divisive hierarchical method, which is referred to as PCA-Part (principal components analysis partitioning) for initialization and the method brings about faster convergence of the *K*-means clustering with fewer number of iterations compared to the random initialization methods. Arai and Barakbah [7] propose a new approach to optimize the initial centers for *K*-means. This technique makes use of all the clustering results of *K*-means in most instances, in spite of the fact that some of them get to the local optima and then, transform the result by combining it with the hierarchical algorithm in order to determine the initial center points for *K*-means. The experimental results improved the clustering quality as compared to some other clustering methods. In both of these studies, it can be observed that they are limited in that there is no mention of the dimensionality of the data and the computation time needed in sorting the centers is quite high.

Karthikeyani and Thangavel [8] extended *K*-means clustering algorithm by applying global normalization before performing the clustering on distributed datasets, without necessarily putting the whole data into one location. The effectiveness of the proposed normalization-based distributed *K*-means clustering algorithm was compared against distributed *K*-means clustering algorithm and normalization-based centralized *K*-means clustering algorithm. The quality of clustering was also compared by three normalization procedures for the proposed distributed clustering algorithm. The comparison study implies that the distributed clustering results depend on the type of normalization procedure.

It is clearly observed with the methods found in our exploratory study that most methods presented do not work well with high dimensional datasets, and all the methods were evaluated with uncorrelated datasets. Although this is ideal, but it is useful to test the effectiveness of the methods also with correlated datasets before its application to a real life data. Werner [9] used two types of correlated datasets namely: fixed and controlled correlations to test and compare the performance of his method of identification of multivariate outliers in large datasets. The author further said, "there is no guarantee that actual datasets have zero correlation and sometimes correlated data can influence the statistical analysis in an undesirable manner".

Furthermore, researchers need to consider carefully which assumptions underlying the different normalization methods appear most reasonable in their experimental setting and possibly consider more than one standardization approach to determine the sensitivity of their results to standardization method used [10]. Bearing these points in mind, this research worked along these lines in the selection of the dimensionality of the data set used as well as in the standardized method employed.

11.2 Materials and Methods

The proposed technique is a hybrid of the basic *K*-means method. First, *z*-score standardization is used to rescale the data to fall within a specified range of values, so that any variable with larger domain will not dominate the variable with smaller domain. Second, principal component analysis used the singular value decomposition to obtain a reduced data set containing possibly uncorrelated variables. Third, the resulting reduced data set will be applied to the *K*-means clustering algorithm. All the analysis is carried out using MATLAB M-File 7.6 (R2008a) package. The steps for the technique are as follows:

11.2.1 Step 1

Consider the *z*-score method of data standardization. For convenience, let $\boldsymbol{X} = (\boldsymbol{X}'_1, \boldsymbol{X}'_2, \ldots, \boldsymbol{X}'_n)$ is the *d*-dimensional raw data set. This results in a $n \times d$ data matrix given by:

$$\boldsymbol{X} = \left(\boldsymbol{X}'_1, \boldsymbol{X}'_2, \ldots, \boldsymbol{X}'_n\right) = \begin{bmatrix} x_{11} & x_{12} & \ldots & x_{1d} \\ x_{21} & x_{22} & \ldots & x_{2d} \\ \vdots & \vdots & \ldots & \vdots \\ x_{n1} & x_{n2} & \ldots & x_{nd} \end{bmatrix}$$

The *z*-score of $\boldsymbol{X}_i$ is given as

$$Z(\boldsymbol{X}_i) = \frac{x_i - \bar{x}_i}{S_{x_i}} \tag{11.1}$$

where x_i, is the raw scores to be standardized. $\bar{x}_i$ and S_{x_i} are the sample mean and sample standard deviation, respectively.

11.2.2 Step 2

In step 2, we proceed to find the principal components of the standardized data set in Eq. 11.1. After finding the principal components of the data, a reduced projected data set will also be obtained by multiplying the original and principal components data matrices and retaining 80 percent of the total variance. Consider the standardized data matrix, $\boldsymbol{X}$, with zero mean and variance 1. Mathematically, the transformation is defined by a set of p-dimensional vectors of loadings (loadings here means the weight by which each standardized original variable should be multiplied to get the component score). Though, we add in SVD method to obtain the PCA as follows:

$$\boldsymbol{X}_{n\times d} = \boldsymbol{U}_{n\times d}\boldsymbol{D}_{d\times d}\boldsymbol{V}'_{q\times d} \tag{11.2}$$

11.2.3 Step 3

Given a set of observations, $\hat{\boldsymbol{X}} = (\hat{\boldsymbol{X}}'_1, \hat{\boldsymbol{X}}'_2, \ldots, \hat{\boldsymbol{X}}'_{\boldsymbol{n}})$ where each observation is a p-dimensional real vector, to partition the n observations into k sets $(k \le n)$, $G = (g_1, g_2, \ldots, g_k)$ compute

$$d_{\text{euclidean}}(\boldsymbol{XY}) = \sqrt{(x_i - y_i)^2} = [(x-y)(x-y)']^{\frac{1}{2}} \tag{11.3}$$

The algorithm proceeds by alternating between two steps

$$G_i = \left\{x_p : ||x_p - \mu_i||^2 \le ||x_p - \mu_j||^2 \forall j, 1 \le j \le k\right\} \tag{11.4}$$

where, each x_p is assign to exactly one G. Then update the process by calculating the new centers in the new clusters. The algorithm converges when this assignment no longer changes. Then calculate the total sum of squares error (SSE) that is:

$$\text{SSE} = \arg_g \min \sum_{i=1}^{k} \sum_{x_j \epsilon C_j} ||x_j - \mu_j||^2 \tag{11.5}$$

where $\mu_i = \frac{1}{n}\sum_{X_i \epsilon C_j} \boldsymbol{X}_i$ denotes the centroid of a cluster c_j and n denotes the number of instances in c_j.

11.3 Results and Discussions

Preprocessing [11] is actually essential before using any data exploration algorithms to enhance the results' performance. Two methods of data preprocessing namely, data standardization and principal component analysis using singular value decomposition are applied to infectious diseases datasets, consisting of seven variables and sample size 20 as shown in Table 11.1. The seven variables are Malaria, Typhoid fever, Cholera, Measles, Chicken pox, Tuberculosis, and Tetanus which are denoted by X_1 to X_7, respectively. The numbers in Table 11.1 represents the number of occurrences of each disease.

The cluster formation and the centroids obtained by the basic *K*-means for this data set are shown in Fig. 11.1.

It can be observed that six points are out of the cluster formation. These six points are on the borders, which are marked by the coordinates (2, 2), (0, 9), (0, 13), (3, 0), (6, 0) and (7, 0). These are in cluster 2. This is one of the basic *K*-means drawbacks, which is that the method does not capture all the points within the cluster formation.

It can be observed that, all the points fall inside the cluster formation (that is none of the points is on the border) when *z*-score standardization is used before clustering the points. This is one of the desired targets for cluster formation in our proposed hybrid *K*-means method. These desired targets consist of three things: the

Table 11.1 The Rafindadi Clinic data

	X_1	X_2	X_3	X_4	X_5	X_6	X_7
Day 1	9	6	4	9	2	5	2
Day 2	7	5	5	7	1	1	1
Day 3	7	2	3	7	2	3	2
Day 4	0	0	2	6	0	5	2
Day 5	3	5	0	2	3	11	1
Day 6	12	5	2	0	5	6	1
Day 7	3	0	0	7	1	8	3
Day 8	6	2	3	5	1	7	2
Day 9	5	3	11	2	7	2	3
Day 10	3	7	6	2	5	3	1
Day 11	0	5	7	9	1	2	1
Day 12	2	9	5	4	0	2	5
Day 13	7	3	0	3	6	7	1
Day 14	3	2	3	5	2	0	2
Day 15	2	13	1	2	0	2	3
Day 16	9	7	3	4	4	1	1
Day 17	9	8	1	0	2	3	1
Day 18	6	10	2	2	0	1	2
Day 19	2	2	3	8	4	2	2
Day 20	5	5	7	1	9	2	1

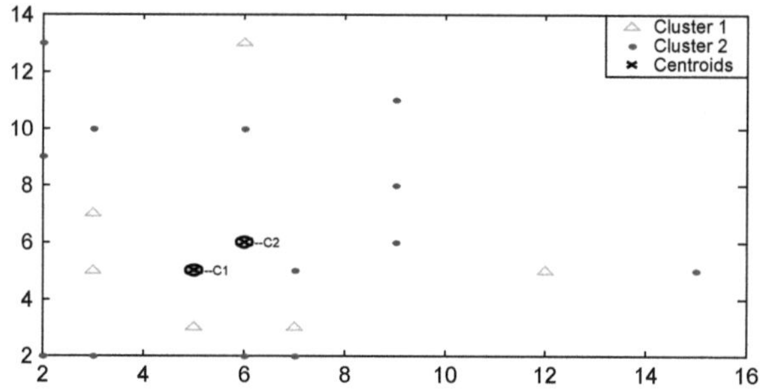

Fig. 11.1 Basic *K*-means centroids

first target is that all the clustered points are expected to fall within the cluster formation as in Fig. 11.2, the second target is that the inter-cluster similarities should be as large as possible and the third target is that the intra-cluster similarities should be very small. Therefore, being that all points fall within the formation, we infer that the *z*-score standardization is a good preprocessing method.

Table 11.2 shows the variances, the percentage of the variances and cumulative percentage of variances, which corresponds to the principal components of the original data. The cumulative percentage variance of the first four components reveals the number of PCs to be retained as it shows 88.4963 % of the total variance. That is PC1 to PC4 would be retained and used for our further data analysis.

Table 11.3 shows the reduced principal components obtained from the original data. The number of principal components obtained is the same with the number of variables in the original data. However, the reduced principal components (PC1, PC2, PC3, and PC4) will be used in obtaining the new projected data by multiplying the original data in Table 11.1 with the reduced principal components in Table 11.3. The computation below is an example to illustrate how this new projected data is obtained.

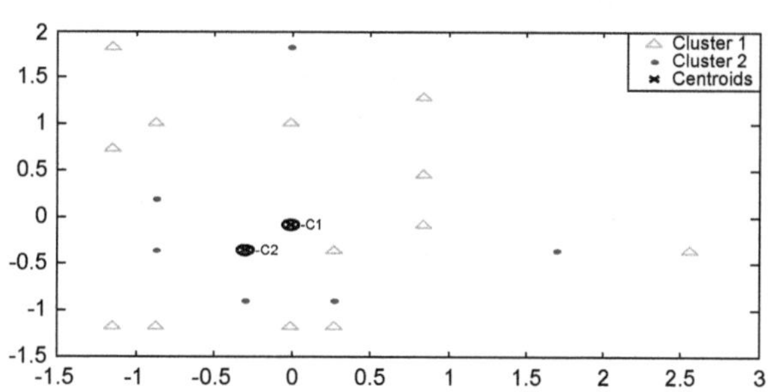

Fig. 11.2 *Z*-score basic *K*-means centroids

Table 11.2 The PCs variances and cumulative percentage

	Variances	Percentage variances	Cumulative % variances
PC1	17.0108	30.2768	30.2768
PC2	14.5370	25.8738	56.1506
PC3	11.8918	21.1658	77.3164
PC4	6.2813	11.1799	88.4963
PC5	4.5518	8.1016	96.5979
PC6	1.3865	2.4678	99.0657
PC7	0.5249	0.9343	100.0000

Table 11.3 The reduced principal components of the original data

PC1	PC2	PC3	PC4
−0.4686	0.4818	−0.1972	0.6848
−0.6273	−0.3548	0.4586	−0.0399
−0.0486	−0.4416	−0.5933	−0.0505
0.5495	−0.1848	0.0116	0.5545
−0.2000	0.1750	−0.5995	−0.3070
0.2020	0.6087	0.1825	−0.3520
0.0428	−0.1080	0.0768	−0.0374

$$\begin{aligned} y_{11} &= 9(-0.4686) + 6(-0.6273) + 4(-0.0486) + 9(0.5495) + 2(-0.2000) \\ &\quad + 5(0.2020) + 2(0.0428) \\ &= -2.5350 \\ y_{21} &= 7(-0.4686) + 5(-0.6273) + 5(-0.0486) + 7(0.5495) + 1(-0.2000) \\ &\quad + 1(0.2020) + 1(0.0428) \\ &= -2.7686 \end{aligned}$$

It can be seen from y_{11} and y_{21} that the dimension of the data is reduced to a reasonable level compared to y_{21} and y_{11} in the original data set in Table 11.1. This is as a result of the transformation carried out using PCA/SVD. Following this transformation, we then applied the basic *K*-means to the projected data set obtained.

Figure 11.3 shows the centroids of the basic *K*-means when applied to the projected data obtained by PCA/SVD. It can be observed that there is no points that are (farthest to the cluster center) outside the cluster formation line and the inter distances between clusters are preserved to be as large as possible using PCA/SVD before clustering the points.

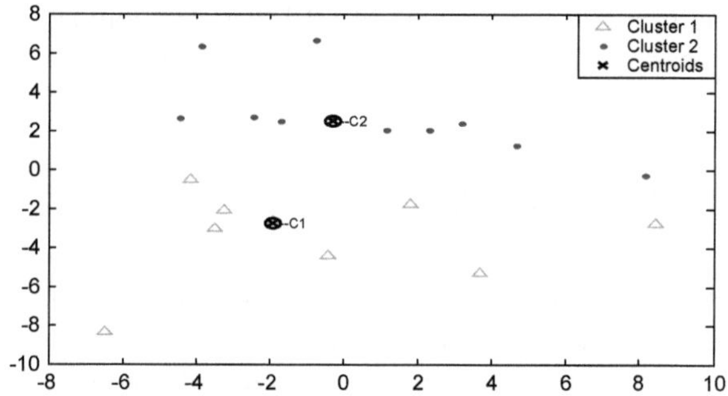

Fig. 11.3 PCA/SVD basic *K*-means centroids

11.3.1 Evaluation of the Hybrid K*-Means Method*

In this section, a simulation experiment is conducted to evaluate the hybrid *K*-means method. In order to show the advantage of this method, we use small and large p in our simulation experiment. That is (p, n) = (5, 500), (10, 500), (20, 500), (50, 500), (100, 500) (200, 500), (500, 1500) and (1000, 1500), where p refers to the number of variables and n is the sample size. The data was generated from multivariate normal distribution $N_p(0, I_p)$ with covariance matrix $\mathbf{\Sigma} = b\Gamma, b > 0$, and Γ is a symmetric matrix of size $(p \times p)$ with all diagonal elements equal 1 and all off diagonal elements equal ρ where $\rho = 0$, and $b = 1.2$ as in [11]. The CPU running time required by each experiment and their error sum of squares for the two approaches are presented in Table 11.4.

Table 11.4 gives the total sum of squares error and their respective CPU time for the basic *K*-means clustering method and the hybrid *K*-means clustering method for simulated datasets. The experimental results implies that the hybrid *K*-means method performs very well with high dimension data set providing better total sum of squares error and reduced CPU time taken for the execution. It was also observed that the clusters are well separated and compact as revealed in Fig. 11.3. This

Table 11.4 The total SSE and time taken for simulated datasets

		Basic *K*-means method		Hybrid *K*-means method	
n	*p*	SSE	CPU time (Sec)	SSE	CPU time (Sec)
500	5	45367.50	28.3610	2736.61	19.2217
500	10	93239.00	29.6870	7184.19	20.9428
500	20	190670.00	35.5526	10753.10	24.0136
500	50	484065.00	40.1859	26535.40	27.7203
500	100	967731.00	64.2412	57352.90	28.2841
500	200	1930896.00	119.1536	116419.53	34.0622
1500	500	14526307.00	466.7896	702725.29	197.6904
1500	1000	28991307.00	914.9216	1268061.05	372.2197

agrees with the findings of [10] that say, compactness and separation are used to measure the significance of clustering results.

11.4 Conclusion

K-means clustering is one of the most popular clustering algorithms. It has become an important multivariate analysis tool in many exploratory studies; however, the quality of the final clusters relies heavily on the initial centers, which are selected randomly and sometimes converges to a local minima of the criterion function especially for high dimensional data set. Hence, this research focuses on two steps in clustering high dimension data set: data preprocessing techniques and developing a good technique for center initialization. The dimensionality reduction was performed using Principal Component Analysis (PCA) by applying Singular Value Decomposition (SVD) as a preprocessing step to data clustering. Thereafter, based on the outcomes of the preprocessing carried out on the data a novel center point initialization method for *K*-means cluster analysis is developed for producing optimum quality clusters and avoiding random selection of initial values which makes the algorithm more efficient. The results and findings were validated with simulation experiments. From this experiments it was observed that, the sum of the total clustering errors was reduced as much as possible whereas inter distances between clusters are preserved to be as large as possible for better identification of clusters.

References

1. Tsai, C.Y., Chiu, C.C.: Developing a feature weight self-adjustment mechanism for a K-means clustering algorithm. Comput. Stat. Data Anal. **52**, 4658–4672 (2008)
2. Zhu, Y., Yu, J., Jia, C.: Initializing K-means clustering using affinity propagation. In: Ninth International Conference on Hybrid Intelligent Systems, vol. 1, pp. 338–343 (2009)
3. Chandrasekhar, T., Thangavel, K., Elayaraja, E.: Effective clustering algorithms for gene expression data. Int. J. Comput. Appl. **32**(4), 25–29 (2011)
4. Chris, D., Xiaofeng, H.: K-means clustering via principal component analysis. In: Proceeding of the 21st International Conference on Machine Learning. Banff, Canada (2006)
5. Rana, S., Jasola, S., Kumar, R.: A hybrid sequential approach for data clustering using K-means and particle swarm optimization algorithm. Int. J. Eng. Sci. Technol. **2**(6), 167–176 (2010)
6. Su, T., Dy, J.: A deterministic method for initializing K-means clustering. In: 16th IEEE International Conference on Tools with Artificial Intelligence, ICTAI 2004, pp. 784–786 (2004)
7. Arai, K., Barakbah, A.R.: Hierarchical K-means: an algorithm for centroids initialization for K-means. Rep. Fac. Sci. Eng. **36**(1), 25–31 (2007). Saga University
8. Karthikeyani, V.N., Thangavel, K.: Impact of normalization in distributed K-means clustering. Int. J. Soft Comput. **4**(4), 168–172 (2009)

9. Werner, M.: Identification of multivariate outliers in large data sets. Doctor Philosophy, University of Colorado, Denver (2003)
10. Zhao, Y., Wang, E., Liu, H., Rotunno, M., Koshiol, J., Marincola, F.M., Teresa, M.L., McShane, M.L.: Evaluation of normalization methods for two channel MicroRNA microarrays. J. Transl. Med. **8**, 62–69 (2010)
11. Berry, M.J.A., Linoff, G.S.: Data mining techniques for marketing, sales and customer support. Wiley, New York (1997)

Chapter 12
A New Framework of Smoothed Location Model with Multiple Correspondence Analysis

Hashibah binti Hamid

Abstract The implication of a considering large binary variables into the smoothed location model will create too many multinomial cells or lead to high multinomial cells and more worrying is that it will cause most of them are empty. We refer this situation as large sparsity problem. When large sparsity of multinomial cells occurs, the smoothed estimators of location model will be greatly biased, hence creating frustrating performance. At worst, the classification rules cannot be constructed. This issue has attracted this paper to further investigate and propose a new approach of the smoothed location model when facing with large sparsity problem.

Keywords Location model · Classification rule · Empty cells · Multiple correspondence analysis · Large binary variables

12.1 Introduction

The focus of this paper is to investigate the statistical methods that are capable of handling many types of variables simultaneously. However, different type of variables needs to be treated differently. As a matter of fact, most of the existing statistical methods are designed to manage single type variables and few methods are able to handle mixture of variables. Thus, one has to be careful in choosing the right method when dealing with mixed variables.

Extensive studies have been devoted to address the impact of different type of variables in a single model. Technically, the inclusion of all mixed variables in discriminant analysis may lead to some complications. The inclusion of too many variables in an analysis may lead to interaction between variables and engage large number of parameters that need to be estimated [1]. Besides, the choice of method is critically depends on the structure of the underlying data and the type of variables [2].

H. Hamid (✉)
School of Quantitative Sciences, College of Arts and Sciences,
Universiti Utara Malaysia, Sintok, Malaysia
e-mail: hashibah@uum.edu.my

A.-R. Ahmad et al. (eds.), *Proceedings of the International Conference on Computing, Mathematics and Statistics (iCMS 2015)*,
DOI 10.1007/978-981-10-2772-7_12

Past literatures have shown three possible strategies to construct a classification rule comprising variables with mixed type.

(i) Transform variables so that they are all in the same type, then construct classification rule fit to this type.
(ii) Construct separate classification rules for each type of variable, then combine the results as the overall classification.
(iii) Develop a model that can manage different type of variables, then derive classification rule from this model.

However, strategy (i) entails possible loss of information [3–5] and strategy (ii) has had limited study [2, 6]. Due to these unsatisfactory consequences, many studies have opened up the possibility of combining different types of variables more optimally via strategy (iii). As in the context of mixed-variables discrimination, a potential approach is based on the location model because it is appropriate to be used in such situation [7]. Recently, the location model has been successfully applied in discriminant analysis problems involving mixture of variables [8–12].

Innovation in the location model has led to the smoothed location model by [13] that is capable to handle the issue of empty cell due to greater number of cell (s) compared to size of sample (n). Thus, the smoothed location model is rectifying the problem of some empty cells of the location model as we can estimate parameters even for the cells that are empty. However, this model still cannot be coped if we are facing with large number of variables especially the binary because it will create too many multinomial cells [8, 14, 15]. Too many multinomial cells may cause many of them become empty which make trouble for the practitioners to handle it as the constructed rule will be greatly biased and result in a very poor performance. Therefore, this paper aims to design a new framework of the smoothed location model to address this problem.

12.2 The Smoothed Location Model as a Classification Rule

A parametric discrimination approach that focuses on mixed variables was proposed by [16], based on the location model introduced by [7]. The discrimination based on the location model assumes without loss of generality that b categorical variables are all binary, each taking on either zero or one values. The combination of zero and one from the vector of binary variables give rise to $s = 2^b$ different multinomial cells. Such realization gives conditional distribution of the location model as

$$f(\mathbf{x}, \mathbf{y}|\pi_i) = \frac{p_{im}}{(2\pi)^{c/2}|\boldsymbol{\Sigma}|^{1/2}} \exp\left\{-\frac{1}{2}(\mathbf{y} - \boldsymbol{\mu}_{im})^T \Sigma^{-1} (\mathbf{y} - \boldsymbol{\mu}_{im})\right\}. \tag{12.1}$$

Location model acts as a rule that distinguishes the groups of objects based on the information obtained from c continuous variables subjected to the underlying multinomial cells created from b binary variables [3, 4, 8, 11, 13]. The structure of the location model grows according to the number of binary variables. For example, the number of multinomial cells is four for each group if the number of measured binary is two $(\mathbf{X}_1, \mathbf{X}_2)$. Further if number of binary variable increases to four $(\mathbf{X}_1, \mathbf{X}_2, \mathbf{X}_3, \mathbf{X}_4)$, then the number of generated multinomial cells growth to 16 per group. This demonstrates that the size of multinomial cell (s) of the location model expanded by the number of binary variable (b) following $s = 2^b$.

In order to have a clear picture about the location model, suppose that a vector $\mathbf{z}^T = (\mathbf{x}^T, \mathbf{y}^T)$ is observed on each object where $\mathbf{x}^T = (x_1, x_2, \ldots, x_b)$ is a vector of b binary variables and $\mathbf{y}^T = (y_1, y_2, \ldots, y_c)$ is a vector of c continuous variables. The binary variables can be treated as a single multinomial cell $\mathbf{m} = \{m_1, m_2, \ldots, m_s\}$ where s $= 2^b$, and each distinct pattern of $\mathbf{x}$ defines a multinomial cell uniquely, with $\mathbf{x}$ falling in cell $m = 1 + \sum_{q=1}^{b} x_q 2^{q-1}$. The probability of obtaining an object in cell m of group π_i is p_{im} where $i = 1, 2$. Next, we assume that $\mathbf{y}$ to have a multivariate normal distribution with mean $\boldsymbol{\mu}_{im}$ in cell m of π_i and a homogeneous covariance matrix across cells and populations, $\boldsymbol{\Sigma}$. Hence, the conditional distribution of $\mathbf{y}$ given $\mathbf{x}$ is $(\mathbf{y}|\mathbf{x}) = m \sim \text{MVN}(\boldsymbol{\mu}_{im}, \boldsymbol{\Sigma})$ for π_i. The general density function of the location model is as Eq. (12.1).

Based on the density function of Eq. (12.1) and if parameters $\boldsymbol{\mu}_{im}$, $\boldsymbol{\Sigma}$ and p_{im} are known, then the optimal rule for allocating a future object $\mathbf{z}^t = (\mathbf{x}^t, \mathbf{y}^t)$ can be derived easily [17]. In a case of two groups problem, the $\mathbf{z}^t = (\mathbf{x}^t, \mathbf{y}^t)$ will be inserted into Eq. (12.1) for each π_1 and π_2, hence gives $f(\mathbf{x}^t, \mathbf{y}^t|\pi_1)$ and $f(\mathbf{x}^t, \mathbf{y}^t|\pi_2)$. Then, $\mathbf{z}^t$ will be allocated to group with greater density. A straightforward strategy is to take a ratio of density of two groups, $f(\mathbf{x}^t, \mathbf{y}^t|\pi_1)/f(\mathbf{x}^t, \mathbf{y}^t|\pi_2)$.

By defining a as the cost of misclassifying an object of π_i as an object of π_j where $i \neq j$. Then, the ratio of the two densities is

$$\frac{f(\mathbf{x}, \mathbf{y}|\pi_1)}{f(\mathbf{x}, \mathbf{y}|\pi_2)} = \frac{\frac{p_{1m}}{(2\pi)^{c/2}|\Sigma|^{1/2}} \exp\{-\frac{1}{2}(\mathbf{y} - \boldsymbol{\mu}_{1m})^T \Sigma^{-1}(\mathbf{y} - \boldsymbol{\mu}_{1m})\}}{\frac{p_{2m}}{(2\pi)^{c/2}|\Sigma|^{1/2}} \exp\{-\frac{1}{2}(\mathbf{y} - \boldsymbol{\mu}_{2m})^T \Sigma^{-1}(\mathbf{y} - \boldsymbol{\mu}_{2m})\}} \geq a. \tag{12.2}$$

Next, we take the logarithm on both sides of Eq. (12.2) as $\log\left[\frac{f(\mathbf{x},\mathbf{y}|\pi_1)}{f(\mathbf{x},\mathbf{y}|\pi_2)}\right] \geq \log(a)$, hence gives

$$\log\left[\frac{p_{1m}}{p_{2m}}\right] + \left[-\frac{1}{2}(\mathbf{y} - \boldsymbol{\mu}_{1m})^T \Sigma^{-1}(\mathbf{y} - \boldsymbol{\mu}_{1m})\right] - \left[-\frac{1}{2}(\mathbf{y} - \boldsymbol{\mu}_{2m})^T \Sigma^{-1}(\mathbf{y} - \boldsymbol{\mu}_{2m})\right] \geq \log(a). \tag{12.3}$$

We make further mathematical simplification and expand the Eq. (12.3). Finally, we obtain the following optimal rule which will allocates $\mathbf{z}^t$ to π_1 if

$$(\boldsymbol{\mu}_{1m} - \boldsymbol{\mu}_{2m})^T \Sigma^{-1} \left[\mathbf{y} - \frac{1}{2}(\boldsymbol{\mu}_{1m} + \boldsymbol{\mu}_{2m}) \right] \geq \log\left(\frac{p_{2m}}{p_{1m}}\right) + \log(a) \tag{12.4}$$

otherwise allocates to π_2. The parameter values $\boldsymbol{\mu}_{im}$, $\boldsymbol{\Sigma}$ and p_{im} are often unknown and will be replaced by the estimators obtained from the samples as discussed next.

12.2.1 Nonparametric Smoothing Estimation

The ideas of using linear model estimations and MANOVA-log-linear formulation provide reasonable justification on parameters estimation, but the issue of cells without objects (empty cells) is still remains unsolved. In earlier work of [13], they imposed the idea of using smoothing approach to estimate the probability of a cell in attempt to avoid obtaining a zero value. The smoothing idea is simple as assigns each cell with a weight that represents the relative closeness between the observed cells to its neighbour. Then, all estimators are obtained from the cells that take some contributions from their neighbour in the same group. This smoothing technique successfully overcomes the problem of empty cell.

In the smoothing approach, the mean $\boldsymbol{\mu}_{im}$ of each cell is fitted by a weighted average of all continuous variables from the data in the relevant group π_i. The vector of mean of the jth continuous variables $\mathbf{y}$ for cell m of π_i is estimated by

$$\hat{\mu}_{imj} = \left\{ \sum_{k=1}^{s} n_{ik} w_{ij}(m,k) \right\}^{-1} \sum_{k=1}^{s} \left\{ w_{ij}(m,k) \sum_{r=1}^{n_{ik}} y_{rijk} \right\} \tag{12.5}$$

under conditions; $0 \leq w_{ij}(m,k) \leq 1$ and $\left\{ \sum_{k=1}^{s} n_{ik} w_{ij}(m,k) \right\} > 0$

where

m, k = 1, 2, …, s; i = 1, 2 and j = 1, 2, …, c
n_{ik} = the number of objects of π_i in cell k
y_{rijk} = the jth continuous variable of the rth object in cell k of π_i
$w_{ij}(m, k)$ = the weights with respect to the variables j and cell m of all objects fall in cell k

If the value of λ was obtained and the vector of the cell means $\hat{\boldsymbol{\mu}}_{1m}$ and $\hat{\boldsymbol{\mu}}_{2m}$ have been estimated, then the smoothed pooled covariance matrix is defined by

$$\hat{\boldsymbol{\Sigma}} = \frac{1}{(n_1 + n_2 - g_1 - g_2)} \sum_{i=1}^{2} \sum_{m=1}^{s} \sum_{r=1}^{n_{im}} (\mathbf{y}_{rim} - \hat{\boldsymbol{\mu}}_{im})(\mathbf{y}_{rim} - \hat{\boldsymbol{\mu}}_{im})^T \tag{12.6}$$

where

n_i = the number of training objects of π_i
n_{im} = the number of objects in cell m of π_i
y_{rim} = the vector of continuous variables of the rth object in cell m of π_i
g_i = the number of non-empty cells in the training set of π_i

Finally, the estimation of the cell probabilities (p_{im}) can be obtained through any of the following four possible strategies namely (i) exponential smoothing, (ii) kernel smoothing, (iii) nearest neighbour smoothing and (iv) standardized exponential smoothing.

(i) The exponential smoothing [13] is defined as

$$\hat{p}_{im} = \frac{\sum_{k=1}^{s} w_{ij}(m,k)n_{im}}{\sum\limits_{m=1}^{s} \sum_{k=1}^{s} w_{ij}(m,k)n_{im}}. \tag{12.7}$$

The Eq. (12.7) is exactly a maximum likelihood estimator if λ approaches zero and it closes to 1 when λ approaches 1. However, the log ratio of the cell probabilities of the two groups will be zero if both groups have the same weight hence hinder the classification rule to be constructed.

(ii) The kernel smoothing [18] estimates the cells probability using

$$\hat{p}_{im} = n_i^{-1}\lambda_i^b \sum_{j=0}^{b} N_{ij}(m,k)\left\{\frac{1-\lambda_i}{\lambda_i}\right\}^j; \frac{1}{2} \le \lambda_i \le 1 \tag{12.8}$$

where
b = the number of binary variables
$s = 2^b$ = the number of multinomial cells
n_i = the number of training objects of π_i
$N_{ij}(m, k)$ = the number of training objects that fall in cell k of π_i whose binary vector **x** is đ binary variables distant from the cell m, $d(\mathbf{x}_m, \mathbf{x}_k)$ = đ for $m, k = 1, 2, \ldots, s$ and $i = 1, 2$
λ_i = a smoothing parameter for π_i

(iii) The nearest neighbour smoothing [19] states that

$$\hat{p}_{im} = n_i^{-1} \sum_{j=0}^{L} w_{ij}N_{ij}(m,k); 0 \le L \le b-1 \tag{12.9}$$

here b, n_i and $N_{ij}(m, k)$ have the same interpretation with the kernel smoothing estimator. The weights are chosen to minimize the mean squared error and the expectation being with respect to repeated sampling from a multinomial distribution.

$$\Delta_i = (w_{i0}, w_{i1}, \ldots, w_{iL}) = \sum_{k=1}^{s} E(\hat{p}_{ik} - p_{ik})^2. \quad (12.10)$$

(iv) The standardize exponential smoothing is simply as by taking a standardize function of exponential smoothing of Eq. (12.7) equally

$$\hat{p}_{im(\text{std})} = \hat{p}_{im} / \sum_{m=1}^{s} \hat{p}_{im}. \quad (12.11)$$

This Eq. (12.11) rectifies the problem of zero log ratio cell probabilities and provides convincing probability of groups under study. Standardize exponential smoothing is an easy method and suitable because both groups can be smoothed by a single parameter [9].

12.3 The Variables Reduction Techniques on Large Number of Measured Variables

The incorporation of too many variables may lead to unstable and bias classification rules [20] which then produce inaccuracy and unreliable decision [21]. The difficulty arises in dealing with big matrices (e.g. computing the eigenvalues and covariance matrix) [22] and at worst the study is burden with a singular matrix. Direct solution to this problem is infeasible showing that the classifier cannot be constructed directly with too many variables [23] or it will lead to disappointing performance. Simple strategy to overcome this problem is either to limit the number of binary (categorical) variables or require a big size of training set in the study [4].

As discussed by [24], if we have too many variables for a given sample size, we need either to increase the size of samples or to decrease the number of variables. Increasing the size of sample is more desirable, but it is usually impossible to be carried out in any study. Besides, in real applications, sometimes we rather have some limited size of sample. As a result, most researchers will try to reduce the number of variables in their analysis. Li [25] addressed that reducing the dimension of the appearance of objects helps to improve not only the recognition accuracy but also efficiency. It is desirable to keep the dimensionality of input variables as small as possible because we can obtain the most significant features to describe important phenomenon of data and simultaneously eliminating the redundant information [26, 27].

Align with this, [8] tried to use variable selection strategy to overcome large consideration of the binary in the smoothed location model, but the process and computation time for selecting the best set of variables are expensive and sometimes infeasible even with only six binary variables. Zhu [28] also concluded that the variable selection techniques handle the variables one by one, it is therefore not a genuine multivariate method in spirit. This paper will adopt a more direct multivariate approach which considers all the variables simultaneously. Therefore, this paper will offer an alternative to variable selection strategy in constructing the smoothed location model purposely for handling too many empty cells created from the consideration of large binary variables using variable extraction technique namely multiple corresponding analyses.

12.3.1 Multiple Correspondence Analyses for Reducing Large Binary Variables

There are many statistical techniques and their number is growing. One of the most confounding hurdles for anyone beginning to use suitable techniques is to know which techniques are possible for a given data type and which are most suitable for a given research question. This section outlines multiple correspondence analysis (MCA) that is comparable to the PCA which offers information and strategy when confronted with a large number of binary variables.

In order to improve the PCA technique, MCA was designed in the 1960s and 1970s when the former loses its parametric estimation optimal properties and to provide more powerful tools of the hidden structure in a set of categorical variables [29]. In the original formulation of [30], the MCA technique was described as a PCA of qualitative (nominal) variables. As pointed out by [31], MCA can be understood as PCA with categorical data as in survey research, more or less all the information is categorical. Research in social and behavioural science areas often results in data that are non-numerical which typically consist of qualitative or categorical variables in a limited number of categories [32]. Also as highlighted by [33] that MCA is a popular technique for analyzing multivariate categorical data, which is usually used to reveal systematic patterns among categories of variables of interest.

In short, MCA can be seen as a way of analyzing an object by variable matrix with categorical variables or an object by item matrix of multiple-choice data or a multi-way contingency table [34]. It is a popular technique to explore the relationships among multiple categorical variables [35–39]. The advantages of using MCA are to examine the association of categorical data to obtain a simplified representation of the multiple associations characterizing attributes as well as to remove noise and redundancy in the data [40]. Glynn [41] admitted that MCA offers a simple but powerful tool for identifying patterns in multi-factorial data. The options for visualization of its results can be difficult to explain, but extremely rich in the information displayed.

Using the notation of [34] and also [42], suppose that an original data matrix of n objects is described by a set of m categorical variables $X_1, X_2, \ldots, X_m$ with $k_1, k_2, \ldots, k_m$ is a category number in the mth variable. We define jl as category l of variable j which then are coded into a binary matrix $\mathbf{Z}$. The idea of replacing a categorical variable with a binary matrix can be found in [30]. Thus, the general entries of **Z** are define as

$$\mathbf{Z}_{ijl} = \begin{cases} 1 & \text{if object } i \text{ is in category } l \text{ of variable } j \\ 0 & \text{otherwise} \end{cases} \tag{12.12}$$

By merging the matrices **Z**, we will obtain a complete disjunctive table **Z** = [Z_1, Z_2, ..., Z_m..., Z_d] with n rows and d columns [$d = \sum_{j=1}^{m} k_j$]. It described the d positions for n objects of **X** through the binary coding. Then, a matrix **B** = $Z^T Z$ called *Burt* table is built where Z^T is the transpose matrix of **Z.** *Burt* table **B** is a (d, d) symmetric matrix which contains all the category marginal on the main diagonal and all possible cross-tables in the off-diagonal. Let **X** be a (d, d) diagonal matrix which has the same diagonal elements of **B** and zeros otherwise. We constructed a new matrix **S** from **Z** and **X** using

$$\mathbf{S} = \frac{1}{d} Z^T Z X^{-1} = \frac{1}{d} B X^{-1}. \tag{12.13}$$

By diagonalizing **S**, we obtain the diagonal elements called eigenvalues denoted as λ_j. Each eigenvalue λ_j is associated to eigenvector u_j where

$$\mathbf{S} u_j = \lambda_j u_j. \tag{12.14}$$

An eigenvalue represents the amount of inertia (variance) that reflects the relative importance of the dimension [43]. The first constructed dimension always has the largest inertia indicating that most of the variations [44] exist in such dimension. Due to that, practitioners often keep only a few dimensions which are able to summarize maximum information contained in the original data set. This is interesting because the reduced dimensions reflecting the relationship and concentrate only on significant information. Nonetheless, it is up to the users to choose the number of reduced dimensions to be kept by examining the eigenvalue or some other criteria [42, 45].

12.4 Methodology and Framework for Constructing the Smoothed Location Model with MCA Technique

The discrimination process is straight forward; we organize the steps in a leave-one-out fashion to avoid producing a bias evaluation. In the leave-one-our procedure, we take an object as a test set and treat the remaining objects of $n - 1$ as a training set. The training set is used to extract variables and to construct the smoothed location model (smoothed LM) rule, while the test set is used to evaluate the constructed rule. The following algorithm will be performed to obtain new approach of the location model. We hope that the new approach produced can be used as an alternative method to the classification problems, primarily when practitioners fronting with large binary variables in the study.

Algorithm
Construction of the Smoothed LM with MCA Technique

Step 1: Omit an object k from the sample, $k = 1, 2, \ldots, n$ where $n = n_1 + n_2$ and let the remaining $n - 1$ as a training set.
Step 2: Based on the training set, perform MCA to extract and reduce the large number of binary variables based on eigenvalues greater than 1.0.
Step 3: Find an optimized smoothing parameter (λ_{opt}) based on new variables (from Step 2) using an optimization approach.
Step 4: Compute the smoothing estimators $\hat{\boldsymbol{\mu}}_{im}$, $\hat{\boldsymbol{\Sigma}}$ and $\hat{p}_{im}$ using the λ_{opt} obtained in Step 3 by nonparametric smoothing technique.
Step 5: Construct the smoothed LM rule based on the smoothing estimators computed in Step 4, producing new approach.
Step 6: Predict the group of the omitted object k using the new constructed rule, if the prediction is correct then assign

$$\text{error}(\varepsilon_k) = 0 \,\text{otherwise}\, \varepsilon_k = 1.$$

Step 7: Repeat Steps 1 to 6 for all objects in turn.
Step 8: Compute the error rate using the omitted object by

$$\sum_{k=1}^{n} \text{error}_k / n.$$

Noticed that the new constructed approach is evaluated using a test set by computing error rates that represent the proportion of objects misclassified by the approach.

12.5 Expected Findings

A new mechanism of discriminant analysis of the smoothed LM for handling large number of binary (categorical) variables will be produced. This new approach provides options for practitioners in this field which can be used as an alternative method for classification purposes.

Acknowledgment Author would like to thank to Universiti Utara Malaysia, Malaysia for financial support.

References

1. Krzanowski, W.J.: Mixtures of continuous and categorical variables in discriminant analysis. Biometrics **36**, 493–499 (1980)
2. Wernecke, K.D.: A coupling procedure for the discrimination of mixed data. Biometrics **48** (2), 497–506 (1992)
3. Krzanowski, W.J.: Discrimination and classification using both binary and continuous variables. J. Am. Stat. Assoc. **70**(352), 782–790 (1975)
4. Krzanowski, W.J.: The location model for mixtures of categorical and continuous variables. J. Classif. **10**, 25–49 (1993)
5. Hand, D.J.: Construction and Assessment of Classification Rules: Wiley Series in Probability and Statistics. Wiley, Chichester (1997)
6. Xu, L., Krzyżak, A., Suen, C.Y.: Methods of combining multiple classifiers and their applications to handwriting recognition. IEEE Trans. Syst. Man Cybern. **22**(3), 418–435 (1992)
7. Olkin, I., Tate, R.F.: Multivariate correlation models with mixed discrete and continuous variables. Ann. Math. Stat. **32**(2), 448–465 (1961)
8. Mahat, N.I., Krzanowski, W.J., Hernandez, A.: Variable selection in discriminant analysis based on the location model for mixed variables. Adv. Data Anal. Classif. **1**(2), 105–122 (2007)
9. Mahat, N.I., Krzanowski, W.J., Hernandez, A.: Strategies for non-parametric smoothing of the location model in mixed-variable discriminant analysis. Mod. Appl. Sci. **3**(1), 151–163 (2009)
10. Hamid, H.: A new approach for classifying large number of mixed variables. In: International Conference on Computer and Applied Mathematics, pp. 156–161. World Academy of Science, Engineering and Technology (WASET), France (2010)
11. Leon, A.R., Soo, A., Williamson, T.: Classification with discrete and continuous variables via general mixed-data models. J. Appl. Stat. **38**(5), 1021–1032 (2011)
12. Hamid, H., Mahat, N.I.: Using principal component analysis to extract mixed variables for smoothed location model. Far East J. Math. Sci. (FJMS) **80**(1), 33–54 (2013)
13. Asparoukhov, O., Krzanowski, W.J.: Non-parametric smoothing of the location model in mixed variable discrimination. Stat. Comput. **10**(4), 289–297 (2000)
14. Vlachonikolis, I.G., Marriott, F.H.C.: Discrimination with mixed binary and continuous data. Appl. Stat. **31**(1), 23–31 (1982)
15. Krzanowski, W.J.: Stepwise location model choice in mixed-variable discrimination. Appl. Stat. **32**(3), 260–266 (1983)
16. Chang, P.C., Afifi, A.A.: Classification based on dichotomous and continuous variables. J. Am. Stat. Assoc. **69**(346), 336–339 (1974)
17. Moussa, M.A.: Discrimination and allocation using a mixture of discrete and continuous variables with some empty states. Comput. Programs Biomed. **12**(2–3), 161–171 (1980)
18. Aitchison, J., Aitken, C.G.G.: Multivariate binary discrimination by Kernel method. Biometrika **63**, 413–420 (1976)
19. Hall, P.: Optimal near neighbour estimator for use in discriminant analysis. Biometrika **68**(2), 572–575 (1981)
20. Wang, X., Tang, X.: Experimental study on multiple LDA classifier combination for high dimensional data classification. In: Roli, F., Kittler, J., Windeatt, T. (eds.) Proceedings of the 5th International Workshop on Multiple Classifier Systems, 9–11 June 2004, Cagliari, Italy, pp. 344–353. Springer, Heidelberg (2004)
21. Lukibisi, F.B., Lanyasunya, T.: Using principal component analysis to analyze mineral composition data. In: 12th Biennial KARI (Kenya Agricultural Research Institute) Scientific Conference on Socio Economics and Biometrics, pp. 1258–1268. Kenya Agricultural Research Institute, Kenya (2010)

22. Yu, H., Yang, J.: A Direct LDA algorithm for high-dimensional data with application to face recognition. Pattern Recogn. **34**(10), 2067–2070 (2001)
23. Das, K., Osechinskiy, S., Nenadic, Z.: A Classwise PCA-based recognition of neural data for brain-computer interfaces. In: Proceedings of the 29th IEEE Annual International Conference of Engineering in Medicine and Biology Society, pp. 6519–6522. IEEE Press, France (2007)
24. Katz, M.H.: Multivariate Analysis : A Practical Guide for Clinicians, 2nd edn. Cambridge University Press, Cambridge (2006)
25. Li, Q.: An integrated framework of feature selection and extraction for appearance-based recognition. Unpublished doctoral dissertation, University of Delaware Newark, USA (2006)
26. Ping, H.: Classification methods and applications to mass spectral data. Unpublished doctoral dissertation, Hong Kong Baptist University, Hong Kong (2005)
27. Young, P.D.: Dimension reduction and missing data in statistical discrimination. Doctoral dissertation, Baylor University, USA (2009)
28. Zhu, M.: Feature extraction and dimension reduction with applications to classification and analysis of co-occurrence data. Doctoral dissertation, Stanford University (2001)
29. LouisMarie, A.: Analysis of Multidimensional Poverty : Theory and Case Studies. Springer, New York (2009)
30. Guttman, L.: The quantification of a class of attributes : a theory and method of scale construction. In: Horst, P., Wallin, P., Guttman, L. (eds.) The Prediction of Personal Adjustment, pp. 319–348. Social Science Research Council, New York, NY (1941)
31. de Leeuw, J.: Here's looking at multi-variables. In: Blasius, J., Greenacre, M.J. (eds.) Visualization of Categorical Data, pp. 1–11. Academic Press, San Diego (1998)
32. Meulman, J.J., van Der Kooij, A.J., Heiser, W.J.: Principal components analysis with nonlinear optimal scaling transformations for ordinal and nominal data. In: Kaplan, D. (ed.) The SAGE Handbook of Quantitative Methodology for the Social Sciences, pp. 49–70. Sage, Thousand Oaks (2004)
33. van Buuren, S., de Leeuw, J.: Equality constraints in multiple correspondenc analysis. Multivar. Behav. Res. **27**(4), 567–583 (1992)
34. Tenenhaus, M., Young, F.W.: An analysis and synthesis of multiple correspondence analysis, optimal scaling, dual scaling, homogeneity analysis and other methods for quantifying categorical multivariate data. Psychometrika **50**(1), 91–119 (1985)
35. Benzécri, J.P.: L'analyse des Données : l'analyse des Correspondances [Data Analysis : Correspndence Analysis]. Dunod, Paris (1973)
36. Nishisato, S.: Analysis of Categorical Data : Dual Scaling and Its Applications. University of Toronto Press, Toronto (1980)
37. Greenacre, M.J.: Theory and Applications of Correspondence Analysis. Academic Press, London (1984)
38. Lebart, L., Morineau, A., Warwick, K.M.: Multivariate Descriptive Statistical Analysis : Correspondence Analysis and Related Techniques for Large Matrices. Wiley, New York (1984)
39. Gifi, A.: Nonlinear Multivariate Analysis. Wiley, Chichester (1990)
40. D'Enza, A.I., Greenacre, M.J.: Multiple correspondence analysis for the quantification and visualization of large categorical data sets. In: Di Ciaccio, A., Coli, M., Ibañez, J.M.A. (eds.) Advanced Statistical Methods for the Analysis of Large Data-Sets : Studies in Theoretical and Applied Statistics, pp. 453–463. Springer, Heidelberg (2012)
41. Glynn, D.: Correspondence analysis: exploring data and identifying patterns. In: Glynn, D., Robinson, J. (eds.) Polysemy and Synonymy: Corpus Methods and Applications in Cognitive Linguistics, pp. 133–179. John Benjamins, Amsterdam (2012)
42. Messaoud, R.B., Boussaid, O., Rabaséda, S.L.: A multiple correspondence analysis to organize data cubes. In: Vasilecas, O., Eder, J., Caplinskas, A. (eds.) Databases and Information Systems IV : Frontiers in Artificial Intelligence and Applications, pp. 133–146. IOS Press, Amsterdam (2007)

43. Hoffman, D.L., Franke, G.R.: Corresponding analysis: graphical representation of categorical data in marketing research. J. Mark. Res. **23**(3), 213–227 (1986)
44. Beh, E.J.: Simple correspondence analysis: a bibliographic review. Int. Stat. Rev. **72**(2), 257–284 (2004)
45. Hwang, H., Tomiuk, M.A., Takane, Y.: Correspondence analysis, multiple correspondence analysis and recent developments. In: Millsap, R.E., Maydeu-Olivares, A. (eds.) The SAGE Handbook of Quantitative Methods in Psychology, pp. 243–263. Sage, Thousand Oaks (2009)

Chapter 13
Comparing Methods for Testing Association in Tables with Zero Cell Counts Using Logistic Regression

Nurin Dureh, Chamnein Choonpradub and Hilary Green

Abstract Logistic regression is one of the most useful methods used to describe the relationship between a binary dependent variable and a set of independent variables. However, when any of the counts are zero, a nonconvergence problem will occur. A procedure for solving such a problem has been proposed by Firth (Biometrika 80:27–38, 1993 [4]), and provides finite parameter estimates based on penalized maximum likelihood. This study suggests a simpler method which involves modifying the data by replacing the zero count by one and doubling the corresponding nonzero count. Results show that this simple data modification method gives similar results to those from the Firth's procedure.

Keywords Logistic regression · Zero counts · Firth's procedure · Nonconvergence

13.1 Introduction

Logistic regression is one of the most used methods for modeling and testing the association between a binary outcome and one or more determinants. However, when one of the four cells in a 2 by 2 table of counts is equal to zero, maximum likelihood estimates of the model parameters fail to converge [1–3]. A solution to this problem was proposed by Firth [4], giving finite parameter estimates based on penalized maximum likelihood. This method is available in statistical software

N. Dureh (✉) · C. Choonpradub
Department of Mathematics and Computer Sciences,
Prince of Songkla University, Pattani, Thailand
e-mail: dnurin@gmail.com

C. Choonpradub
e-mail: cchamnein.c@gmail.com

H. Green
Department of Statistics, Macquarie University, Sydney, NSW, Australia
e-mail: hilary.green@mq.edu.ac

A.-R. Ahmad et al. (eds.), *Proceedings of the International Conference on Computing, Mathematics and Statistics (iCMS 2015)*,
DOI 10.1007/978-981-10-2772-7_13

packages such as SAS, S-PLUS, and R [5, 6]. Since this method requires special software we considered the possibility of simply modifying the data rather than using Firth's method. Lunn and McNeil [7] used a similar simple approach for modeling competing risks in survival analysis. Agresti [8] and Little and Rubin [9], also recommended modification in preference to new methodology when cell counts are small or the data are incomplete.

13.2 Method

The data modification (DM) method proposed in this paper is an improvement from the standard approach suggested by Agresti [8] where 0.5 is added to each cell in a 2 by 2 table. To deal with zero count problem in logistic regression, we introduce a new simple method for which the statistical significance determined by conventional procedures aligns closely with Firth's method.

Suppose the 2 by 2 table contains counts a, b, c, d with all possible cases of zero count as in Table 13.1 (Panel A). The DM method simply replaces zero by 1 and double the count in the corresponding cell. The modified tables contain count with asterisk as in Table 13.1 (Panel B).

Table 13.1 The general counts of a 2 by 2 table with a zero count (A) and modified table (B)

A				B		
y	x		→	y	x	
	1	0			1	0
1	$a = 0$	b		1	$a^* = 1$	b
0	c	d		0	$c^* = c + c$	d
y	x		→	y	x	
	1	0			1	0
1	a	$b = 0$		1	a	$b^* = 1$
0	c	d		0	c	$d^* = d + d$
y	x		→	y	x	
	1	0			1	0
1	a	b		1	$a^* = a + a$	b
0	$c = 0$	d		0	$c^* = 1$	d
y	x		→	y	x	
	1	0			1	0
1	a	b		1	a	$b^* = b + b$
0	c	$d = 0$		0	c	$d^* = 1$

13.2.1 Simulation Data for 2 by 2 Tables with Zero Count

The 2 by 2 tables were obtained by generating binomial random numbers using R software. Suppose we have a binary response variable where $Y = 1$ denotes a success and $Y = 0$ otherwise. We also have a binary covariate X, also with values 0 or 1. If p_{ij} is the probability of a successful outcome, $P(Y|X)$, the logistic regression model is given by:

$$P(Y|X) = \frac{e^{\beta_0 + \beta_1 X}}{1 + e^{\beta_0 + \beta_1 X}} \qquad (13.1)$$

And

$$\text{Logit}(p_{ij}) = \beta_0 + \beta_1 X \qquad (13.2)$$

The logistic regression model for a 2 by 2 table can be shown as in Table 13.2.

If $\beta_1 = 0$ in Eq. 13.2, then the rows and columns of the 2 by 2 table are independent. In the following we simulate data based on condition $\beta_1 = 0$ and $\beta_0 = -3$, for example. Using Eq. 13.2, the p_{ij} for this independent model may represented by Table 13.3.

Moreover, we also simulate data tables where the row and columns are dependent/nearly independent. According to the logistic regression model, we calculate the p_{ij} using $\beta_1 = 0.5$, $\beta_0 = -3$ to generate another set of tables. Table 13.4 is in accordance with this model.

Table 13.2 The general probabilities given by logistic regression model

Y	*X*	
	1	0
1	$P(Y = 1\|X = 1)$	$P(Y = 1\|X = 0)$
0	$1 - P(Y = 1\|X = 1)$	$1 - P(Y = 1\|X = 0)$

Table 13.3 The probabilities given by the logistic regression model, using $\beta_1 = 0$, $\beta_0 = -3$

Y	*X*	
	1	0
1	0.0474	0.9526
0	0.9526	0.0474

Table 13.4 The probabilities given by logistic regression model using $\beta_1 = 0.5$, $\beta_0 = -3$

Y	*X*	
	1	0
1	0.0759	0.9526
0	0.9241	0.0474

Table 13.5 The counts in simulated 2 by 2 tables which include zero counts (A) and corresponding tables (B) modified according to the DM method

A					B				
Table ID	a	b	c	d	Table ID	a	b	c	d
1	10	10	0	0	1	20	20	1	1
2	9	10	1	0	2	9	20	1	1
3	10	9	0	1	3	20	9	1	1
4	8	10	2	0	4	8	20	2	1
:					:				
:					:				
103	100	96	0	4	103	200	96	1	4
104	96	100	4	0	104	96	200	4	1
105	100	97	0	3	105	200	97	1	3
106	94	100	6	0	106	94	200	6	1
107	100	95	0	5	107	200	95	1	5
108	97	100	3	0	108	97	200	3	1

The sample sizes of the simulated data sets vary from 10 to 100 with equal sizes for groups $X = 0$ and $X = 1$. For the purposes of this study, only the data from tables which include zero counts are selected. An example of data tables is shown in Table 13.5 (Panel A) with different sample sizes for $X = 1$ and $X = 0$ and the corresponding data tables after DM are shown in Table 13.5 (Panel B).

13.3 Results

13.3.1 Comparison the Percentages of Correctly Identified p-Values

Figure 13.1 shows the percentage of times each of three methods correctly identified that the explanatory variable (X) and outcome (Y) variable are dependant, that is, that the test produced a p-value less than 0.05. While the results from our simulation study yielded the highest percentage (approximately 46 %) of correct identification of dependence, Lancaster's mid-P test and Firth's method correctly identified dependence in 38 and 41 % of cases, respectively.

When these methods were applied to data where the explanatory variable (X) and outcome (Y) variable are independent, that is, that the test produced a p-value greater than 0.05, we found that Firth's method correctly identified the highest percentage, approximately 81 % of independent cases the p-values, while the DM method correctly identified only 63 % as shown in Fig. 13.2.

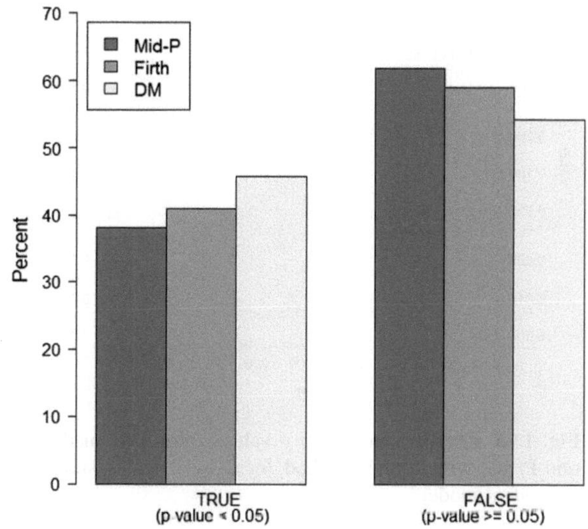

Fig. 13.1 Percentages of times the methods correctly identified a dependent model

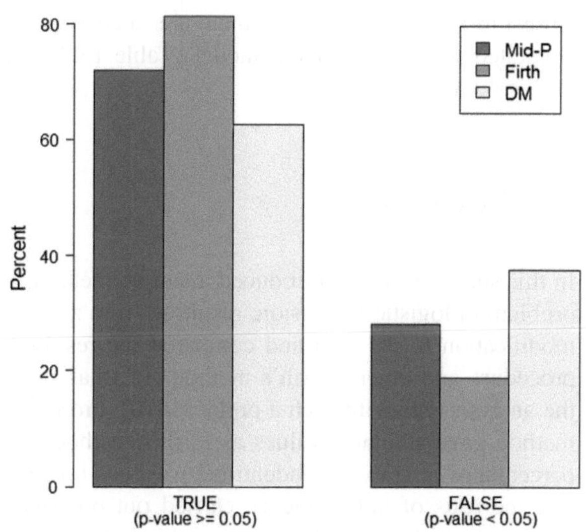

Fig. 13.2 Percentage times the methods correctly identified an independent model

13.3.2 Comparison of the p*-Values from DM, Fisher and Lancaster's mid-P Test to Firth's*

Figure 13.3 shows a comparison of the p-values from DM, Fisher's exact test and Lancaster's mid-P test to Firth's method. The red diagonal line indicates that p-values from the first three methods are in agreement with those obtained using

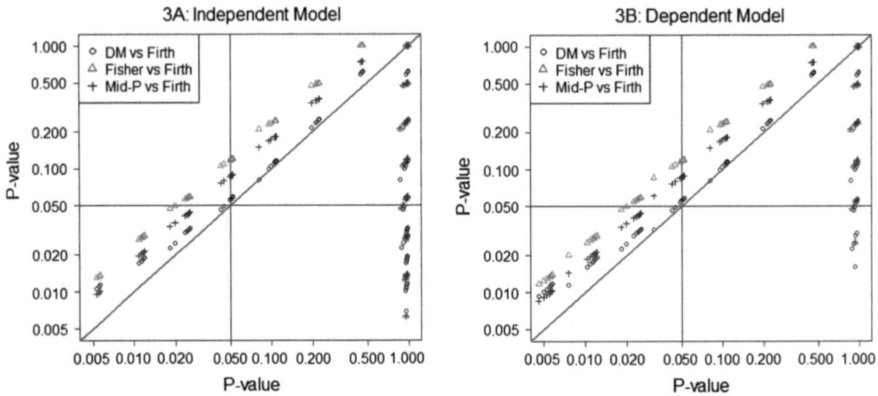

Fig. 13.3 Comparison of the *p*-values from DM method, Fisher's exact test and Lancaster's mid-P test with Firth's method for 2 by 2 tables simulated from the independent model and dependent model

Firth's method. The finding shows that the *p*-values from DM method appear closest to Firth's and some of them are in complete agreement both on the data for generated form a dependent model (Table 13.3) and from an independent model (Table 13.4).

13.4 Conclusion

In this study, we have introduced a simple method for solving the nonconvergence problem in logistic regression, resulting from zero counts. We have made a simple modification to the data and compared the results to those obtained from Firth's procedure and others. Firth's method (FL) has previously been recommended for the analysis data with such a problem [10]. From this study, we found that the DM method gave similar *p*-values as Firth's method. In addition, when compared the percentages of correctly indentified *p*-values for logistic regression, we also found that the tests of independence carried out on data modified by the DM method, produced *p*-values which were more likely to be in agreement with those produced by the same tests applied to data modified using Firth's method, compared to *p*-values obtained by Lancaster's mid-P test carried out on the data. Furthermore, the *p*-values from the tests on the DM data were closer in value to those from the tests on the data modified using Firth's method, than those obtained from both Fisher's exact text and Lancaster's mid-P test.

References

1. Aitkin M., Chadwick T.: Bayesian analysis of 2 × 2 contingency tables from comparative trials, School of Mathematics and Statistics, University of Newcastle UK (2003)
2. Bester C.L., Hansen C.: Bias reduction for Bayesian and frequentist estimators, University Chicago, Chicago (2005)
3. Eyduran, E.: Usage of penalized maximum likelihood estimation method in medical research: an alternative to maximum likelihood estimation method. J. Res. Med. Sci. **13**, 325–330 (2008)
4. Firth, D.: Bias reduction of maximum likelihood estimates. Biometrika **80**, 27–38 (1993)
5. Heinze, G., Ploner, M.: Fixing the nonconvergence bug in logistic regression with SPLUS and SAS. Comput. Meth. Prog. Bio. **71**, 181–187 (2003)
6. Heinze G., Ploner M.: A SAS macro, S-PLUS library and R package to perform logistic regression without convergence problems. Medical University of Vienna, Wien (2004)
7. Lunn, M., McNeil, D: Applying cox regression to competing risks. Biometrics **51**, 524–532 (1995)
8. Agresti, A.: Categorical Data Analysis. Wiley, New York, pp. 70–71 (2002)
9. Little, R.J., Rubin, D.B.: Statistical Analysis with Missing Data. Wiley, New York (2002)
10. Heinze, G., Schemper, M.: A solution to the problem of separation in logistic regression. Stat. Med. **21**, 2409–2419 (2002)

Chapter 14
Effect of Tumor Microenvironmental Factors on the Stability of Tumor Growth Dynamics with Nonzero Correlation Time

Ibrahim Mu'awiyya Idris and Mohd Rizam Abu Bakar

Abstract The effect of tumor microenvironmental factors on tumor growth dynamics modeled by multiplicative colored noise is investigated. Using the Novikov theorem and Fox approach, an approximate Fokker–Planck equation for the non-Markovian stochastic equation is obtained and an analytic expression for the steady-state probability distribution $P_{st}(x)$ is derived. We found that the strength of the microenvironmental factors θ have a negative effect on the stability of tumor growth at weak correlation time τ and at strong correlation time, the effect of θ is opposed and instead a positive growth stability is noticed for the tumor growth dynamics which in other words corresponds to growth effect. The result indicated that the growth effect exerted by the non-immunogenic components of tumor microenvironmental depend on the strength of the correlation time τ.

Keywords Langevin equation · Fokker–Planck equation · Colored noise · Tumor growth dynamics

14.1 Introduction

Study of systems driven by noise have been of growing interest in the last three decades, and stochastic approach is proving to be a reliable approach with its application cutting across many fields such as in biology, physics, and chemistry. A more general situation of studying system affected by noise especially from external source formulated as Langevin equation is that which involved colored noise, where the correlation at different times are not proportional to the delta function $\delta(t - t')$. Colored noise problem is non-Markovian with associated correlation time and therefore no longer characterized by the Fokker–Planck equation

I.M. Idris (✉) · M.R.A. Bakar
Universiti Putra Malaysia, Serdang, Malaysia
e-mail: Talk2midris@yahoo.com

M.R.A. Bakar
e-mail: drrizam@gmail.com

A.-R. Ahmad et al. (eds.), *Proceedings of the International Conference on Computing, Mathematics and Statistics (iCMS 2015)*,
DOI 10.1007/978-981-10-2772-7_14

derived under the Markovian assumption. However, to the leading order in the correlation time, an approximate approach to the Markovian Fokker–Planck equation can be found [1, 2]. Consideration of the effect of colored noise on tumor growth system has been the subject of many communications in the recent decades [3–5]. Moreover stochastic dynamics for systems driven by correlated colored noise are discussed in [6–8]. Transitions in a logistic growth model induced by noise coupling and noise color shows some interesting behavior [9], and the influence of colored correlated noises on the steady-state probability distribution and the mean of tumor cell number are analyzed and discussed [10]. The effect of multiplicative colored noise on the steady-state distribution of bacterium growth system was investigated and the result shows that, the strength of the multiplicative colored noise influences the bacterium number and even bacterium extinction [11].

Tumor growth is an open biological system of growth that interacts constantly with its internal surrounding microenvironment and as well as responds to the external environmental factors such as the effects from therapy [12, 13]. In addition, the internal tumor microenvironment contain processes that can either promotes or inhibits tumor growth depending on some uncertain factors, and such microenvironmental processes includes tumor interaction with extracellular matrix and fibroblast cells, signal transduction in cellular activity, nutrients among others. In the study of tumor cell growth, the logistic growth equation is mostly considered as the deterministic model equation, this is due to the fact that the logistic equation describes the general features of tumor growth. In this paper, we model the tumor response to the microenvironmental factors using multiplicative colored noise and logistic growth equation as the deterministic evolution equation. In addition, the tumor microenvironmental factors considered in this research are non-immunogenic but their natural biological functions within the tumor site are capable of supporting or undermining tumor growth [14]. This paper is organized as follows, Sect. 14.2 present the model formulation, Sect. 14.3 present the steady-state analysis based on the approximate Fokker–Planck equation, Sect. 14.4 present the numerical results and discussions and Sect. 14.5 concludes the paper.

14.2 Model Formulation

The phenomenological model equation is the Langevin dynamical equation given below

$$\dot{x}(t) = f(x) + g(x)\zeta, \tag{14.1}$$

where $g(x) = x$, and the function $f(x)$ is the deterministic logistic equation given as

$$f(x) = ax - bx^2 \quad a > 0, b > 0, \tag{14.2}$$

To account for the effect of the second term in Eq. (14.1), we consider the following equation

$$\dot{\zeta}(t) = -\frac{1}{\tau}\zeta + \Gamma(t), \tag{14.3}$$

where $\Gamma(t)$ is a Gaussian white noise with properties:

$$\langle \Gamma(t) \rangle = 0 \tag{14.4}$$

$$\langle \Gamma(t)\Gamma(t') \rangle = 2\theta\delta(t - t') \tag{14.5}$$

Equation (14.3) has a formal solution given by

$$\zeta(t) = \zeta(0)\exp(-t/\tau) + \int_0^t \exp[-(t - t')/\tau]\Gamma(t')\mathrm{d}t'. \tag{14.6}$$

For which the mean and the correlation function are given as

$$\langle \zeta(t) \rangle = 0 \tag{14.7}$$

$$\langle \zeta(t)\zeta(t') \rangle = \frac{2\theta}{\tau}\exp\left[-\frac{|t - t'|}{\tau}\right] \tag{14.8}$$

where in Eq. (14.8), as $\tau \to 0$, we recover the white noise case in Eq. (14.5). Moreover, the non-immunogenic tumor microenvironmental fluctuations are modeled by colored multiplicative noise $\zeta(t)$ with statistical properties given in Eqs. (14.7) and (14.8).

14.3 Steady-State Analysis

The stochastic Liouville equation corresponding to Eq. (14.1) for the time evolution of probability density is given by

$$\partial_t \rho(x,\ t) = -\partial_x\{f(x) + g(x)\zeta(t)\}\rho(x,t), \tag{14.9}$$

where $\rho(x,\ t)$ is the probability distribution for the realization of $\zeta(t)$. By averaging Eq. (14.10) we have a non-stochastic form with an averaged probability $P(x,\ t)$. This trivial fact is known as the Van Kampen lemma [15]

$$P(x,\, t) = \langle \rho(x,\, t) \rangle, \tag{14.10}$$

where

$$\rho(x,\, t) = \delta(x(t) - x). \tag{14.11}$$

The general equation satisfying the probability distribution $P(x,\, t)$ which is the average taking over the ensemble of many $\rho(x,\, t)$ for the possible realizations of $\zeta(t)$ is given as

$$\partial_t P(x,\, t) = -\partial_x f(x) P(x,\, t) - \partial_x g(x) \langle \zeta(t) \delta(x(t) - x) \rangle. \tag{14.12}$$

For the average in Eq. (14.12), we follow the approach given in [1, 2, 16] that collectively yield an approximate Fokker–Planck equation in steady-state regime.

$$\partial_t P(x,t) = -\partial_x f(x) P(x,t) + \frac{\theta}{M} \partial_x g(x) \partial_x g(x) P(x,t), \tag{14.13}$$

where

$$M = 1 - \tau \left[f'(x_s) - \frac{g'(x_s)}{g(x_s)} f(x_s) \right]. \tag{14.14}$$

In Eq. (14.14), $x_s = a/b$ is the steady-state value for the process $x(t)$ and τ is the correlation time. In addition, Eq. (14.13) is valid under the condition that $M > 0$, and interpreted in the sense of Stratonovich we have

$$\partial_t P(x,\, t) = -\partial_x A(x) P(x,\, t) + \partial_{xx} B(x) P(x,\, t), \tag{14.15}$$

where $A(x)$ and $B(x)$ are the modified drift and diffusion terms, respectively

$$A(x) = ax - bx^2 + \frac{\theta x}{1 + a\tau}, \tag{14.16}$$

$$B(x) = \frac{\theta x^2}{1 + a\tau}. \tag{14.17}$$

Rewriting Eq. (14.15) in terms of the conservation equation for probability we have

$$\partial_t P(x,\, t) + \partial_x J(x) = 0, \tag{14.18}$$

$J(x)$ is the probability current density for the system. At steady state it is required that $\partial_t P(x,\, t) = 0$, and we have the steady-state current

$$J_{st} = A(x)P_{st}(x) - B(x)\,\partial_x P_{st}(x). \tag{14.19}$$

Integrating Eq. (14.19) with reflecting boundary condition, the steady-state probability distribution $P_{st}(x)$ for the system is obtained [17, 18],

$$P_{st}(x) = \mathrm{N}B(x)^{-1/2}\exp[-U(x)/\theta], \tag{14.20}$$

where N is the normalization constant and $U(x)$ is given as

$$U(x) = b(1 + a\tau)x - [a(1 + a\tau) + \theta]\ln(x). \tag{14.21}$$

The stationary mean $\langle x\rangle_{st}$ for the tumor population is determined from

$$\langle x\rangle_{st} = \frac{\int_0^\infty xP_{st}(x)\mathrm{d}x}{\int_0^\infty P_{st}(x)\mathrm{d}x}. \tag{14.22}$$

14.4 Results and Discussion

Figure 14.1 shows the behavior of the steady-state distribution $P_{st}(x)$ against the tumor population with intervening parameter τ and θ. Keeping the correlation time strength fixed at $\tau = 0.3$, it is observed that increasing the microenvironmental factors strength θ result to decreasing the value of the steady-state distribution $P_{st}(x)$ which indicates that θ have a negative effect on the stability of tumor growth dynamics at weak τ. Figure 14.2 shows the same result as in Fig. 14.1 but with increased correlation time $\tau = 3.0$, and a sharp increase in the value of the steady-state distribution $P_{st}(x)$ is observed which in other words corresponds to positive stability for the tumor growth dynamics. This is evident in Fig. 14.3, where the effect of the correlation time τ on the steady-state distribution $P_{st}(x)$ is depicted

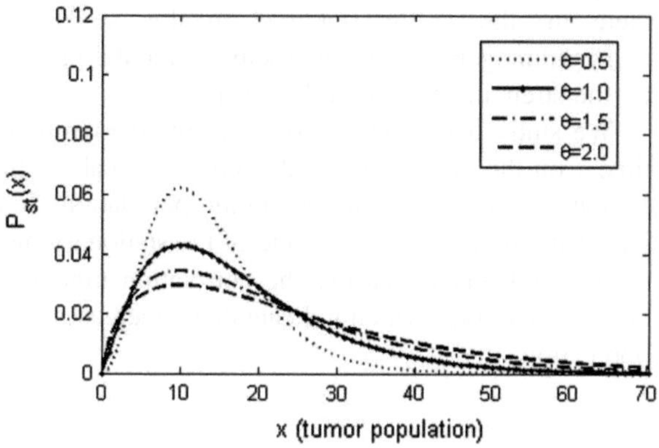

Fig. 14.1 Plot of the steady-state distribution $P_{st}(x)$ against the tumor population x at different microenvironmental noise strength θ. Other parameter values remain fixed τ = 0.3, a = 1.0, b = 0.1 (units are arbitrary)

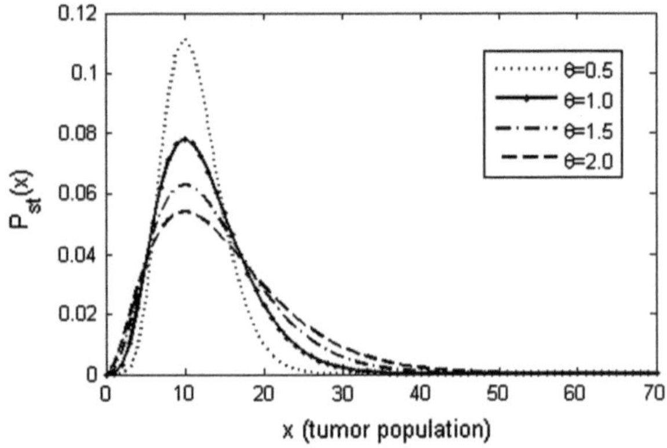

Fig. 14.2 Plot of the steady-state distribution $P_{st}(x)$ against the tumor population x at different microenvironmental noise strength θ. Other parameter values remain fixed $\tau = 3.0$, $a = 1.0$, $b = 0.1$ (units are arbitrary)

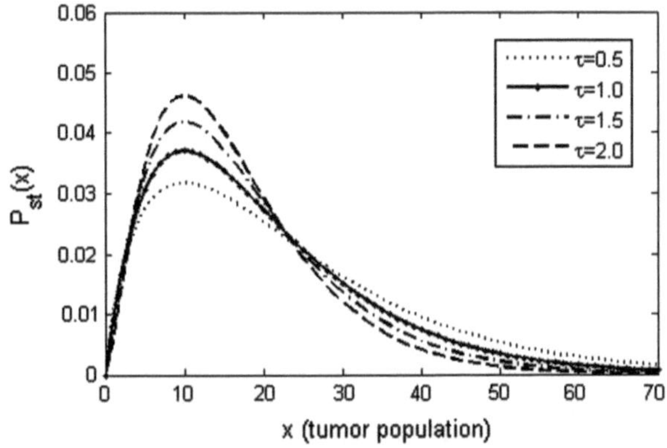

Fig. 14.3 Plot of the stationary distribution $P_{st}(x)$ against the tumor population x at different correlation time τ. Other parameter values remain fixed $\theta = 2.0$, $a = 1.0$, $b = 0.1$ (units are arbitrary)

at fixed microenvironmental factors strength θ, it is observed that increasing the correlation time strength τ increases the value of the steady-state distribution $P_{st}(x)$, this corresponds to positive stability for the tumor growth dynamics. In other words, tumor microenvironment plays a critical role in tumor progression as reported in many clinical and experimental investigations as discussed in [14] and references there in, the results obtained in this study show that the ability of the non-immunogenic tumor microenvironmental factors to exert growth effect depend on the strength of the correlation time τ.

We study the effect of both the microenvironmental factors θ and correlation time τ on the tumor growth dynamics by making the numerical simulation of the stationary mean $\langle x \rangle_{st}$ of the tumor population. It is observed in Fig. 14.4 that increasing θ decreases the value of the stationary mean $\langle x \rangle_{st}$. While in Fig. 14.5, it is observed that increasing the correlation time τ increases the value of the stationary mean $\langle x \rangle_{st}$ which indicate that τ increases the expected number of the tumor population.

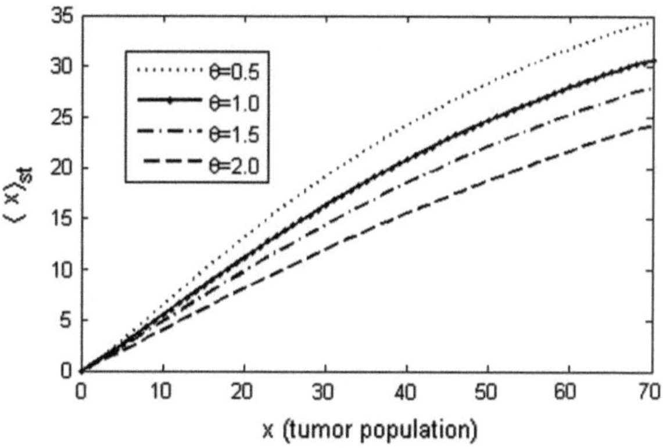

Fig. 14.4 Plot of the stationary mean $\langle x \rangle_{st}$ against the tumor population x at different microenvironmental strength θ. Other parameter values remain fixed $\tau = 0.1$, $a = 1.0$, $b = 0.1$ (units are arbitrary)

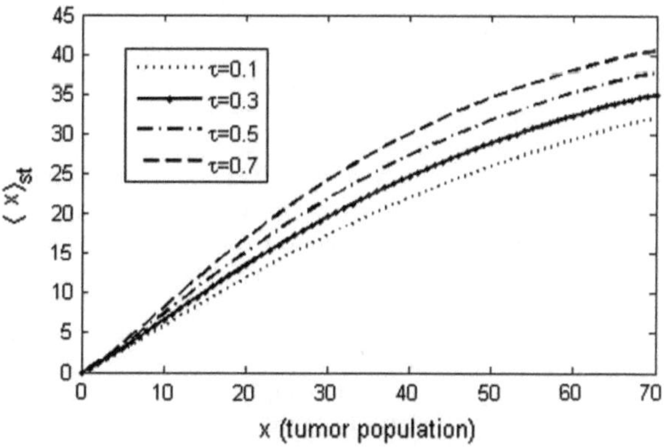

Fig. 14.5 Plot of the stationary mean $\langle x \rangle_{st}$ against the tumor population x at different correlation time τ. Other parameter values remain fixed $\theta = 5.0$, $a = 1.0$, $b = 0.1$ (units are arbitrary)

14.5 Conclusion

We make the steady-state analysis for the effect of non-immunogenic tumor microenvironmental factors on tumor growth dynamics modeled by multiplicative colored noise. The numerical result for the steady-state distribution $P_{st}(x)$ and the stationary mean $\langle x \rangle_{st}$ of the tumor population are obtained. Our result shows that the microenvironmental factors have a negative effect on the stability of tumor growth dynamics at weak correlation time τ, and at strong correlation time τ a positive growth stability is noticed for the tumor growth dynamics, in other words the positive stability corresponds to promoting tumor growth. Biologically, this means that tumor is an adaptive process that utilize the surrounding biological processes toward its growth and aggression, and the stronger the biological processes the more the chances of growth, and the weaker the biological processes the less the chances of growth. The result indicated that the growth effect exerted by the non-immunogenic components of tumor microenvironmental depend on the strength of the correlation time τ.

Acknowledgments One of the Authors (Ibrahim Mu'awiyya Idris) acknowledges financial support from Umaru Musa Yar'adua University Katsina Nigeria.

References

1. Novikov, E.A.: Functionals and the random force method in turbulence theory. Sov. Phys. JETP **20**(5), 1290–1294 (1965)
2. Fox, R.F.: Functional-calculus approach to stochastic differential equations. Phys. Rev. A **33** (1), 467–476 (1986)
3. Wang, C.-J., Di, L., Mei D.-C.: Pure multiplicative noises induced population extinction in an anti-tumor model under immune surveillance. Commun. Theor. Phys. **52**(3), 463–467 (2009)
4. Wang, C.-J., Wei, Q., Mei, D.-C.: Mean first-passage time of a tumor cell growth system subjected to a colored multiplicative noise and a white additive noise with colored cross-correlated noises. Mod. Phys. Lett. B **21**(13), 789–797 (2007)
5. Wang, C.-J., Wei, Q., Mei, D.-C.: Associated relaxation time and the correlation function for a tumor cell growth system subjected to color noises. Phys. Lett. A **372**, 2176–2182 (2008)
6. Wei, X., Cao, L., Wu, D.: Stochastic dynamics for systems driven by correlated colored noise. Phys. Lett. A **207**, 338–341 (1995)
7. Zeng, C., Wang, H.: Colored noise enhanced stability in a tumor cell growth system under immune response. J. Stat. Phys. **141**, 889–908 (2010)
8. Wang, C.-Y., Gao, Y., Wang, X.-W., Song, Y., Zhou, P., Yang, H.: Dynamical properties of a logistic growth model with cross-correlated noises. Phys. A **390**, 1–7 (2011)
9. Jin, S., Zhu S.-Q.: Transitions in a logistic growth model induced by noise coupling and noise color. Commun. Theor. Phys. **46**(1), 175–182 (2006)
10. Li-Bo, H., Xiao-Long, G., Li, C., Da-Jin, W.: Influence of colored correlated noises on probability distribution and mean of tumor cell number in the logistic growth model. Chin. Phys. Lett. **24**(3), 632–635 (2007)
11. Liao, H.-Y., Ai, B.-Q., Lian, H.: Effects of multiplicative colored noise on bacteria growth. Braz. J. Phys. **37**(3B), 1125–1128 (2007)
12. Ai, B.-Q., Wang, X.-J., Liu, G.-T., Liu, L.-G.: Correlated noise in a logistic growth model. Phys. Rev. E **67**, 022903 (2003)
13. Bose, T., Trimper, S.: Stochastic model for tumor growth with immunization. Phys. Rev. E **79**, 051903 (2009)
14. Witz, I.P.: Yin-yang activities and vicious cycles in the tumor microenvironment. Cancer Res. **68**(1), 9–13 (2008)
15. Van Kampen, N.G.: Stochastic differential equations. Phys. Rep. **24**(3), 171–228 (1976)
16. Hanggi, P., Mroczkowski, T.J., Moss, F., McClintoch, P.V.E.: Bistability driven by colored noise: theory and experiment. Phys. Rev. A **32**(1), 695–698 (1985)
17. Gardiner, C.W.: Handbook of stochastic methods for physics, chemistry and the natural sciences, 2nd edn. Springer, Berlin (1985)
18. Risken, H.: The Fokker Planck equation method of solution and application. Springer, Berlin (1996)

Chapter 15
Explorative Spatial Analysis of Crime Rates Among the District of Peninsular Malaysia: Geographically Weighted Regression

Syerrina Zakaria and Nuzlinda Abdul Rahman

Abstract In crime study, regression analysis can be used to test the relationship between crime rates and factors that are believed to be statistically significant. Regression model also provides the statistically measurable level for each unit change in the independent variables that affect crime rate. However, this global model does not take into account the spatial effects. It is believed that when spatial effects are included, it provides the more accurate coefficient estimates and standard errors for variables of interest. By taking into account the spatial effects, each study location will have unique coefficient estimate which is also known as the local estimates. The objective of this study is to analyze the spatial relationship between crime cases and social, environment and economic status for the districts in Peninsular Malaysia by using the Geographically Weighted Regression (GWR). For comparison purposes, OLS regression, known as global measure model, was used to measure the relationship between violent crime rates with factors that influence it. The results suggest that GWR model fitted better than OLS model.

Keywords Violent · Property · OLS · GWR

15.1 Introduction

Criminal activities are the inevitable social problem that has been faced by most of the countries in this world. Various plans and suitable strategies have been taken by law enforcement and local government in addressing this problem. These problems

S. Zakaria (✉)
School of Informatics and Applied Mathematics, Universiti Malaysia Terengganu, Kuala Terengganu, Malaysia
e-mail: syerrina@umt.edu.my

N.A. Rahman
School of Mathematical Sciences, Universiti Sains Malaysia, Penang, Malaysia
e-mail: nuzlinda@usm.my

A.-R. Ahmad et al. (eds.), *Proceedings of the International Conference on Computing, Mathematics and Statistics (iCMS 2015)*,
DOI 10.1007/978-981-10-2772-7_15

can give large impact on countries around the world; be it developed countries, developing countries and also undeveloped countries. In order to take any appropriate action in dealing with crime problems, the study on the crime situation and the distribution should be carried out.

There are a number of previous studies related to crime issue in various countries. Dutt and Venugopal [1] studied the crime patterns in India by using data of 14 different types of crime from 99 cities in India. They used standard correlation methods to analyze the relationships between the types of crime. They were found that crime rates are higher in cities than in rural area. Meanwhile, Tsushima [2] studied the impact of economic structure on crime (1986–1988) in 47 prefectures of Japan. In order to examine the impact, multiple regression analysis was employed. They reported that unemployment rates have a positive relation with homicide and robbery and the degree of inequality has positive correlated with larceny. Appiahene-Gyamfi [3] was study to investigate the robbery trends and patterns in Ghana from 1982 to 1993. He found that many robbery cases occurred in a central of socio-economic activities.

In Malaysia, Sidhu [4] studied the crime situation from 1980 to 2004 by looking at the overall pattern of crime and also property crime and violent crime. He analyzed the crime situation and discussed the factors that are believed to give impact on crime situation. Furthermore, in 2006, Sidhu [5] extended his work in studying the crime situation using the same data set. He also found that, crime situation also affected by global economic situation, i.e. when the economic downturn or recession happened, the unemployment rate increased and people faced difficulties in finding higher paying jobs. This scenario increased the crime rate. Another study was done by Cole and Gramajo [6]. They explore the relationship between homicide rates in 91 countries and several types of factors such as socio-economic, cultural and institutional factors using regression analysis. They found that the higher level of cultural and ethnic heterogeneity will increase the rate of homicide, and education was significantly associated with homicide rates.

The pattern of crime problem can be detected differently based on specific types of environments. It is believed that crime was higher in the most developed and densely populated area such as at large cities, town or urban area compared to undeveloped area such as rural area. This situation happened because of several factors such as environment characteristic, economics, social, political and demographics [7, 8].

There are several factors that are believed will give significant impact on crime cases. Appiahene-Gyamfi [9] reported that the factor of gender and demographic characteristics influenced the types and locations of some criminal types. The demographic characteristics that the author considered are population growth, unemployment and also education level. Meanwhile, Kanyo and Norizan [10] reported that other factors that influenced the crime cases are increasing cost of life, urbanization and poverty.

In spatial analysis, the location where an event occurs may provide an indication of the reason why that particular event occurs. In the beginning, spatial analysis involves mapping methods, reviews and geographic location without formal techniques. Starting from twenty-first century, spatial modern analyses were widely developed and focus on specific use of computer-based. Mapping is one of the most widely used techniques in spatial analysis [11, 12].

One of the common methods used to measure the relationship between the independent with the dependent variable is Ordinary Least Square (OLS). The assumption of this regression method is that the relationship between the dependent and independent variables are constant over regions that known as global estimate model. Regression model also provides the statistically measureable level for each unit change in the independent variables that affect crime rate. However, this global model does not take into account the spatial effects. Because regression-based approaches carried out on the average relationship between a dependent variable and other variable of interest, they often fail to examine the unique characteristics of places. As a result, utilizing a combination of complimentary approaches, including regression, may explain structural theories of crime and the environmental characteristics of a community. It is believed that when include spatial effects, it provides the more accurate coefficient estimates and standard errors for variables of interest. By taking into account the spatial effects, each study location will have unique coefficient estimate which is also known as the local estimates. For the purpose of spatial regression analysis, there are many methods that can be used such as spatial lag regression, spatial error, spatial autoregressive, spatial panel and geographically weighted regression (GWR).

The usage of spatial regression methods grow and it can be applied in various fields of study such as crime study [13, 14, 15], traffic accident and death [16], mortality analysis [17], analysis of coca crops [18], childhood obesity analysis in Athens and Greece [19]. These studies used the GWR method as a spatial model.

However, GWR should only be used as an exploratory tool to visualize the relationships between variables change in the entire study area. Obviously, this is a major departure from OLS, which is confirmatory in nature and able to test hypotheses about the impact of independent variables on a dependent variable. However, GWR is largely limited to exploratory purposes because has a weakness of multiple significance testing for parameters estimated [20, 13].

In crime studies, one of the important issues needed to be investigated are the factors that influence the incidence of crime. In order to understand the crimc trend or pattern, two theories often been used which are social disorganization theory and routine activity theory [21, 22, 23]. For this purpose, one of the ways that can be used is through regression methods. Due to the reason that location factors are considered, spatial regression methods should be used. In Malaysia, limited studies have been done on crime data using statistical method. Most of the studies been done were mainly focused on expert opinion or knowledge and discussed about causes and effects of crime. For example, studies done by Kanyo and Norizan [10], Sidhu [4, 5]. The investigation of the crime situation will become more significant by using the combination of qualitative and quantitative techniques. The objective of this study is to analyze the spatial relationship between violent crime cases and social, environment and economic status in Peninsular Malaysia by using the GWR. For comparison purposes, OLS is used to measure the relationship between violent crime rates with factors that influence, which is known as global measure model.

15.2 Data

The data employed in this study were obtained from Royal Malaysia Police (PDRM), which is the number of crime cases. There are two categories of crime cases included in index crime statistics known as violent and property crime. The definition of index crime statistics is the crimes that are reported with sufficient regularity and with sufficient significance to be meaningful as an index to the crime situation [4].

The violent crime included murder, gang robbery with firearm, gang robbery without firearm, robbery with firearm, robbery without firearm, rape and voluntarily causing hurt. The data crime for the year 2001 according to districts reported in Peninsular Malaysia will be analyzed in this study. Meanwhile, the property crime including housebreaking and theft by day, housebreaking and theft by night, theft of lorries and van, theft motor car, theft of motorcycles and scooters, snatch theft and other form of theft.

Six independent variables were considered in this study known as population density (popden), unemployment rate (unemp), percentage of non-citizen population (nonciti), percentage of single mother (singlemom), BASIC index and MIDDLE index. These variables are obtained from population and housing census report in Malaysia for the year 2000. BASIC and MIDDLE index were household-based socio-economic index that was built using factor analysis method [24].

15.3 Methodology

15.3.1 Ordinary Least Square, OLS

The *OLS* regression model with one variable as follows:

$$\mathbf{Y} = \beta_0 + \beta_1 \mathbf{X}_1 + \varepsilon$$

where $\mathbf{Y}$ is the dependent variable, $\mathbf{X}_1$ is the independent variable, β_0 is a constant and β_1 are parameters to be estimated, and ε is random error that normally distributed. β_0 and β_1 is taking a constant value without considering a different location or place. So, for purposes of analysis in spatial, few changes were made to allow regression carried out to meet the needs of different locations or space research.

An ordinary linear regression model can be written as below

$$y_i = \beta_0 + \sum_{r=1}^{R} \beta_r x_{ri} + u_i \quad i = 1, 2, \ldots, N$$

where y_i are the observation of the dependent variable Y, β_r $(r = 1, 2, \ldots, R)$ are the regression coefficients, x_{ri} is the ith value of x_r and u_i are the independent normally distributed error terms with zero mean,$E(u_i) = 0$ and constant variance, $\mathrm{Var}(\varepsilon_i) = \sigma^2$.

In matrix notation;

$$\mathbf{Y} = \beta_0 + \sum_{r=1}^{R} \beta_r \mathbf{X}_r + \mathbf{U}$$

where **Y** as vector of dependent variable, $\mathbf{X}_r$ as vector of the rth independent variable and **U** as vector of the error term. The OLS estimator takes the form:

$$\hat{\beta} = (\mathbf{X}'\mathbf{X})^{-1}\mathbf{X}'\mathbf{Y}$$

where $\hat{\beta}$ is the vector of estimated coefficients, **X** is the matrix of the independent variables with a columns, **Y** is the vector of dependent value, and $(\mathbf{X}'\mathbf{X})^{-1}$ is the inverse of the variance–covariance matrix.

15.3.2 Geographically Weighted Regression, GWR

The spatial regression is also one of the spatial analyses. This analysis is the result of modification of the method of *Ordinary Least Square* regression *(OLS)*. However, the spatial regression was able to resolve the problem of unstable parameters and unreliable significant testing. This method measures the spatial dependency where it measures the spatial relationship among the variables that were considered.

Instead of OLS, GWR estimates model coefficients at each location in study area, and produces more accurate predictions for the response variable. It is also known as local estimates model, instead of OLS which known as global estimate model. Furthermore, the residuals of the GWR model have more desirable spatial distributions than the ones derived from the OLS and other models.

Let us say there are several locations in the study assumed the (u, v) coordinates is at a point in space. The GWR model with an independent variable can be written as follows [17]:

$$\mathbf{Y}(u, v) = \beta_0(u, v) + \beta_1(u, v)\mathbf{X}_1 + \varepsilon(u, v)$$

where $\beta_0(u, v)$ and $\beta_1(u, v)$ is a continuous function for (u, v) at one point. In GWR, the global regression coefficients are replaced by local parameter.

$$y_i = \beta_0(u_i, v_i) + \sum_{r=1}^{R} \beta_r(u_i, v_i)x_{ir} + \varepsilon_i \quad i = 1, 2, \ldots, N, \; r = 1, 2, \ldots R$$

where β_{ri} are the regression coefficients for each location i and each variable r. The ordinary linear regression model is actually a special case of the GWR-function where the β_{ri} are constant for all $i = 1, 2, \ldots, N$.

The GWR equation recognizes that there may exist the spatial effect and variations in relationships and provides a way in which they can be measured [25]. In the GWR-function, there are more unknown parameters than degrees of freedom. Thus, the local estimates are made using weighted regression, with weights assigned to observations being function of the distance from point i. The GWR estimator is

$$\hat{\beta}_i = (\mathbf{X}'\mathbf{W}_i\mathbf{X})^{-1}\mathbf{X}'\mathbf{W}_i\mathbf{Y}$$

with $\hat{\beta}_i = (\hat{\beta}_{i0}\,\hat{\beta}_{i1}\ldots\hat{\beta}_{iM})'$ and $\mathbf{W}_i = \begin{bmatrix} w_{i1} & \ldots & 0 & 0 \\ 0 & w_{i2} & \ldots & 0 \\ \vdots & \vdots & \ddots & \vdots \\ 0 & 0 & \ldots & w_{iN} \end{bmatrix}$ where $\mathbf{W}_i = w_{ij}$ represents the weight of the data at point j on the calibration of the model around point i. These weights will vary with i, which distinguishes GWR from traditional Weighted Least Squares where the weighting matrix is fixed.

15.4 Result and Discussion

15.4.1 Ordinary Least Square, OLS

The independent variable nonciti was positively correlated with the violent crime rate whereas other variables unemp, popden, middle and singlemom were negatively correlated. However, only three variables have statistically significant correlation with violent crime rate which are nonciti, BASIC index and singlemom. The nonciti was has statistically significant positive correlation and others variable have statistically negative correlation with violent crime rate in 2001.

In 2005, the independent variable popdens, nonciti and BASIC index were positively correlated with the violent crime rate whereas other variables unemp and singlemom were negatively correlated. However, only two variables have statistically significant correlation with violent crime rate. The nonciti factors were statistically significant at 0.05 significant level and BASIC index factor was

Table 15.1 Model summary for the year 2001, 2005 and 2009

Model	*R*	*R* Square	Adjusted *R* Square
2001	0.472	0.223	0.160
2005	0.351	0.123	0.066
2000	0.389	0.151	0.083

statistically significant at 0.1 significant level. In 2009, the independent variable nonciti, BASIC and MIDDLE index were positively correlated with the violent crime rate whereas other variables unemp, popdens and singlemom were negatively correlated. The nonciti and singlemom factors were statistically significant at 0.05 significant level and MIDDLE factor was statistically significant at 0.1 significant level.

The result from this 3 years study showed that the nonciti factor give the stable result for these 3 years. In order to measure the value of relationship between the violent crime rates with the factors of interest, OLS regression will be performed. Table 15.1 showed the OLS model fitted for the year 2001, 2005 and 2009.

The model with six covariates explains 16.0 % (R^2 adj = 0.16) of the variance in the rate of violent crime for the year 2001, 6.6 % for the year 2005 and 8.3 % for the year 2009 as shown in Table 15.4. Due to the relatively low percentage of the selected variables explain the variation in the model; it is believed that there are other factors not taken into consideration in this model.

The Analysis of Variance was conducted to check whether OLS regression model was statistically significant or not. The *F* statistics ($F = 3.580$; *p*-value = 0.004) showed that the overall regression model was highly significant in 2001. Similar results were shown in 2005 ($F = 2.138$; *p*-value = 0.070) and 2009 ($F = 2.224$; *p*-value = 0.050) which overall regression was significant at 0.1 significant level.

Table 15.2 shows statistical results of fitted OLS model for the year 2001, 2005 and 2009. For the year 2001, the regression coefficients are statistically significant at the level of 5 % level for the BASIC index variables. The violent crime rates are negatively related to the BASIC index. For every unit increase in BASIC index, the

Table 15.2 OLS model for violent crime rates in Peninsular Malaysia (2001, 2005 and 2009)

Model	Coefficients value (**p-value**)		
	2001	2005	2009
unemp			
popden			
nonciti		8.501670 (**0.009**)	6.476166 (**0.025**)
basic	−0.920874 (**0.003**)		
singlemomper			
middle			

violent crime rate is decreasing by approximately 1 unit. BASIC index represent the household with basic amenities. The results suggest that violent crime rate become increase when the Basic index is decrease. That means, the area where have less access to household basic amenities would cause an increase in violent crime rates. This situation can be attributed to the basic needs cannot be met. Due to lack of basic needs, many social problems will exist and one of them is this violent crime rates.

For the year 2005, the regression coefficient of nonciti was statistically significant at 5 % and significance level (p-value = 0.00898) and had positive contribution on the response variable where the increase in the variable leads to increase in the violent crime rate. Meanwhile, the regression coefficient of nonciti factor was statistically significant at 5 % of significance level (p-value = 0.02535) and had positive contribution on the response variable where the increase in the variable leads to increase in the violent crime rate on 2009. The nonciti variable also can be referred to the foreigner variable. In this study, the result showed that the higher nonciti rate is connected to the higher violent crime rate. This result parallel with the result obtained by Kakamu et al. [26]. The result for property crime rate can be referred to Table 15.3.

Since the global model with six variables explains only 16.0, 6.6 and 8.3 % (29.28, 8.00 and 13.55 %) of the variance in the violent crime rate (property crime rate) for 3 years, it is indicating that there are any other factors influencing the risk of violent crime and property crime unaccounted for in the model. Thus, some of the unexplained variance may be associated with the assumption of spatial stationary. It is reasonable to argue that the global regression model do not explain the local effect in the relationships of violent crime rate and the explanatory variables. This is because the parameter estimates of OLS are constant across the study area without distinguishing between affluent locations surrounded by an affluent neighbourhood or deprived neighbourhood. It is believed that OLS model misspecification of the actual relationship between the violent crime rate and explanatory variables. In order to analyze the local effect is to employ the GWR model [25, 27].

Table 15.3 OLS model for property crime rates in Peninsular Malaysia (2001, 2005 and 2009)

Model	Coefficients value **(p-value)**		
	2001	2005	2009
unemp			
popden			
nonciti		−1.00610 (**0.017**)	−1.37740 (**0.018**)
basic	0.19945 (**0.000**)	−0.06069 (**0.048**)	
singlemomper			
middle			

15.4.2 Geographically Weighted Regression, GWR and the Comparison with OLS Model

The standard error will be compared to the interquartile range of the local estimates from GWR result to determine the non-stationary relationship across space. If the interquartile range of the local estimates is greater than that ±1 standard deviation of the equivalent global parameter, then we are believed that the parameter under investigation might be non-stationary. In contrast to a single set of constant estimated values over the study area generated by the OLS, GWR produced set of parameter estimates and model statistics at each sampled point. For a general understanding of non-stationarity relationship across space, Fotheringham et al. [25] suggested comparing the range of the GWR local parameter estimates with confidence intervals (CI) around OLS global estimate of the equivalent parameter. If the interquartile (i.e., 25–75 % quartiles) range of the local parameter estimated is greater than that of ±1 standard deviation of the equivalent global parameter (i.e., 2 × S.E. of each global estimate), then the parameter under investigation might be non-stationary [25].

In 2001, the local variation in the all factors parameter estimates does not appear to be very unusual whereas the local variation in the nonciti, BASIC index and singlemom estimates on 2005 showed that spatial variation of relationships exist. Meanwhile, the result obtained in 2009 showed that the interquartile values for some factors (i.e., popdens, BASIC, MIDDLE and singlemom) were greater that their corresponding 2 × S.E. of the global estimates, indicating the presence of non-stationarity of relationship over the study area. It is believed that there is an advantage to use GWR model over the simplified OLS model.

One question arises in this regression analysis whether GWR model is statistically significant better than OLS model. It is possible to evaluate whether the rate of change for any specific variable alters significantly across the study region. As explanation given before, there will be non-stationary exist of relationship over the study area. It is believed that the GWR model is better than OLS model. One of the ways to compare performance of these two models is by looking model diagnostic such as the value of R^2 and AIC as shown in Table 15.4. Generally, the best model is the model with higher value of R^2 and lower value of AIC.

As expected, the results suggest that GWR model fitted better than OLS model for both crimes. For violent crime that shown in Table 15.4, the AIC value reduced from 96.6845 for the OLS model to 92.5394 for the GWR model for the year 2001. For the year 2005 and 2009, the AIC value was decreased from 89.8384 to 68.6990

Table 15.4 Model diagnostic for the violent crime rates

Parameter	OLS			GWR		
	2001	2005	2009	2001	2005	2009
R^2 Adj	0.1604	0.0656	0.0831	0.2798	0.4119	0.5268
Akaike information criterion (AIC)	96.6845	89.8384	69.9512	92.5694	68.6990	36.6215

Table 15.5 Model diagnostic for the property crime rates

Parameter	OLS			GWR		
	2001	2005	2009	2001	2005	2009
R^2 adj	0.1604	0.0656	0.0831	0.2798	0.4119	0.5268
Akaike information criterion (AIC)	96.6845	89.8384	69.9512	92.5694	68.6990	36.6215

and from 69.9512 to 36.6215. In addition, Adjusted R^2 value is higher for GWR than it was for our best OLS model which is OLS was 16.0 %; GWR is almost 28.0 % in 2001. This value increased from 6.6 % for OLS model to 41.2 % for GWR model for the year 2005 and increased from 8.3 to 52.7 % in 2009. The results obtained for property crime rate was more or less similar to violent crime rate (Table 15.5).

So, it is interesting to evaluate the factors that give impact on violent crime rate and property crime rate by incorporating the spatial locations of data. It can be concluded that GWR seem able to produce better model than that produced by OLS by accounting for spatial non-stationarity in the modelling process.

15.5 Conclusion

This study has reported empirical finding about spatial patterns in rate of violent crime and its relationship with several explanatory variables of districts in Peninsular Malaysia. These relationships have been measured using OLS and GWR method. The GWR results have shown that there are significant spatial variations in the relationships between the violent crime rate and the explanatory variables. GWR results are potentially useful in targeting priority areas for crime prevention and for informing local planning and policy development.

The results suggest that violent crime rate become increase when the Basic index is decrease in 2001. That means, the area where have less access to household basic amenities would cause an increase in violent crime rates. This situation can be attributed to the basic needs cannot be met. Due to lack of basic needs, many social problems will exist and one of them is this violent crime rates. Meanwhile, the result showed that the higher nonciti rate is connected to the higher violent crime rate in 2005 and 2009. As expected, the results suggest that GWR model fitted better than OLS model in all these 3 years study. It can be concluded that GWR seem able to produce better model than that produced by OLS by accounting for spatial non-stationarity in the modelling process.

Therefore, some actions or strategic planning in controlling and addressing the crime problem in the high risk areas not only has to consider the factors that influence crime in that particular area, but also the situation of neighbouring areas. The authorities should pay more attention to those areas in taking necessary actions to overcome this problem. As far as possible, we want to reduce both crimes because they affect not only the people in particular area, but also to the nation in general.

Acknowledgments This research was supported by Short Term Grant offered by Universiti Sains Malaysia (304/PMATHS/6310041) and a postgraduate scholarship from the Ministry of Higher Education, Malaysia. This study also funded by the Research Grants NIC NRGS-UMT (NRGS/2015/53131/30). The crime data were provided by Royal Malaysia Police (PDRM).

References

1. Dutt, A.K., Venugopal, G.: Spatial patterns of crime among Indian cities. Geoforum **14**(2), 223–233 (1983)
2. Tsushima, M.: Economic structure and crime: the case of Japan. J Socio-Econ **25**(4), 497–515 (1996)
3. Appiahene-Gyamfi, J.: Violent crime in Ghana: the case of robbery. J. Crim. Justice **26**(5) (1998)
4. Sidhu, A.S.: An academic and statistical analysis. J Kuala Lumpur R Malays Police College **1** (4), 1–28 (2005)
5. Sidhu, A.S.: Crime levels and trends in the next decade. J Kuala Lumpur R Malays Police College **5** (2006)
6. Cole, J.H., Gramajo, A.M.: Homicide rates in a cross-section of countries: evidence and interpretations. Popul. Dev. Rev. **35**(October), 749–776 (2009)
7. Perreault, B.S., Savoie, J.: Crime and justice research paper series neighbourhood characteristics and the distribution of crime on the Island of Montréal. Additional **85** (2008)
8. Savoie, B.J., Bédard, F., Collins, K.: Crime and justice research paper series neighbourhood characteristics and the distribution of crime on the Island of Montréal (2006)
9. Appiahene-Gyamfi, J.: An analyses of the broad crime trends and patterns in Ghana. J. Crim. Justice **30**(3), 229–243 (2002)
10. Kanyo, N., Norizan, M.N.: Pelakuan Jenayah dari Perspektif Geografi: Satu Kajian Kes di Daerah Timur Laut Pulau Pinang. Persidangan Geografi 2007 Memperkasa Geografi di Malaysia: Isu dan Cabaran (2007)
11. Berke, O.: Choropleth mapping of regional count data of Echinococcus multilocularis among red foxes in Lower Saxony. Prev. Vet. Med. **52**, 119–131 (2001)
12. Chaikaew, N., Tripathi, N. K., Souris, M.: Exploring spatial patterns and hotspots of diarrhea in Chiang Mai, Thailand. Int. J. Health Geographics **8**, 36 (2009)
13. Grubesic, T. H., Mack, E. a, & Kaylen, M. T.: Comparative modeling approaches for understanding urban violence. Social science research, 41(1), 92–109 (2012)
14. Huang, Y., Leung, Y.: Analysing regional industrialisation in Jiangsu province using geographically weighted regression. J. Geogr. Syst. **4**(2), 233–249 (2002)
15. Ye, X., Wu, L.: Analyzing the dynamics of homicide patterns in Chicago: ESDA and spatial panel approaches. Appl. Geogr. **31**, 800–8007 (2011)
16. Erdogan, S.: Explorative spatial analysis of traffic accident statistics and road mortality among the provinces of Turkey. Journal of Safety Research **40**(5), 341–351 (2009)
17. Holt, J.B., Lo, C.P.: The geography of mortality in the Atlanta metropolitan area. Comput. Environ. Urban Syst. **32**(2), 149–164 (2008)
18. Rincón-Ruiz, A., Pascual, U., Flantua, S.: Examining spatially varying relationships between coca crops and associated factors in Colombia, using geographically weight regression. Appl. Geogr. **37**, 23–33 (2013)

19. Chalkias, C., Papadopoulos, A.G., Kalogeropoulos, K., Tambalis, K., Psarra, G., Sidossis, L.: Geographical heterogeneity of the relationship between childhood obesity and socio-environmental status: empirical evidence from Athens, Greece. Appl. Geogr. **37**, 34–43 (2013)
20. Brunsdon, C., Fotheringham, A.S., Charlton, M.E.: Geographically weighted regression: a method for exploring spatial nonstationarity. Geogr. Anal. **28**(4) (1996)
21. Andresen, M.A.: Estimating the probability of local crime clusters: the impact of immediate spatial neighbor. J. Crim. Justice **39**, 394–404 (2011)
22. Ceccato, V., Dolmen, L.: Crime in rural Sweden. Appl. Geogr. **31**(1), 119–135 (2011)
23. Ceccato, V., Oberwittler, D.: Comparing spatial patterns of robbery: Evidence from a Western and an Eastern European city. Cities **25**(4), 185–196 (2008)
24. Nuzlinda, A.R., Syerrina, Z.: The household-based socio-economic index for every districts in Peninsular Malaysia. In: Proceeding of World Academy of Science, Engineering and Technology (2012)
25. Fortheringham, A.S., Brunsdon, C., Charlton, M.: Geographically Weighted Regression the analysis of spatially varying relationships. John Wiley & Sons, LTD (2002)
26. Kakamu, K., Polasek, W., Wago, H.: Spatial interaction of crime incidents in Japan. Math. Comput. Simul. **78**(2–3), 276–282 (2008)
27. Poetz, A.: Residential burglaries and neighborhood socioeconomic context in London, Ontario: global and local regression analysis*. Prof. Geogr. **57**, 516–529 (2005)

Chapter 16
Hybridizing Wavelet and Multiple Linear Regression Model for Crude Oil Price Forecasting

Ani Shabri and Ruhaidah Samsudin

Abstract Crude oil prices play a significant role in the global economy and contribute an important factor affecting government's plans and commercial sectors. In this paper, the accuracy of the simple wavelet multiple linear regression (WMLR) model in crude oil prices forecasting was investigated. The WMLR model was improved by combining two methods: discrete wavelet transform (DWT) and a multiple linear regression (MLR) model. To assess the effectiveness of this model, daily crude oil market-West Texas Intermediate (WTI) was used as the case study. Time series prediction capability performance of the WMLR model is compared with the Artificial neural network (ANN), autocorrelation integrated moving average (ARIMA), MLR and Generalized Autoregressive Conditional Heteroscedasticity (GARCH) models using various statistics measures. The results show that the hybrid WMLR is more accurate and perform better than of any individual model in the prediction of crude oil prices series.

Keywords Wavelet · Multiple linear regression · GARCH, ARIMA · Crude oil · Forecasting

16.1 Introduction

The international crude oil prices play an important part in the economy as the trends in changing oil prices will affect the financial markets. Crude oil prices do play a significant role in the global economy and constitute an important factor

A. Shabri (✉)
Department of Science Mathematic, Faculty of Science, Universiti Teknologi Malaysia, Johor Bahru, Malaysia
e-mail: ani@utm.my

R. Samsudin
Department of Software Engineering, Faculty of Computing, Universiti Teknologi Malaysia, Johor Bahru, Malaysia
e-mail: ruhaidah@utm.my

A.-R. Ahmad et al. (eds.), *Proceedings of the International Conference on Computing, Mathematics and Statistics (iCMS 2015)*,
DOI 10.1007/978-981-10-2772-7_16

affecting government's plans and commercial sectors. Therefore, proactive knowledge of its future fluctuations can lead to better decisions in several managerial levels. Due to the importance of crude oil price in future, there have been abundant studies on analysis and forecasting of crude oil price.

Crude oil price forecasting has been a challenging topic in the field of energy market research. The application of classical time series models such as Autoregressive Moving Average (ARMA) [1, 2] and Generalized Autoregressive Conditional Heteroscedasticity (GARCH) type models [3–5] for crude oil forecasting has received much attention in the last decade. However, they are basically linear models and have a limited ability to capture non-linearities and nonstationary in crude oil forecasting.

In 2014, wavelet transforms has become a useful method for analyzing variations, periodicities and trends in time series. Recently, new hybrid models on wavelet transform processes have been improved for forecasting. For example the wavelet-neural network [6–8], wavelet-least square support vector machines (LSVM) [9] and wavelet-fuzzy neural network [10] have been employed recently in numerous studies for crude oil forecasting. It was observed that the wavelet transform performs slightly better in improving forecasting accuracy.

In this paper, a hybrid wavelet multiple linear regression (WMLR) model integrating wavelet and multiple linear regression (MLR) was proposed to forecast crude oil prices. The sequence of implementation consists of the initial stage of decomposition of the original price using the Mallat algorithm of the wavelet transform. The decomposed levels were then used as the input layer of MLR model. For verification purpose, the daily crude oil market-West Texas Intermediate (WTI) was used for one-daily-ahead forecasting. Finally, to evaluate the model ability, the proposed model was compared with individual MLR, ARIMA and GARCH models studied by Yu et al. [11] and Ani and Samsudin [12].

16.2 Methodology

16.2.1 Discrete Wavelet Transformation

Wavelet transformations provide useful decomposition of original time series by capturing useful information on various decomposition levels. Discrete wavelet transformation (DWT) is preferred in most forecasting problems because of its simplicity and ability to compute with less time. The DWT can be written as

$$\psi_{m,n}(t) = \frac{1}{\sqrt{s_0^m}}\psi\left(\frac{t - n\tau_0 s_0^m}{s_0^m}\right) \tag{16.1}$$

where t is time, $\psi(t)$ called the mother wavelet, m and n are integers that control the scale and time. The most common choices for the parameters $s_0 = 2$ and $\tau_0 = 1$. For a discrete time series $x(t)$, the DWT can be defined as

$$W_{m,n} = 2^{-m/2} \sum_{t=0}^{N-1} \psi(2^{-m}t - n)\, x(t) \tag{16.2}$$

According to Mallat's theory, the original discrete time series $x(t)$ can be decomposed into a series of linearity independent approximation and detail signals by using inverse DWT. The inverse DWT is given by [13]

$$x(t) = T + \sum_{m=1}^{M} \sum_{t=0}^{2^{M-m-1}} W_{m,n} 2^{-m/2} \psi(2^{-m}t - n). \tag{16.3}$$

or in a simpler form

$$x(t) = A_M(t) + \sum_{m=1}^{M} D_m(t) \tag{16.4}$$

in which $A_M(t)$ is the approximation sub-series or residual term at levels M and $D_m(t)$ (m = 1, 2,…, M) are detail sub-series which can capture small features of interpretational value in the data.

16.2.2 Multiple Linear Regressions

MLR model is one of the modelling techniques to investigate the relationship between a dependent variable and several independent variables. Let the MLR have p independent variables with n observations. Thus the MLR can be written as

$$Y = w_0 + w_1 x_i + w_2 x_2 + \cdots + w_p x_p + \varepsilon_t \tag{16.5}$$

where w are regression coefficients, Y is dependent variable, x_i are independent variables and ε_t is fitting errors. The method of least squares is used to estimate the coefficients model.

16.3 An Application

In this study, the WTI crude oil price series was the experimental sample selected for forecasting. The WTI crude oil is the most famous benchmark prices, which is widely used as the basis of many crude oil Price Formulae. The daily data dated

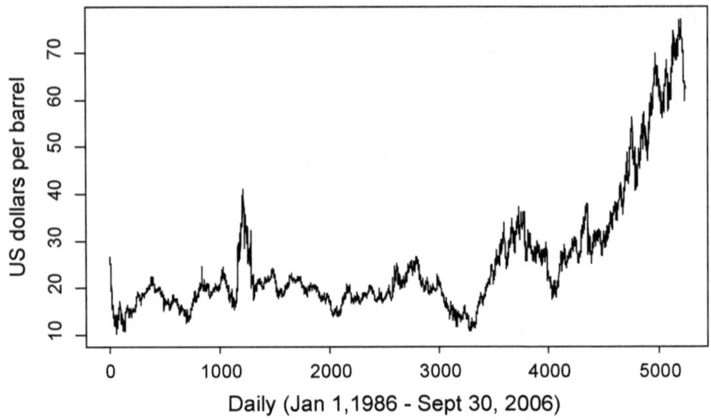

Fig. 16.1 Daily crude oil prices

from January 1, 1986 to September 30, 2006, excluding public holidays, with a total of 5237 was employed as experimental data. For convenience of WLR modelling, the data from January 1, 1986 to December 31, 2000 was used for training set (3800 observations), and the remainder was used as testing set (1437 observations). Figure 16.1 shows the daily crude oil prices from January 1, 1986 to September 30. In practice, short-term forecasting results are more useful as they provide timely information for the correction of forecasting value. In this study, two performance criteria such as RMSE and MAE were used to evaluate the accuracy of the models. The criteria are given below

$$\text{RMSE} = \sqrt{\frac{1}{n}\sum_{t=1}^{n}(x_t - \hat{x}_t)^2} \quad \text{and} \quad \text{MAE} = \frac{1}{n}\sum_{t=1}^{n}|x_t - \hat{x}_t|$$

where x_t is the actual and $\hat{x}_t$ is the forecasted value of period t, and n is the number of total observation.

16.4 Fitting Hybrid Wavelet-Regression Model to the Data

The WLR model was obtained by combining two methods, the wavelet transform and the MLR model. The WMLR model is a MLR model, which uses sub-time series components obtained using DWT on original log return data. The observed series was decomposed into a number of wavelet components, depending on the selected decomposition levels. Deciding the optimal decomposition level of the time series data in wavelet analysis plays a significant role in preserving the information and reducing the distortion of the datasets. Several researchers have

Table 16.1 The correlation coefficients between each of sub-time series and original machinery data series

Wavelet	Coefficient of determination (R^2) between Q_t and						Mean
Components	D_{t-1}	D_{t-2}	D_{t-3}	D_{t-4}	D_{t-5}	D_{t-6}	R^2
D1	0.0000	0.0000	0.0000	0.0000	0.0000	0.0001	0.0000
D2	0.0001	0.0000	0.0000	0.0003	0.0001	0.0001	0.0001
D3	0.0007	0.0006	0.0003	0.0000	0.0003	0.0010	0.0005
A3	0.9896	0.9917	0.9937	0.9953	0.9965	0.9973	0.9940

used an empirical equation to select the number of decomposition levels [14–16]. In this study, the following empirical equation is used to determine the decomposition level

$$M = \text{int}[\log(N)] \tag{16.6}$$

where N is length of the time series, int[.] is the integer-part function and M is decomposition level. In this study, $N = 5237$ daily data are used, which gives $M = 3$ decomposition levels. Thus, three decomposition levels (2–4–8 months) were considered for this study.

The effectiveness of wavelet components is determined using the coefficient of determination (R^2) between each D sub-time series and original data was given in Table 16.1. It can be seen that the D1 component ($R^2 = 0$ %) shows that no variance in D1 is shared with the original data. However, the wavelet component D2 and D3 displayed significantly higher R^2 compared to the D1. According to the R^2 analyzes, the effective components D2 and D3 were selected as the dominant wavelet components. Thereafter, the significant wavelet components D2, D3 and approximation (A3) component were added to each other to constitute the new series to the LR model. Figure 16.1 shows the original crude oil prices and their Ds, that is the time series of 2-month mode (D1), 4-month mode (D2), 8-month mode (D3) and approximate mode (A3). Figure 16.2 shows the structure of the WLR model (Fig. 16.3).

Six input combinations based on previous log return of daily oil prices are evaluated to estimate current prices value. The input combinations evaluated in this study are (i) r_{t-1}, (ii) r_{t-1}, r_{t-2}, (iii) $r_{t-1}, r_{t-2}, r_{t-3}$, (iv) $r_{t-1}, r_{t-2}, r_{t-3}, r_{t-4}$, (v) $r_{t-1}, r_{t-2}, r_{t-3}, r_{t-4}, r_{t-5}$ and (vi) $r_{t-1}, r_{t-2}, r_{t-3}, r_{t-4}, r_{t-5}, r_{t-6}$ where $r_t = \ln(x_t/x_{t-1})$ is log return.

A program code including wavelet toolbox was written in MATLAB language for the development of MLR and WMLR model. The forecasting performances of the MLR and WMLR models in terms of the RMSE and MAE at testing phase are compared and shown in Table 16.2. Table 16.2 shows the MLR model, the M6 with six lags obtained the best MSE statistics of 0.9706, while the M1 with 1 lag obtained the best MAE statistics of 0.7123. For WMLR, model M6 with lags 6 obtained the best RMSE and MAE statistics of 0.4830 and 0.3363, respectively.

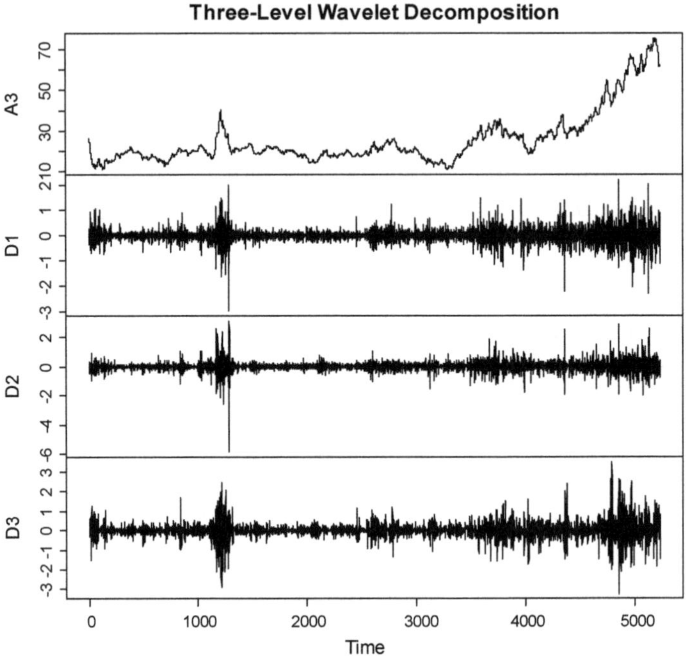

Fig. 16.2 Decomposed wavelet sub-time series components and approximation of crude oil

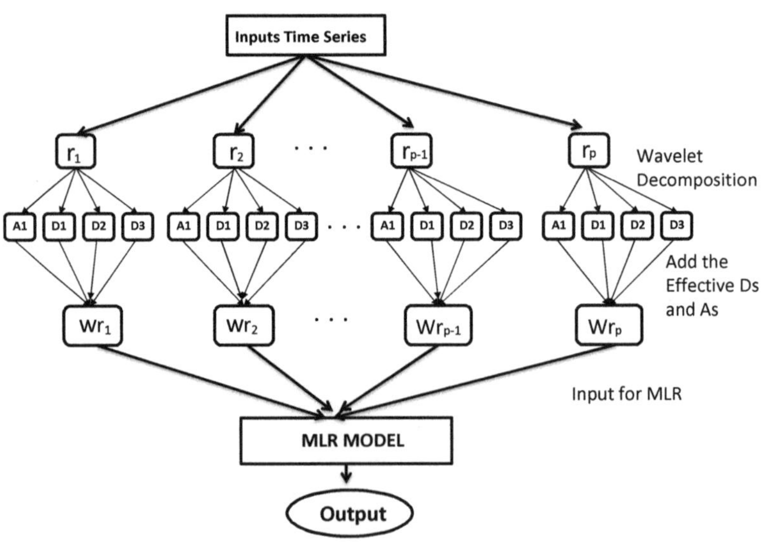

Fig. 16.3 The structure of WMLR model

Table 16.2 Forecasting performance indices of MLR and WMLR

Model	Lag	MLR		WMLR	
		RMSE	MAE	RMSE	MAE
M1	1	0.9708	**0.7123**	0.7592	0.5642
M2	1,2	0.9709	0.7124	0.6712	0.4909
M3	1,2,3	0.9697	0.7136	0.5861	0.4262
M4	1,2,3,4	0.9710	0.7134	0.5381	0.3875
M5	1,2,3,4,5,	0.9715	0.7134	0.5001	0.3524
M6	1,2,3,4,5,6	**0.9706**	0.7127	**0.4830**	**0.3363**

Table 16.3 The RMSE and MAE comparisons for different models

Model	RMSE	MAE
ARIMA(2,1,5) [12]	0.9573	0.6970
GARCH(1,1) [12]	0.9513	0.6947
WMLR + PCA [12]	0.6572	0.4834
MLR	0.9708	0.7123
WMLR (proposed model)	**0.4830**	**0.3363**
Yu' ARIMA [11]	2.0350	–
Yu' FNN [11]	0.8410	–

For further analysis, the best performance of WMLR was compared with the best results of ARIMA and forward neural network (FNN) studied by Yu et al. [11], GARCH(1,1) and WMLR using PCA studied by Ani and Samsudin [12]. In Table 16.3, it shows that the proposed WMLR model has good performance where they outperformed all models in terms of all the standard statistical measures. This results show that the new series (DWT) have significant positive effect on MLR model results.

16.5 Conclusions

A hybrid wavelet-WMLR model to predict world crude oil prices was proposed. The performance of the proposed WMLR model was compared to regular MLR and several models from previous studies for crude oil forecasting. Results of the study show that the WMLR model was substantially more accurate than other models. The study concludes that the forecasting abilities of the MLR model are found to be improved when the wavelet transformation technique was adopted for the data pre-processing. The decomposed periodic components obtained from the DWT technique were found to be most effective in yielding accurate forecast when used as inputs in the MLR model. The WMLR model presented in this study is a simple explicit mathematical formulation. The accurate forecasting results indicate that WMLR model provides a superior alternative to other models and a potentially very useful new method for crude oil forecasting.

Acknowledgments The authors thankfully acknowledged the financial support afforded by MOE, UTM and GUP Grant (VOT 4F681).

References

1. Mohammadi, H., Su, L.: International evidence on crude oil price dynamics: applications of ARIMA-GARCH models. Energy Econ. **32**, 1001–1008 (2010)
2. Ahmad, M.I.: Modelling and forecasting Oman crude oil prices using Box-Jenkins techniques. Int. J. Trade Global Markets **5**, 24–30 (2012)
3. Agnolucci, P.: Volatility in crude oil futures: a comparison of the predictive ability of GARCH and implied volatility models. Energy Econ. **31**, 316–321 (2009)
4. Wei, Y., Wang, Y., Huang, D.: Forecasting crude oil market volatility: further evidence using GARCH-class models. Energy Econ. **32**, 1477–1484 (2010)
5. Liu, L., Wan, J.: A study of Shanghai fuel oil futures price volatility based on high frequency data: long-range dependence, modeling and forecasting. Econ. Model. **29**, 2245–2253 (2012)
6. Jammazi, R., Aloui, C.: Crude oil forecasting: experimental evidence from wavelet decomposition and neural network modeling. Energy Econ. **3**, 828–841 (2012)
7. Qunli, W., Ge, H., Xiaodong, C.: Crude oil price forecasting with an improved model based on wavelet transform and RBF neural network. In: International Forum on Information Technology and Applications, pp. 231–234 (2009)
8. Yousefi, S., Weinreich, I., Reinarz, D.: Wavelet-based prediction of oil prices. Chaos, Solitons Fractals **25**, 265–275 (2005)
9. Bao, Y., Zhang, X., Yu, L., Lai, K.K., Wang, S.: Hybridizing wavelet and least squares support vector machines for crude oil price forecasting. In: Proceedings of IWIF-II, Chengdu, China (2007)
10. Liu, J., Bai, Y., Li, B.: A new approach to forecast crude oil price based on fuzzy neural network. In: Fourth International Conference on Fuzzy Systems and Knowledge Discovery, Haikou (2007)
11. Yu, L., Wang, S., Lai, K.K.: Forecasting crude oil price with an EMD-based neural network ensemble learning paradigm. Energy Econ. **30**, 2623–2635 (2008)
12. Ani, S., Samsudin, R.: Crude oil price forecasting based on hybridizing wavelet multiple linear regression model, particle swarm optimization techniques, and principal component analysis. Sci. World J. 1–8 (2014)
13. Mallat, S.G.: A theory for multi decomposition signal decomposition: the wavelet representation, IEEE Trans. Pattern Anal. Mach. Intell **11**(7), 674–693(1989)
14. Nejad, F.H., Nourani, V.: Elevation of wavelet denoising performance via an ANN-based streamflow forecasting model. Int. J. Comput. Sci. Manage. Res. **1**(4), 764–770 (2012)
15. Belayneh, A., Adamowski, J.: Standard precipitation index drought forecasting using neural networks, wavelet neural networks, and support vector regression. Appl. Comput. Intell. Soft. Comput. 6. http://dx.doi.org/10.1155/2012/794061 (2012)
16. Seo, Y., Kim, S., Kisi, O., Singh, V.P.: Daily water level forecasting using wavelet decomposition and artificial intelligence techniques. J. Hydrol. **520**, 224–243 (2015)

Chapter 17
Modeling Relationship Between Stock Market of UK and MENA Countries: A Wavelet Transform and Markov Switching Vector Error Correction Model Approach

Amel Abdoullah Dghais and Mohd Tahir Ismail

Abstract This study combines wavelet filtering and Markov-Switching Vector Error Correction model (MSVECM) in order to investigate the dynamic relationship among financial time series. The UK stock market and four stock markets of the Middle East and North Africa (MENA) region namely, Egypt, Morocco, Saudi Arabia, and Istanbul are used and the monthly data are from May 1, 2001 to December 30, 2011. The series generated by the discrete wavelet transform (DWT) is then analyzed to determine the long-run and short-run relationships between the stock markets using a cointegration test and a MSVECM model. The comparison between the proposed and traditional models demonstrates that the former dominates the latter in performance and fitting the financial stock market series. The cointegration test affirms the existence of long-term relationship between the studied series. The proposed model also shows the existence of a short-term relationship between stock market in the UK and all other stock.

Keywords Discrete wavelet transform · MSVECM model · MENA stock market, cointegration

17.1 Introduction

Recently, increasing economic integration of the global economy has become a popular research topic in the economy which has encouraged the scholars and investors to focus on the issue of studying a dynamic relationship among the stock

A.A. Dghais (✉) · M.T. Ismail
School of Mathematical Sciences, Universiti Sains Malaysia, Penang, Malaysia
e-mail: Amel_usm@yahoo.com

M.T. Ismail
e-mail: m.tahir@usm.my

A.-R. Ahmad et al. (eds.), *Proceedings of the International Conference on Computing, Mathematics and Statistics (iCMS 2015)*,
DOI 10.1007/978-981-10-2772-7_17

market indices around the globe. Nevertheless, most of the researchers investigated the relationship among the stock markets by applying linear models in return to stock markets, such as a vector autoregressive model (VAR) or vector error correction model (VECM), however these models has a weakness, include the inability to capture asymmetries in the series, and thus, nonlinear models have been progressively applied such as Markov switching models.

The main advantages of Markov switching model, as frequently advocated in the literature, include their capability to handle several crucial features of a time series (such as nonlinear phenomena) and to model temporal asymmetries and the persistence of a macroeconomic time series [1]. These features are crucial in analyzing the relationship among global stock market returns. The Markov switching autoregressive model has become increasingly popular since [2] applied this model to measure business cycle in the US. Later, [3] extend to Markov switching vector autoregressive model (MSVAR) and Markov switching vector error correction model (MSVECM). Some of the papers that used these MSVAR and MSVECM models are as follows.

Ismail and Isa [4] used monthly data (1990–2004) to investigate the relationship among the return of three markets of Southeast Asian region, Malaysia, Singapore, and Thailand, based on cointegration test, VAR model, and MSVAR model. The study showed that there was no long-run relationship between the stock markets of ASEAN region, while in short term, the MSVAR had showed co-movements among the three returns stock markets. Additionally, they explored that the performance and fitting data of MSVAR model are much better than that of VAR model.

While, [5] used the weekly closing price for stock market indices of the US, UK, Japan, Hong Kong, and Mainland China regions to study the long- and short-run relationships between the China and International main stock markets from June 5, 1992 to December 26, 2008. Their study was based on cointegration test, VECM, and MSVECM models. They found a long-run relationship between the China and International main stock markets since 1999. In case of short-term relationship, they observed that the International main stock markets have been effecting on China stock market directly or indirectly. In comparison between VECM and MSVECM models, it was studied that the MSVEC model has less AIC, HQ, and SC. Therefore MSVEC model is better to provide effective information about these relationships among these stock markets.

Then, [6] investigated the relationships among the stock markets of the US, Australia, and New Zealand via the MSVAR and VAR models using weekly stock market returns. The results revealed that the MSVAR is more appropriate than the linear VAR in modeling the relationship among these stock markets. The correlations among the three markets were significantly higher in the bear regime than in the bull regime. In addition, the responses of these markets to shocks in other stock markets were stronger in the bear regime.

Later, [7] used the MSVAR model to examine the relationship of stock market returns between the International stock markets (specifically, US, UK, Singapore, and China) and the stock market of the Philippines from 2000 until 2010. Weekly stock market returns were used for this study. The findings revealed that, the Philippines stock market is most correlated with stock markets of Hong Kong and Singapore compared to stock markets of US and UK in both regimes. These results indicated that there is stronger relationship between economies in the same region.

In the present study, the aim is to develop an economic model using a new technique, based on the combination of the DWT and MSVECM model. It is a novel technique as to the limit of the author knowledge there does not exist any work in the literature using the proposed model (DWT with VECM model) to study the relationship between stock market of UK and selected stock markets of the MENA region. It also compares the proposed and traditional model in terms of their performance and fit with the financial stock market series.

17.2 Research Methodology

17.2.1 Discrete Wavelet Transform (DWT)

DWT aims to decompose the discrete time signal to basic functions called the wavelets to provide a clear analytic view of the analyzed signal. Proposed by [8], the DWT was implemented through pyramid algorithm to find the wavelet coefficients of a discrete series. The DWT can be defined in terms of two filters, namely, high-pass filter denoted as $H = \{h_l\}$ and low-pass filter denoted as $G = \{g_l\}$. The $\{h_l\}$ and $\{g_l\}$ are the coefficients of the filters and are defined as follows:

$$
\begin{aligned}
g_l &= \frac{1}{\sqrt{2}} \int \varphi(t)\varphi(2t - l)\mathrm{d}t \\
h_l &= \frac{1}{\sqrt{2}} \int \psi(t)\varphi(2t - l)\mathrm{d}t
\end{aligned}
\tag{17.1}
$$

where the father wavelet $\varphi(t)$ and mother $\psi(t)$ functions are defined as:

$$
\begin{aligned}
\varphi(t) &= \sqrt{2}\sum g_l\varphi(2t - l) \\
\psi(t) &= \sqrt{2}\sum h_l\varphi(2t - l)
\end{aligned}
\tag{17.2}
$$

Figure 17.1 shows a flow diagram for the first step of the pyramid algorithm where the symbol denotes $\downarrow 2$ downsampling (elimination of every other value of input vector) by 2.

Where x is a length vector $(N = 2^J)$ of observations. At each iteration of the pyramid algorithm, the filtering requires three objects: the data x, the high-pass filter, and low-pass filter. The first step is filtering the data with each filter to produce detail coefficients and scaling coefficients, subsample both filter coefficients to half their original lengths, keep the subsampled output from h_1 filter and repeat the filtering procedure for the subsampled output from g_1 filter as follows:

Fig. 17.1 Schematic illustrating of DWT decomposition using the pyramid algorithm

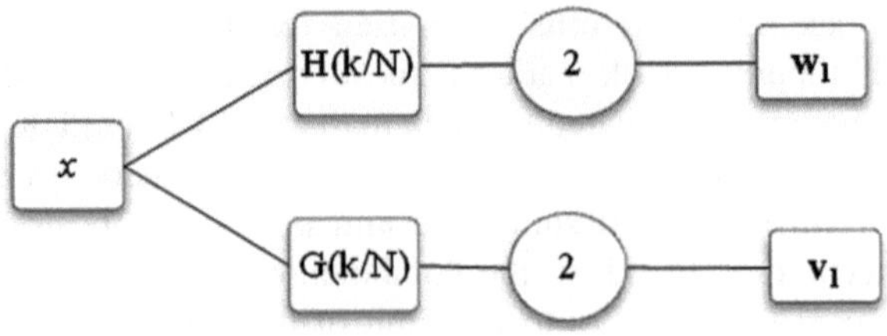

$$\begin{aligned} w_{1,t} &= \sum_{l=0}^{L-1} h_l x_{2t+1-l \bmod N} \\ v_{1,t} &= \sum_{l=0}^{L-1} g_l x_{2t+1-l \bmod N} \end{aligned} \tag{17.3}$$

where the $t = 0, 1, \ldots, N/2 - 1$ and the downsampling operation has been included in the filtering step through the subscript x_t. The x series has been filtered high as well as low pass to produce $N/2$ coefficients associated with this information. Repeating this procedure up to J times, the number of multi-resolution levels such that $J = \log_2(N)$, the DWT coefficients for scales can be computed as

$$\begin{aligned} w_{j,t} &= \sum_{l=0}^{L-1} h_l v_{j-1,2t+1-l \bmod N} \\ v_{j,t} &= \sum_{l=0}^{L-1} g_l x_{j-1,2t+1-l \bmod N} \end{aligned} \tag{17.4}$$

17.2.2 Wavelet with MSVECM

In this section, the DWT and MSVECM model are combined to study the relationship among the stock markets. As a primary step, the DWT is used to filter the data and then MSVECM model is applied to obtain the objective. A MSVECM model is an equilibrium model with shifts in some of the parameters are given by

$$\begin{aligned} &\Delta \mathbf{x_t} = \mathbf{v}(\mathbf{s_t}) + \alpha(s_t)[\beta' \mathbf{x_{t-1}}] + \sum_{k=1}^{p-1} \Gamma_i(s_t) \mathbf{\Delta x_{t-k}} + u_t \\ &u_t \sim N\left(0, \sum (s_t)\right) \end{aligned} \tag{17.5}$$

where α and β are $(2 \times r)$ matrices. The cointegrating rank = r as the number of stationary long-run relations while the cointegrating vector β is determined by solving an eigenvalue problem. Thus, $\beta'\mathbf{x_t}$ is a stationary long-run relation with the adjustment towards the equilibrium driven by the vector of loading matrix [9]. Therefore, the intercept terms, $\mathbf{v}(\mathbf{s_t})$, the error correction terms $\alpha(s_t)$, the stationary dynamics $\Gamma_i(s_t)$, and the variance–covariance terms $\sum(s_t)$ of the innovations of this MSVECM model are conditioned on the realization of the state variables.

The MSVECM model is often related to the notion of multiple equilibria in dynamic economic theory. Therefore, each regime is characterized by an attractor of the system defined by the long-run equilibrium $\mu(s_t)$ and the drift $\delta(s_t)$;

$$\Delta\mathbf{x_t} - \delta(s_t) = \alpha(s_t)[\beta'\mathbf{x_{t-1}} - \mu(s_t)] + \sum_{k=1}^{p-1} \Gamma_i(s_t)[\Delta\mathbf{x_{t-k}} - \delta(s_t)] + u_t \qquad (17.6)$$
$$u_t \sim NID\left(0, \sum(s_t)\right)$$

Both $\Delta\mathbf{x_t}$ and $\beta'\mathbf{x_t}$ are expressed as deviations about their regime and time dependent means, $\delta(s_t)$ and $\mu(s_t)$.

17.3 Empirical Results and Discussion

17.3.1 Data and Wavelet Decomposition

The data in the current study consist of the monthly price of two stock market indices of the UK and four stock markets of the MENA region (Egypt, Istanbul, Morocco, and Saudi Arabia), taken from DataStream which covers the period from May 1, 2001 to December 30, 2011. The stock market indices were decomposed using Daubechies (db2) basic function based on DWT, which is better than the MODWT (Maximal Overlap DWT) for producing a less noisy series for a given financial dataset as shown by [10]. The level of the wavelet decomposition corresponds to a level-2 decomposition of each stock market index that results are reasonable smoothing and produces a signal that identifies its identity without losing the basic characteristics of the original series.

17.3.2 Stationarity Test

Stationarity tests were performed on the level-2 approximation series to test whether or not the data have a unit root. The results in Table 17.1 indicate a p-value > 0.05. Thus, the level-2 approximation series of all stock market indices are non-stationary time-series processes. The first difference in the series is stationary and the order of integration is at $I(1)$.

Table 17.1 Unit root test results

Country	Augmented Dickey Fuller test (ADF)		Phillips-Perron test (PP)	
	p-value at level	p-value at first difference	p-value at level	p-value at first difference
UK	0.3513	0.000*	0.4455	0.000*
Egypt	0.4169	0.000*	0.5761	0.000*
Istanbul	0.3840	0.000*	0.5094	0.000*
Morocco	0.6763	0.000*	0.5518	0.000*
Saudi Arabia	0.2993	0.000*	0.3816	0.000*

*Denotes rejection of the hypothesis at the 0.05 level

17.3.3 Cointegration Test

For level-2 series, shown in Table 17.1, the logarithm price sequences are $I(1)$. Therefore, Johansen–Juselius test can be used to test the cointegration relationship among the variables. The results thus obtained show that the optimal lag order for the return level-2 series is 2. The results for the cointegration test are given in Table 17.2 where there are two cointegration vectors between UK stock market and stock markets of MENA. Thus, DWT-MSVECM and MSVECM models will be employed to model the relationship.

17.3.4 Model Performance Comparison

As shown in Table 17.3 for the stock markets of the UK and MENA region, the Akaike information criterion (AIC), Schwarz criterion (SBC), and Hannan-Quinn criterion (HQ) produced a minimum value for the proposed model (DWT-MSVECM). These results indicate that the performance and fit the financial series of the proposed model is better than the MSVECM.

Table 17.2 Results of Johansen-Juselius cointegration Test of UK and MENA

No. of cointegrating relations	Trace statistic	Maximum-Eigen statistic
None	87.651*	34.436*
At most 1	53.215*	25.631
At most 2	34.436	12.733

*Denotes rejection of the hypothesis at the 0.05 level

Table 17.3 Comparative performance of the models

Criterion	DWT-MSVECM	MSVECM
AIC	27.732	33.495
SBC	29.376	35.054
HQ	31.779	37.333

17.3.5 Model Estimation

Figure 17.2 presents the results of DWT-MSVECM on the UK and MENA equity markets and reveals that, the ECT1 coefficients on the Istanbul stock market have a significance level of 5 %, and is positively signed. While, the case of that of the UK and Saudi Arabia stock markets, which have negative sign, and a significance level of 10 and 5 % respectively. The ECT2 coefficients of UK, Istanbul, and Saudi

	UK	Egypt	Istanbul	Morocco	Saudi Arabia
Constant (Reg. 1)	−1.005	−6.776	−1.988	−1.427	−1.516
Constant (Reg.2)	0.512	2.402	2.147	0.872	0.542
UK_1	0.737**	0.504**	0.152	0.103	−0.062
UK_2	0.613	0.480	−3.994**	0.596	1.634**
UK_3	0.329**	0.442*	0.312582*	0.472987**	−0.147301
UK_4	−0.019972	−0.348394	0.222*	−0.452**	0.549**
UK_5	−0.132	0.263	−0.367**	0.277**	−0.509**
Egypt_1	−0.0753**	0.0299	−0.118**	0.004	−0.080
Egypt_2	−0.154	0.152	0.770**	−0.026	−0.161
Egypt_3	−0.033	0.108*	−0.203**	0.079**	−0.120**
Egypt_4	0.0130	0.142**	0.050	0.0138	0.0134
Egypt_5	−0.022	0.112*	−0.116**	0.099**	0.003
Istanbul_1	−0.166**	−0.154*	0.356**	0.112**	−0.051
Istanbul_2	−0.199**	0.475*	0.138*	−0.050	−0.339*
Istanbul_3	0.020	0.102	0.175*	−0.064	0.124
Istanbul_4	0.006	−0.029	−0.532**	0.138**	−0.039
Istanbul_5	0.026	−0.236**	0.407**	−0.028	0.043
Morocco_1	0.173**	0.511**	0.217**	0.245**	0.179**
Morocco_2	−0.119	−0.797**	0.537**	−0.325**	−0.408**
Morocco_3	0.039	−0.184**	0.299**	−0.045	0.158**
Morocco_4	−0.004	−0.139*	−0.042	0.049	−0.036
Morocco_5	0.111**	−0.072	0.374**	−0.107**	−0.010
Saudi Arabia_1	0.076*	0.203*	0.152**	−0.250**	0.703**
Saudi Arabia_2	0.334**	−0.144	−0.829**	0.124	0.498*
Saudi Arabia_3	−0.108**	0.224*	−0.158**	−0.236**	0.163*
Saudi Arabia_4	−0.159**	−0.173	0.117*	−0.040	−0.247**
Saudi Arabia_5	0.115**	−0.225*	−0.079	0.129*	0.127
Variance (Reg.1)	3.325	33.323	4.258	22.850	5.912
Variance (Reg.2)	3.682	4.418	7.990	2.431	4.011
CointEq1	1.000	0.000	0.429*	−0.665**	−0.024
CointEq2	0.000	1.000	2.914**	−2.689**	−1.567**
ECT1	−0.829*	−0.944	3.739**	−0.570	−1.748**
ECT2	0.226**	−0.008	−0.545**	0.062	0.387**

*significance at 10% level, **significance at 5% level.

Fig. 17.2 Estimates of DWT-MSVECM for the UK and MENA stock markets

Arabia stock markets have a significance level of 5 % and are positively signed. In the short term, most of the movements of the stock markets can be explained by certain national factors, as represented by the past performance of their own indices.

Bilateral causal relationships exist among the stock indices, such as those in the stock markets of the UK and Egypt, Istanbul, Morocco, and Saudi Arabia at a significance level of 5%. Also, the causal relationship holds among the four stock markets of the MENA countries at a significance level of 5 %, except stock market of Saudi Arabia is affected by Egypt at a significance level of 10 %, and in the same time is affecting on stock market of Istanbul by a significance level of 10 %.

17.4 Conclusion

This study employs a new technique to investigate the relationship between the stock markets of MENA region and the UK by combining the wavelet transform and MSVECM. Daubechies2 (db2) based on the DWT technique has been used to separate low-frequency signals from high-frequency noise in the UK and MENA index series and thereby eliminate noise in the stock price signal. The low-frequency data at level-2 has been used to examine the long and short-term relationships between the stock market of the UK and four of the stock markets of the MENA region. The results of the comparison between the proposed and traditional model show that the DWT-MSVECM model outperforms the MSVECM model in terms of performance and fit with the financial stock market series. The cointegration test for the level-2 data reveal a long-term relationship between the stock market of the UK and those of the MENA region. In the short-term, the stock market in the UK has relationships with all these stocks.

Acknowledgments We would like to acknowledge support for this project from Universiti Sains Malaysia (USM Short-term grant 304/PMATHS/6313045).

References

1. Diebold, F.: Comment on modeling the persistence of conditional variances. Econometric Rev. **5**, 51–56 (1986)
2. Hamilton, J.D.: A new approach to the economic analysis of nonstationary time series and the business cycle. Econometrica **57**, 357–384 (1989)
3. Krolzig, H.-M.: Markov-Switching Vector Autoregressions: Modelling, Statistical Inference, and Application to Business Cycle Analysis. Springer, Berlin (1997)
4. Ismail, M.T., Isa, Z.: Modelling nonlinear relationship among selected ASEAN stock markets. J. Data Sci. **6**, 533–545 (2008)
5. Fan, K., Lu, Z., Wang, S.: Dynamic linkages between the China and international stock markets. Asia-Pac. Finan. Markets. **16**, 211–230 (2009)

6. Qiao, Z., Li, Y., Wong, W.K.: Regime-dependent relationship among the stock markets of the US, Australia and New Zealand: a Markov-switching VAR approach. Appl. Financ. Econ. **21** (24), 1831–1841 (2011)
7. Tan, T.A.: Stock Market Integration: Case of the Philippines. Philippine Management Review. 19 (2012)
8. Gençay, R., Selçuk, F., Whitcher, B.J.: An Introduction to Wavelets and Other Filtering Methods in Finance and Economics. Academic Press, New York (2001)
9. Tillmann, P.: Inflation regimes in the US term structure of interest rates. Econ. Model. **24**, 203–223 (2007)
10. Ahmed, A.A., Ismail, M.T.: A comparative study between discrete wavelet transform and maximal overlap discrete wavelet transform for testing stationarity. Int. Math. Comput. Sci. Eng. **7**, 1–5 (2013)

Chapter 18
Parameter Estimation of the Exponential Smoothing with Independent Regressors Using R

Ahmad Farid Osman

Abstract As an extension to the state space approach of the exponential smoothing methods, Osman and King recently introduced a new class of exponential smoothing methods by integrating regressors into the model. For the model to be utilized successfully, it requires a proper estimation procedure. The parameter estimation can be done through optimization using the "optim" function available in R statistical software. The objective of this paper is to discuss the effective use of the "optim" function in estimating parameters of the state space model of the exponential smoothing method augmented with regressors. The study started by considering several sets of optimization R codes to be supplied to the "optim" function. These codes use different set of initial values as the starting points for the optimization routine. The other difference between optimization codes is also in terms of restrictions imposed on the optimization routine. The second phase of the study was done by generating a number of simulated time series data with predetermined parameter values. The final phase of the study was then conducted by applying the optimization codes on all simulated series. By analyzing the performance of each of the optimization codes to accurately estimate parameters, a guideline or suggestion on how to effectively execute "optim" function in R with respect to the use of the new forecasting approach then outlined.

Keywords Optimization · Time-varying parameter · State-space model · Parameter space

---

A.F. Osman (✉)
Department of Applied Statistics, Faculty of Economics and Administration,
Universiti Malaya, Kuala Lumpur, Malaysia
e-mail: faridosman@um.edu.my

A.-R. Ahmad et al. (eds.), *Proceedings of the International Conference on Computing, Mathematics and Statistics (iCMS 2015)*,
DOI 10.1007/978-981-10-2772-7_18

18.1 Introduction

The exponential smoothing method is an approach that used to produce future forecasts for a univariate time series. The method is believed to have been around since 1950s. The classical forms of exponential smoothing method are including simple exponential smoothing [1], linear [2] and Holt–Winter's additive and multiplicative [3]. Following the idea of nonlinear state-space framework, characterized by a single source of errors introduced by [4], the exponential smoothing method can now be expressed via state space representation. Hyndman et al. [5] thoroughly explain the state-space approach of the exponential smoothing method. A total of 30 possible models makes this approach to be very useful in producing forecasts for any univariate time series data. The disadvantage of this approach, however, it does not allow for integration of regressor(s) into the model. To mitigate this problem, Osman and King [6] introduced an extended version in which regressor(s) can be included into the model. By taking an example of a basic model that consist of one regressor, $z_{1,t}$, the method with time varying parameter of regressor can be expressed as follows:

$$y_t = \ell_{t-1} + b_{1,t-1}\Delta_{z_{1,t}} + \varepsilon_t \tag{18.1a}$$

$$\ell_t = \ell_{t-1} + b_{1,t-1}\Delta_{z_{1,t}} + \alpha\varepsilon_t \tag{18.1b}$$

$$b_{1,t} = \begin{cases} b_{1,t-1} + \frac{\beta_1\left(\varepsilon^+_{1,t-1} + \varepsilon_t\right)}{\Delta^*_{z_{1,t}}}, & \text{if}\left|\Delta^*_{z_{1,t}}\right| \geq L_{b_1} \\ b_{1,t-1}, & \text{if}\left|\Delta^*_{z_{1,t}}\right| < L_{b_1} \end{cases} \tag{18.1c}$$

$$\varepsilon^+_{1,t} = \begin{cases} 0, & \text{if}\left|\Delta^*_{z_{1,t}}\right| \geq L_{b_1} \\ \varepsilon^+_{1,t-1} + \varepsilon_t, & \text{if }\left|\Delta^*_{z_{1,t}}\right| < L_{b_1}, \end{cases} \tag{18.1d}$$

where

$$\Delta^*_{z_{1,t}} = z_{1,t} - z^*_{1,t-1} \quad \text{and} \quad z^*_{1,t} = \begin{cases} z_{1,t}, & \text{if}\left|\Delta^*_{z_{1,t}}\right| \geq L_{b_1} \\ z^*_{1,t-1}, & \text{if}\left|\Delta^*_{z_{1,t}}\right| < L_{b_1} \end{cases}. \tag{18.2}$$

ℓ_t in Eq. (18.1) denotes the level of the series, $b_{1,t}$ represents a regressor parameter, ε_t denotes errors, while α and β_1 represent the smoothing parameters. Both $\varepsilon^+_{1,t}$ and $z^*_{1,t}$ are representing dummy errors and dummy regressor, respectively. On the other hand, L_{b_1} represents the lower boundary for a "switching procedure." Detailed explanation on this model is given in Osman and King [6].

Given that this is a new developed method, there is no available statistical software that can be used to execute parameter estimation for the method at this moment. In general, the objective of this study is to determine how best of executing "optim" function in R statistical software to estimate parameters of the new models.

18.2 Research Methodology

There are two main issues involved in estimating parameters of Eq. (18.1). First, what constraints need to be imposed on the smoothing constants, α and β_1. Second, what values of starting points (for the two smoothing constants) should be used in optimization routine. In order to find answer for the first issue, two approaches were considered that provide different parameter space for the two smoothing constants. The first approach is to impose parameter space according to the classical boundary of the smoothing constants in exponential smoothing method, that is between 0 and 1 for α and between 0 and α for β_1 as explained by [6] and Osman and King [3]. The second approach is to impose restrictions on eigenvalues of the characteristic equation of the model to be within the unit circle as explained by Osman and King [7]. The second approach is known as forecastability concept which gives wider parameter space for the smoothing constants as compared to the first approach. With regard to the second issue, different sets of starting points for the smoothing constants were considered in this analysis that is either 0.01, 0.1, or 0.5 for each of the smoothing constants.

The analysis which only involves one model as described by Eq. (18.1) was started by producing two optimization R codes for parameter estimation using two different sets of restrictions as explained above. The next step then followed to produce six simulated series with different predetermined values of parameters as listed in Table 18.1. For all cases of simulation, initial level and growth term were set to be $\ell_0 = 500$ and $b_{1,0} = 0.5$. By considering a situation where the "switching procedure" is not required, all simulated series were generated based on the following equations where the error term, ε_t is a normally distributed generated series.

$$y_t = \ell_{t-1} + b_{1,t-1}\Delta_{z_{1,t}} + \varepsilon_t \tag{18.3a}$$

$$\ell_t = \ell_{t-1} + b_{1,t-1}\Delta_{z_{1,t}} + \alpha\varepsilon_t \tag{18.3b}$$

$$b_{1,t} = b_{1,t-1} + \beta_1\varepsilon_t/\Delta_{z_{1,t}} \tag{18.3c}$$

Table 18.1 Predetermined parameter values for six simulated series

	α	β_1		α	β_1
Simulation 1	0.10	0.01	Simulation 4	0.50	0.10
Simulation 2	0.10	0.10	Simulation 5	0.90	0.01
Simulation 3	0.50	0.01	Simulation 6	0.90	0.10

Table 18.2 Starting points used in optimization routine

α	β_1	Possible combination of starting points considered in optimization, (α, β_1)
0.01	0.01	(0.01, 0.01)
0.10	0.01	(0.10, 0.01)
0.10	0.10	(0.10, 0.10)
0.50	0.01	(0.50, 0.01)
0.50	0.10	(0.50, 0.10)
0.50	0.50	(0.50, 0.50)
0.01, 0.10	0.01, 0.10	(0.01, 0.01), (0.01, 0.10), (0.10, 0.01), (0.10, 0.10)
0.01, 0.10, 0.50	0.01, 0.10, 0.50	(0.01, 0.01), (0.01, 0.10), (0.01, 0.50), (0.10, 0.01), (0.10, 0.10), (0.10, 0.50), (0.50, 0.01), (0.50, 0.10), (0.50, 0.50)

The third stage of analysis was done by applying optimization R codes for parameter estimation on all simulated series. A total of eight sets of starting points for smoothing constants were used with six of them that are simply combination between single starting point for both α and β_1. For other two sets of starting points, either two or three different values of starting points were used for each smoothing constant. All possible combinations of starting points are then considered in the optimization routine. Table 18.2 depicts all starting points used in estimation process. Note that, even though multiple starting points were used, only the one that gives the smallest value of the objective function was examined.

The estimation process was performed iteratively by minimizing the objective function $\left(n\log\left(\sum_{t=1}^{n}\varepsilon_t^2\right)\right)$ of the optimization code. This optimization procedure was conducted using "optim" function with the use of Nelder–Mead algorithm. The final step of analysis was then to evaluate the estimation results by comparing the predetermined values and the estimated values of all parameters.

18.3 Estimation Results

Estimation results are given in Tables 18.3, 18.4, 18.5, 18.6, 18.7, and 18.8. It can be seen that in most cases, small values of starting points ($\alpha = 0.01$, $\beta_1 = 0.01$ and $\alpha = 0.1$, $\beta_1 = 0.01$) managed to produce the smallest value of the objective function. This finding, however, is not true for simulation 5 and also not true for simulations 3 and 4 in which restrictions on parameter space were imposed according to the forecastability concept. Another finding is that, when the actual parameters (predetermined values) are small, the use of large starting points has led to the failure of the optimization routine to produce estimation that close to the actual values as evident by the results of analysis in the case of simulation 1 and 2. As mentioned earlier, estimation based on the forecastability concept provides

Table 18.3 Estimation results for simulated series 1

Classical boundary						
Starting points		Estimated coefficients				Objective function
α	β_1	ℓ_0	$b_{1,0}$	α	β_1	
0.01	0.01	490.5707	0.5042	0.0610	0.0080	**882.0783**
0.10	0.01	490.5887	0.5042	0.0609	0.0080	**882.0783**
0.10	0.10	491.3279	0.5037	0.0607	0.0079	882.1475
0.50	0.01	489.1021	0.5033	**0.4089**	0.0135	894.1491
0.50	0.10	492.9763	0.5024	0.0660	0.0079	882.7981
0.50	0.50	490.5314	0.5044	0.0596	0.0080	882.0814
0.01, 0.10		490.5887	0.5042	0.0609	0.0080	**882.0783**
0.01, 0.10, 0.50		490.5887	0.5042	0.0609	0.0080	**882.0783**
Forecastability concept						
Starting points		Estimated coefficients				Objective function
α	β_1	ℓ_0	$b_{1,0}$	α	β_1	
0.01	0.01	489.9522	0.5060	**2.18E-06**	**0.0049**	886.3208
0.10	0.01	490.5844	0.5042	0.0608	0.0080	**882.0783**
0.10	0.10	490.6940	0.5044	0.0601	0.0082	882.1154
0.50	0.01	493.5977	0.5023	0.0696	0.0084	883.2511
0.50	0.10	490.6622	0.5042	0.0602	0.0080	882.0793
0.50	0.50	489.4188	0.5053	0.0590	0.0078	882.3468
0.01, 0.10		490.5844	0.5042	0.0608	0.0080	**882.0783**
0.01, 0.10, 0.50		490.5844	0.5042	0.0608	0.0080	**882.0783**

Table 18.4 Estimation results for simulated series 2

Classical boundary						
Starting points		Estimated coefficients				Objective function
α	β_1	ℓ_0	$b_{1,0}$	α	β_1	
0.01	0.01	499.6713	0.5437	0.0965	0.0965	888.2322
0.10	0.01	496.6999	0.5679	0.0987	0.0986	**885.1003**
0.10	0.10	494.0989	0.6097	0.0986	0.0986	887.5330
0.50	0.01	497.1064	0.5793	0.0968	0.0968	887.3669
0.50	0.10	498.8807	0.5607	0.0961	0.0961	887.9756
0.50	0.50	496.0680	0.5655	**0.2274**	0.0970	899.7527
0.01, 0.10		496.6999	0.5679	0.0987	0.0986	**885.1003**
0.01, 0.10, 0.50		496.6999	0.5679	0.0987	0.0986	**885.1003**
Forecastability concept						
Starting points		Estimated coefficients				Objective function
α	β_1	ℓ_0	$b_{1,0}$	α	β_1	
0.01	0.01	496.8073	0.5991	0.0442	0.0978	**880.7533**
0.10	0.01	496.0696	0.5654	**0.2279**	0.0970	899.7527
0.10	0.10	493.4104	0.6271	0.0551	0.1029	882.9565
0.50	0.01	496.0702	0.5655	**0.2274**	0.0970	899.7527
0.50	0.10	492.9681	0.5885	0.1120	0.1032	891.7968
0.50	0.50	496.6072	0.5614	**0.2319**	0.0970	899.7886
0.01, 0.10		496.8073	0.5991	0.0442	0.0978	**880.7533**
0.01, 0.10, 0.50		496.8073	0.5991	0.0442	0.0978	**880.7533**

Table 18.5 Estimation results for simulated series 3

Classical boundary						
Starting points		Estimated coefficients				Objective function
α	β_1	ℓ_0	$b_{1,0}$	α	β_1	
0.01	0.01	485.8148	0.5060	0.6596	0.0136	**883.3676**
0.10	0.01	478.3717	0.5063	0.6515	0.0133	884.3208
0.10	0.10	485.8522	0.5059	0.6595	0.0136	**883.3676**
0.50	0.01	485.8272	0.5060	0.6591	0.0136	**883.3676**
0.50	0.10	485.8050	0.5059	0.6593	0.0136	**883.3676**
0.50	0.50	478.6677	0.5062	0.6550	0.0133	884.2457
0.01, 0.10		485.8148	0.5060	0.6596	0.0136	**883.3676**
0.01, 0.10, 0.50		485.8148	0.5060	0.6596	0.0136	**883.3676**
Forecastability concept						
Starting points		Estimated coefficients				Objective function
α	β_1	ℓ_0	$b_{1,0}$	α	β_1	
0.01	0.01	485.0704	0.5061	0.6701	0.0131	883.4157
0.10	0.01	476.6314	0.5065	0.6436	0.0132	884.8260
0.10	0.10	478.0573	0.5062	0.6586	0.0133	884.4020
0.50	0.01	485.8229	0.5060	0.6596	0.0136	**883.3676**
0.50	0.10	485.8287	0.5060	0.6599	0.0136	**883.3676**
0.50	0.50	479.5304	0.5062	0.6512	0.0133	884.0483
0.01, 0.10		485.8167	0.5060	0.6597	0.0136	**883.3676**
0.01, 0.10, 0.50		485.8229	0.5060	0.6596	0.0136	**883.3676**

Table 18.6 Estimation results for simulated series 4

Classical boundary						
Starting points		Estimated coefficients				Objective function
α	β_1	ℓ_0	$b_{1,0}$	α	β_1	
0.01	0.01	491.8341	0.5654	0.5788	0.1019	**883.1337**
0.10	0.01	491.8140	0.5654	0.5787	0.1019	883.1338
0.10	0.10	491.8329	0.5655	0.5788	0.1019	883.1338
0.50	0.01	491.9181	0.5669	0.5814	0.1016	883.1428
0.50	0.10	491.8227	0.5655	0.5788	0.1020	883.1338
0.50	0.50	491.8329	0.5654	0.5787	0.1019	**883.1337**
0.01, 0.10		491.8341	0.5654	0.5788	0.1019	**883.1337**
0.01, 0.10, 0.50		491.8341	0.5654	0.5788	0.1019	**883.1337**
Forecastability concept						
Starting points		Estimated coefficients				Objective function
α	β_1	ℓ_0	$b_{1,0}$	α	β_1	
0.01	0.01	491.8447	0.5654	0.5787	0.1020	883.1338
0.10	0.01	491.8128	0.5654	0.5787	0.1019	883.1338
0.10	0.10	491.8210	0.5655	0.5787	0.1019	883.1338
0.50	0.01	491.8337	0.5653	0.5788	0.1019	**883.1337**
0.50	0.10	491.8399	0.5653	0.5786	0.1019	883.1338
0.50	0.50	491.8136	0.5655	0.5788	0.1019	883.1338
0.01, 0.10		491.8128	0.5654	0.5787	0.1019	883.1338
0.01, 0.10, 0.50		491.8337	0.5653	0.5788	0.1019	**883.1337**

Table 18.7 Estimation results for simulated series 5

Classical boundary						
Starting points		Estimated coefficients				Objective function
α	β_1	ℓ_0	$b_{1,0}$	α	β_1	
0.01	0.01	466.0406	0.4934	1.0000	0.0135	889.1787
0.10	0.01	484.8486	0.4929	0.9993	0.0169	883.9131
0.10	0.10	466.5630	0.4942	1.0000	0.0140	888.8960
0.50	0.01	473.3510	0.4972	1.0000	0.0137	885.8410
0.50	0.10	486.2458	0.4965	1.0000	0.0143	**883.3803**
0.50	0.50	466.0930	0.4936	1.0000	0.0137	889.1486
0.01, 0.10		484.8486	0.4929	0.9993	0.0169	883.9131
0.01, 0.10, 0.50		486.2458	0.4965	1.0000	0.0143	**883.3803**
Forecastability concept						
Starting points		Estimated coefficients				Objective function
α	β_1	ℓ_0	$b_{1,0}$	α	β_1	
0.01	0.01	487.7907	0.4972	**1.1450**	0.0137	**881.5762**
0.10	0.01	487.7899	0.4972	**1.1447**	0.0138	**881.5762**
0.10	0.10	487.7884	0.4973	**1.1446**	0.0137	**881.5762**
0.50	0.01	489.8414	0.5011	**1.1408**	0.0150	881.7747
0.50	0.10	487.7656	0.4972	**1.1446**	0.0137	**881.5762**
0.50	0.50	487.8659	0.4972	**1.1493**	0.0137	881.5781
0.01, 0.10		487.7907	0.4972	**1.1450**	0.0137	**881.5762**
0.01, 0.10, 0.50		487.7656	0.4972	**1.1446**	0.0137	**881.5762**

Table 18.8 Estimation results for simulated series 6

Classical boundary						
Starting points		Estimated coefficients				Objective function
α	β_1	ℓ_0	$b_{1,0}$	α	β_1	
0.01	0.01	488.7248	0.5210	0.9672	0.1045	884.9660
0.10	0.01	488.6802	0.5210	0.9670	0.1046	**884.9659**
0.10	0.10	488.6616	0.5211	0.9670	0.1046	884.9659
0.50	0.01	488.6761	0.5210	0.9668	0.1046	884.9660
0.50	0.10	488.6834	0.5210	0.9671	0.1046	**884.9659**
0.50	0.50	488.6856	0.5211	0.9671	0.1046	**884.9659**
0.01, 0.10		488.6802	0.5210	0.9670	0.1046	**884.9659**
0.01, 0.10, 0.50		488.6834	0.5210	0.9671	0.1046	**884.9659**
Forecastability concept						
Starting points		Estimated coefficients				Objective function
α	β_1	ℓ_0	$b_{1,0}$	α	β_1	
0.01	0.01	488.6967	0.5210	0.9671	0.1046	**884.9659**
0.10	0.01	488.6582	0.5210	0.9670	0.1046	**884.9659**
0.10	0.10	488.6894	0.5211	0.9671	0.1046	**884.9659**
0.50	0.01	488.6784	0.5210	0.9671	0.1046	**884.9659**
0.50	0.10	488.6732	0.5211	0.9671	0.1046	**884.9659**
0.50	0.50	488.6926	0.5211	0.9668	0.1045	884.9660
0.01, 0.10		488.6967	0.5210	0.9671	0.1046	**884.9659**
0.01, 0.10, 0.50		488.6784	0.5210	0.9671	0.1046	**884.9659**

wider space for the smoothing constants. If we look at the estimation result of simulation 5, the estimated values of α when estimated based on the forecastabilty concept are much higher than the predetermined value. In this particular case, estimation performed based on the classical boundary method produced more accurate estimated value for α.

18.4 Conclusion

As discussed in the previous section, the use of small values of starting points for smoothing constants in optimization routine is advised as compared to large values of starting points. However, taking into account the better performance of large starting points in some cases, it is encouraged to use multiple starting points that is also include small values of starting points. It is also can be concluded that the use of forecastability concept in imposing restrictions on parameter space does not necessarily will produce better estimate than the use of classical boundary concept. Since it provides wider parameter space for the smoothing constant, it allows for greater chance of estimation error as well.

The use of classical boundary concept in imposing restrictions to the parameter space of the smoothing constant is actually sufficient if we want to update the levels, ℓ_t and the regressor parameter, $b_{1,t}$ based on the weighted average methodology, the concept used in the classical form of the exponential smoothing methods. This is because restricting α to be between 0 and 1 and β_1 to be between 0 and α in the state-space representation is equivalent to restrict both α and β_1 to be between 0 and 1 in the classical form of the exponential smoothing methods.

References

1. Brown, R.G.: Statistical Forecasting for Inventory Control. McGraw-Hill, New York (1959)
2. Holt, C.C.: Forecasting Seasonal and Trends by Exponentially Weighted Moving Averages. Office of Naval Research Memorandum, vol 52 (1957)
3. Winters, P.R.: Forecasting sales by exponentially weighted moving averages. Manage. Sci. **6**, 324–342 (1960)
4. Ord, J.K., Koehler, A.B., Synder, R.D.: Estimation and prediction for a class of dynamic nonlinear statistical models. J. Am. Statist. Assoc. **92**, 1621–1629 (1997)
5. Hyndman, R.J., Koehler, A.B., Ord, J.K., Snyder, R.D.: Forecasting with Exponential Smoothing: The State Space Approach. Springer Verlag, Berlin (2008)
6. Osman, A.F., King, M.L.: A New Approach to Forecasting based on Exponential Smoothing with Independent Regressors. Working paper series, Monash University (2015)
7. Osman, A.F., King, M.L.: Exponential smoothing with regressors: estimation and initialization. Model Assist. Stat. Appl. **10**(3), 253–263 (2015)

Chapter 19
Review on Economy and Livelihood Status in Setiu Minapolitan: A System Dynamics Approach

L. Muhamad Safiih and R. Mohd Noor Afiq

Abstract This paper presents a modeling approach using the system dynamics to simulate and analyze the economic effects of aquaculture in a wetland. Setiu Wetland is a part of Setiu district that upholds tradition in fishery and maritime-based industry but the area is left behind in terms of economy and livelihood. The average household income in the Minapolitan area in Setiu ranges from 200MYR to 700MYR per month and its population growth is 1.93 % per year. This study aims to recreate a model that represents the economy and livelihood of Setiu Minapolitan area. The study engaged Vinsim software to simulate the environment and to perform a better analysis.

Keywords System dynamics · Stimulate · Economic effects · Aquaculture

19.1 Introduction

Setiu is the youngest among a total of eight districts in Terengganu, a state located in the east coast of peninsular Malaysia. Terengganu is rich in agriculture, fisheries, and marine-based industries as well as tourism. Setiu was established in January 1, 1985 upon certification of Administrative Modernization and Management Planning (MAMPU). This particular district consists of 135,905.80 ha (1359.90 km^2) of land or approximately 10.49 % of the state of Terengganu. As part of the seventh district of Terengganu, this area is classified a partial-Minapolitan concept-based region in the Terengganu marine industry

L. Muhamad Safiih (✉) · R. Mohd Noor Afiq
Jabatan Matematik, Pusat Pengajian Informatik Dan Matematik Gunaan,
Universiti Malaysia Terengganu, Kuala Terengganu, Malaysia
e-mail: safiihmd@umt.edu.my

R. Mohd Noor Afiq
e-mail: afiqramlee@yahoo.com

L. Muhamad Safiih
Institut Biologi Marin, Universiti Malaysia Terengganu, Kuala Terengganu, Malaysia

A.-R. Ahmad et al. (eds.), *Proceedings of the International Conference on Computing, Mathematics and Statistics (iCMS 2015)*,
DOI 10.1007/978-981-10-2772-7_19

judging on its performance in small maritime-based industry. A fully Minapolitan concept-based region should rely on fisheries and marine commodities as much as their sea-based economic value [1]. For Setiu to thrive in this current age, an economic boosts are needed to allow the Setiu region to become a main supporter for Terengganu economic development in terms of aquaculture, agriculture, and marine-based industry [2]. Setiu local population's involvement in industry-based business is still small. Whether in aquaculture, tourism, small-medium enterprise, the major player in Setiu area comes from non-originated people or external investor.

Just like any other food-producing sector in the world, aquaculture relies on renewable and non-renewable resources [3]. Hence, a sustainable aquaculture practices and business canvas should be built and practiced in terms to uplift Setiu to become a fully cooperated Minapolitan without neglecting its environment and ecological surplus.

19.1.1 Objective

1. Develop a conceptual model of the economic interaction of local and aquaculture practitioner in Setiu.
2. Implement a dynamic economic model in order to stimulate
    a. The socioeconomic component of aquaculture practices
    b. The economic spill over from aquaculture practices in neighborhood
3. Comparing stimulated scenario with model output.

19.1.2 Study Area

This particular study was conducted in the Setiu Wetland as it serves the purpose of Minapolitan based industry. The area consists of Merang, Pantai, and Guntung area. The area in Setiu covers approximately 31,672.9 ha or around 23.305 % of Setiu district [4]. Although this particular area represents more than 20 % of Setiu district area, the population is only 16,616 people [5]. Majority of the residents in this particular area works as a small-scale fisherman and their relative income average MYR700 during catchment seasons (March–October). For monsoon and rainy seasons, they rely on government subsidies or MYR200 per month [4]. Apart from that, some residents on this particular area work in small and self-sufficient agricultural sector. About 10–15 % of these farmers earn an average of MYR850 per month from their agriculture industries [4].

In terms of tourism, Merang areas are well known for its jetty for transportation of passengers to some local islands in Terengganu such as the Perhentian Island.

Terengganu receive an average 3.98 million tourists per year from 2008 to 2013 [6]. This massive industries and complex economic development in Setiu area needs to be address to fully capture the developmental status. Hence, for this particular scenario system dynamic has a distinct advantage in analyzing, improving, and managing the system characterized by long development cycle and complex feedback effects [7]. Therefore, the systemic dynamics method meets the modeling requirement in our study.

19.2 Methodology

19.2.1 Conceptual Approach

By understanding the objective requirement, we need to develop a dynamic model using system dynamic method to achieve a full understanding in matter that we need to study. The system dynamics method was created by Professor Forrester of Massachusetts Institute of Technology in the mid-1950s [8]. After decades of development and improvement, the systemic dynamics model has been widely used in the study of economy, society, ecology, and many complex systems [9, 10].

But before we proceed to create a system dynamic model, it is essential to understand the behavior of dynamic that should be created. Therefore, economic and livelihood dynamic must be related with a conceptual or logical framework in order to perform a full dynamic model.

Economic and livelihood model used in system dynamic represent certain sector involved directly in Setiu stakeholder. Relation between livelihood and economic exists symbiotically as each end effect another, as shown in Fig. 19.1. Therefore several components that should be listed and account for.

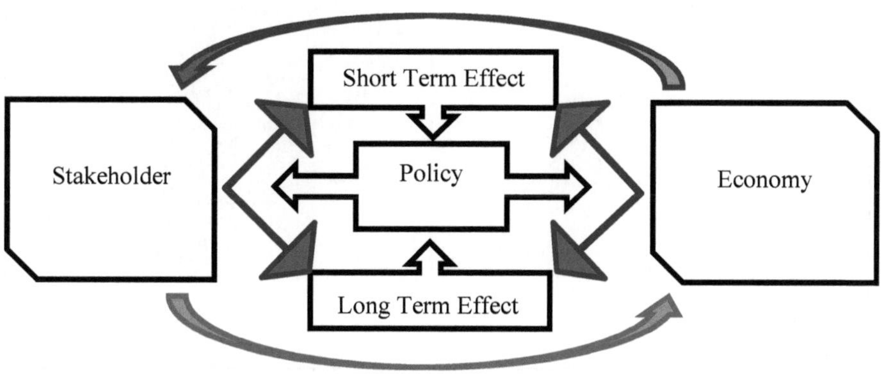

Fig. 19.1 Logic framework for dynamic relation

19.2.2 Policy and Stake Holder

Policy is the heart of framework. By applying a certain policy and guideline, a clear picture of development model could take place within any sector [11]. Hence, without a proper policy one industry of the cases in Sub-Saharan Africa, aquaculture or maritime based industry has such a low profile that institutional structures are very weak and there may be no aquaculture policy at all, or it is incorporated into other policies, such as the fisheries policy. Of the 12 countries reviewed, the Strategy for International Fisheries Research on Aquaculture Development and Research is a type of policy that could drive development towards economic sustainability and greater livelihood.

19.2.3 Case Study: Site and Data Description

The model stimulated relation between economical and livelihood has been applied on residence in Setiu's Minapolitan located at Pantai, Merang, and Chalok costal area (Fig. 19.2) in northeast area in Terengganu [5].

Setiu known for its abundance of tourism spots, marine fresh products, and variety of traditional marine-based products that been offered [4]. According to UPEN, although residence this particular area mainly involve in marine fresh

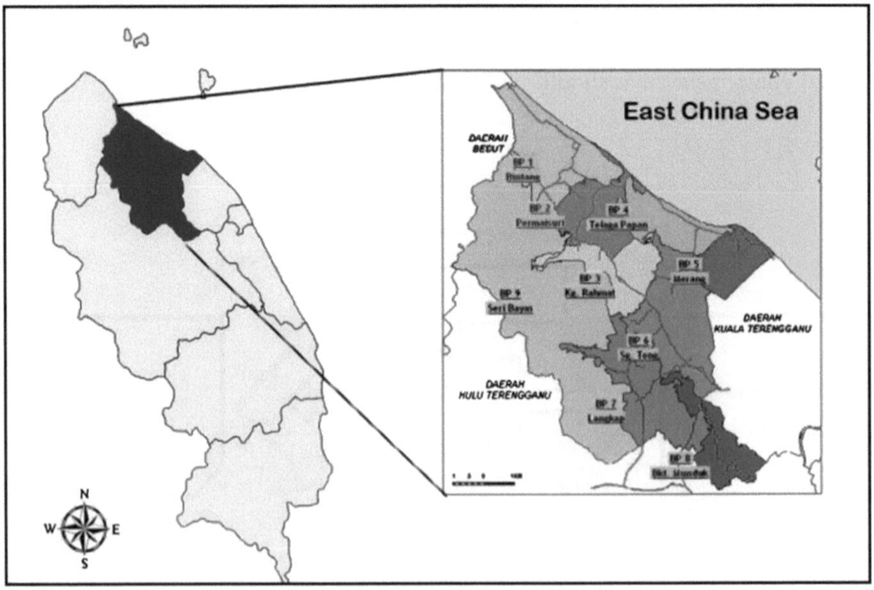

Fig. 19.2 Setiu district map

product catchment, yet the total largest income comes from other sectors such as tourism transportation, farming, and wage earner.

19.2.4 Data Sources

Data sources in this study mostly came from first-hand information and data that has been gathered using interviews and questionnaire with people from local area. For a certain statistical data, we gather the secondary data from reliable sources such as reports and journal of local authorities, universities, and international researcher that have done research in particular field in Setiu namely Universiti Malaysia Terengganu (UMT), Universiti Sultan Zainal Abidin (UniSZA), and World Wide Foundation (WWF). Table 19.1 shows land size and total population of Setiu subdistricts.

19.2.5 System Dynamic Mathematical Design

System Dynamic design consists of stock and flow that represent the mathematical behavior for dynamic simulation. Hence, the formulation for primary stock and flow represent with the following:

Income Dynamic

$$\text{Income}_t = \int (\text{Employability rate} + \text{Business rate})\,\mathrm{d}t \tag{19.1}$$

$$\text{Land use}_t = \int \text{policy}(\text{Land Available})\,\mathrm{d}t \tag{19.2}$$

$$\text{Population}_t = \text{Population}_n(e^{rt}) \tag{19.3}$$

Table 19.1 Land size and total population of Setiu subdistricts

Area	Land (ha)	Total population
Hulu Nerus	54,523.3	12,592
Chalok	20,589.6	14,854
Hulu Setiu	23,292.9	3198
Guntung	16,348.5	7424
Tasik	5827.1	7227
Pantai	8499.4	5994
Merang	6825	3198

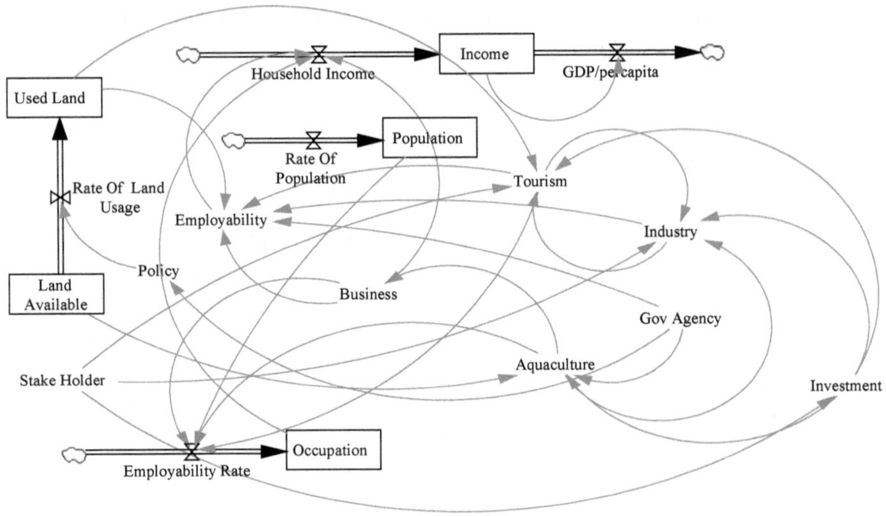

Fig. 19.3 Stock flow diagram for Setiu Economical and livelihood

where

$$t = 0, 1, 2, \ldots, m$$

$$rt = \text{rate over time}$$

Note that the formulae (19.1), (19.2), and (19.3) are affected or affecting with other variable that exists in natural environment in Setiu such as Tourism Activity, Aquaculture Activity, Industry, Investment, and other variable that is visible in Fig. 19.3. Hence, this particular system shows their correlation in Stock Flow Dynamic existing in SD.

19.3 Result

The simulated graph shows indication population growth (Fig. 19.4), aquaculture involvement income, and land uses (Fig. 19.5) in 70 month which is approximately 5.9 years. Economic limitation and low pays restricted the population growth and hinder local aquaculture related business involvement due to certain effect.

For land usage, simulated graph (Fig. 19.6) show significant movement after 35 month marks due to certain policy that has been introduced by local authorities and empowerment of stake holder. This is due to significant rises of research and community based projects managed by universities, private sector and government.

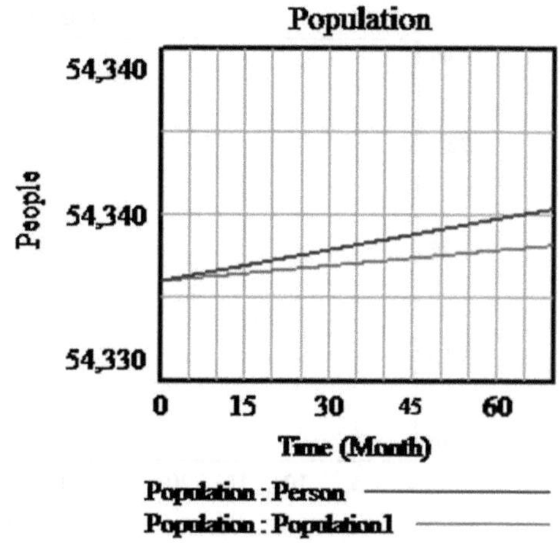

Fig. 19.4 Population growth simulation in Setiu based on authorities statistics

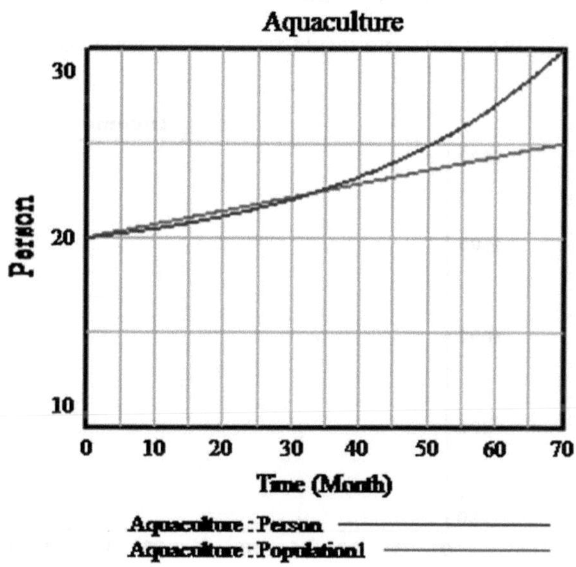

Fig. 19.5 Local involvement in aquaculture related industry for next 70 month

Hence, by strengthen local population with several aspect and promote employability, certain land uses could be maximized and the usage of land could be optimized to be more profitable, as shown in Fig. 19.7. This way, the economical of local could be strengthen and promoting a brand new livelihood towards Setiu district.

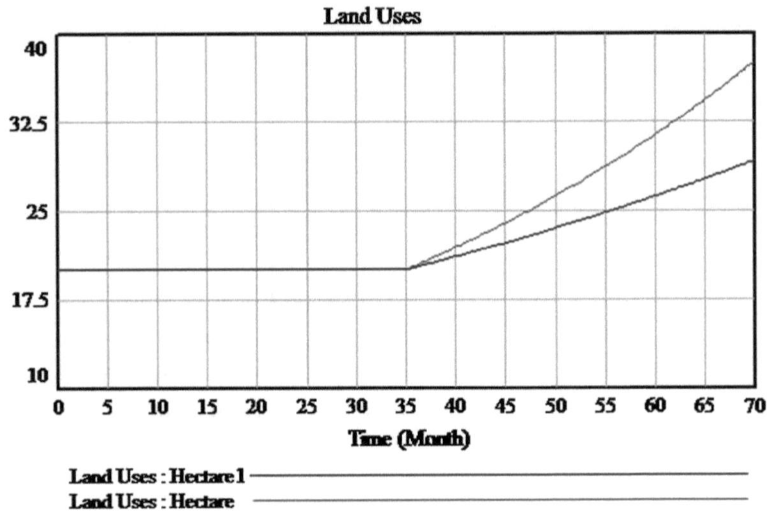

Fig. 19.6 Land conversion In Setiu based on government and private sector intervention

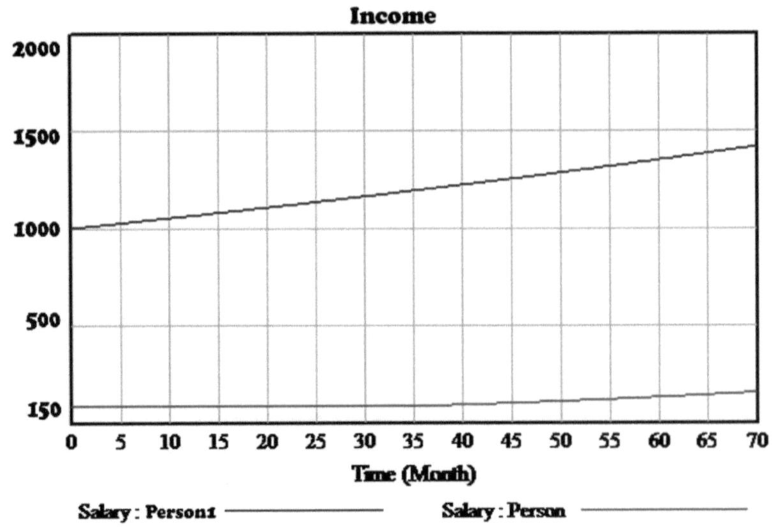

Fig. 19.7 Economic value before and after intervention and policy change

19.4 Conclusion and Discussion

Using system dynamics model for economy and livelihood review, the system in Setiu Wetland Minapolitan could be simulated and analyzed. Some improvement could be applied in the simulated environment to help this particular area to thrive in economical. Hence a few conclusion and recommendation are drawn as follows:

1. System dynamic could be used to simulate supply and demand of local marine products to widen Setiu Minapolitan horizon further.
2. Simulate economic development cost production and its limiter.
3. Simulate interaction of local development model and its effects on environment for a sustainable livelihood.
4. Simulate spatial limiter towards cultural and heritage upon upcoming development.
5. Simulate local tourism industry; as a catalysts of economic development and its impact.
6. Simulate major player in aquaculture industries (iSharp) for its economical spillover or impact towards local residence.

Acknowledgments This study supported by National Research Grant Scheme (NRGS) for Setiu Wetland Development (NRGS/UMT/Vot 1515). Acknowledgement for Terengganu Economic Development Unit (UPEN), YDSM and Setiu District Welfare and Safety Committee (JKKK) for providing insights during interview and questionnaire sessions.

References

1. Mu'tamar, F.M., Eriyatno, M., Soewardi, K.: Dynamic model analysis of raw material supply in minapolitan shrimp agroindustry. Ind Eng Lett **3**: 36–44 (2013)
2. YDSM.: Setiu research and development (2015)
3. Nobre, A., Musango, J., de Wit, M., Ferreira, J.: A dynamic ecological–economic modeling approach for aquaculture management. Ecol Econ **68**: 3007–3017 (2009)
4. UPEN.: Penerangan Pelan Strategik Setiu (2015)
5. Official Portal of Setiu District Council, http://mds.terengganu.gov.my/en/latar-belakang-setiu
6. Auditor General Report, Malaysia: National Audit Department (2012)
7. Tao, Z.: Scenarios of China's oil consumption per capita (OCPC) using a hybrid factor decomposition-system dynamics (SD) simulation. Energy **35**(1), 168–180 (2010)
8. Forrester, J.: Industrial dynamics: a major breakthrough for decision makers. Harvard Bus. Rev. **36**(4), 37–66 (1958)
9. Chang, Y., Hong, F., Lee, M.: A system dynamic based DSS for sustainable coral reef management in Kenting coastal zone, Taiwan. Ecol. Model. **211**(1–2), 153–168 (2008)
10. Wang, Y., Zhang, X.: A dynamic modeling approach to simulating socio-economic effects on landscape changes. Ecol. Model. **140**(1–2), 141–162 (2001)
11. Sterman, J. D.: System Dynamics Modeling for Project Management. MA. System Dynamics Group Sloan, School of Management, Massachusetts Institute of Technology, Cambridge (1992)

Part III
Mathematics

Chapter 20
A Model of HIV-1 Dynamics in Sub-Saharan Africa: An Optimal Control Approach

Mamman Mamuda, Amiru Sule, Jibril Lawal, Usman Nayaya Zurmi, Saidu Abarshi Kanoma and Ganiyatu Aliyu

Abstract In this study a nonlinear deterministic mathematical model was developed on the dynamics of HIV-1 in sub-Saharan Africa with the final goal of implementing an optimal control using two control parameters of treatment and culling. This could minimize the number of infected humans and chimpanzees. Validity of the model was investigated through Invariant region and the model was found to be epidemiologically meaningful. Basic reproduction number which determines the stability of the disease is obtained using next generation operator. Disease free equilibrium was found to be locally asymptotically stable whenever basic reproduction number is greater than one and unstable otherwise. Numerical simulation results reveal that, the two control strategies optimally help in curtailing the spread of the virus.

Keywords Optimal control · Treatment · Culling · Basic reproduction number

---

M. Mamuda (✉) · A. Sule
School of Mathematical Sciences, Universiti Sains Malaysia,
Penang, Malaysia
e-mail: maanty123@gmail.com

A. Sule
e-mail: amirusule@yahoo.com

J. Lawal
Department of Mathematical Sciences, Federal University, Gusau,
Zamfara State, Nigeria
e-mail: jibril.lawal@yahoo.com

U.N. Zurmi · S.A. Kanoma · G. Aliyu
Department of Mathematics, Zamfara State College of Education Maru,
Maru, Nigeria
e-mail: aliyuganiyatu@yahoo.com

A.-R. Ahmad et al. (eds.), *Proceedings of the International Conference on Computing, Mathematics and Statistics (iCMS 2015)*,
DOI 10.1007/978-981-10-2772-7_20

20.1 Introduction

Human immunodeficiency virus (HIV) leads to acquired immune deficiency syndrome (AIDS). It is a global human disaster particularly in sub-Saharan Africa. This may be attributed to poor financial development, poverty, and lack of health services [1]. HIV is now the main cause of years of potential lives lost and the most common cause of death ascribed to many infectious diseases [2]. In 2013, it was estimated that roughly 39 million people so far died of HIV-related illnesses since the virus was first detected in 1981 [3]. And about 35 million people were living with the virus globally [4]. Furthermore, about 2.1 million people were newly infected in 2013. Around 1.5 million people died of AIDS-related illnesses in the year under review. The virus is transmitted through unprotected sex with someone who is infected or transfusion of contaminated blood or blood products. Other ways of transmitting the virus include sharing unsterilized injection equipment and from infected mother to her child. This occurs during pregnancy, at birth, and through breastfeeding [5]. Though, at present there is no cure for the disease, observing various approaches for controlling the spread of the virus in order to reduce the disease prevalence is very essential [6]. Mathematical modeling over the years has been beneficial in examining various diseases, such as Dengue, Ebola, and HIV/AIDS. It also plays an essential role in the well understanding of epidemiological designs for disease control.

Chimpanzees in Sub-Saharan Africa are divided into two main species; the common chimpanzee (Pan Troglodytes) and the bonobos (Pan Paniscus). Pan troglodytes is further divided into four subspecies based on their geographic location in Sub-Saharan Africa; Pan Troglodyte's schweinfurthii, Pan troglodytes troglodytes, Pan Troglodyte's verus, and Pan Troglodyte's vellerosus [7]. Among these four subtypes, pan troglodytes troglodytes (p.t. troglodytes) chimpanzees was believed to harbor SIV that ignites the emergence of HIV-1 subtype in Sub-Saharan Africa. As such, some scientists believed that SIV is the precursor of HIV which is the main virus responsible for high morbidity and mortality worldwide [7–9]. SIV's are believed to spread among primates (chimpanzees) in the wild by sex, through blood to blood contact during a fight and in the case of cross species transmission as a result of predation [10]. Simian immunodeficiency viruses (SIVs) are primate lentiviruses that were believed to be the precursor of HIV 1 and 2. However, scientists believed that certain strains of SIVs bear a very close similarity to two types of HIV, 1 and 2 [11]. Therefore, the aim of this study is to examine the transmission of HIV-1 with the inclusion of chimpanzee population which some scientist believed to be the precursor of the virus, and to suggest possible optimal control of the disease in Sub-Saharan Africa. Two control strategies, treatment and culling [12, 13], using pontryagins maximum are considered.

The paper is organized as follows, in Sect. 20.2, description of the model that consists of ordinary differential equations which describes the dynamics of SIV and HIV/AIDS is given. In Sect. 20.3, basic properties of the model and basic reproduction number are presented. In Sect. 20.4, an optimal control problem is investigated using pontryagins maximum principle. In Sect. 20.5, we have the numerical simulation. And in Sect. 20.6, conclusion is presented.

20.2 Model Formulations

This paper is an improved work of Anderson et al. [14], by incorporating the population of chimpanzees and two time-dependent control strategies. Dynamics of human population at time t, are susceptible human $S_h(t)$, infected human $I_h(t)$, and AIDS human $A_h(t)$. The dynamics of chimpanzee's population at time t are susceptible chimpanzees $S_c(t)$ and infected chimpanzees $I_c(t)$. In this formulation it is assumed that, the flow into the human population is through immigration at a constant rate π_h. A fraction δ_h is infected while the remaining $(1-\delta_h)$ is susceptible human. The susceptible humans proceed to the infected compartment either through sexual contact with the infected humans at a rate $\beta_1 c_1 I_h$, or through contact with infected chimpanzee at rate $\beta_2 c_2 I_c$. The human population suffers a natural death at a rate μ_h and disease-induced death at a rate μ_0. Infected humans move to the AIDS class as a result of low CD4+ counts at a rate α.

It is considered that the population of chimpanzee is a flow into the population that is either infected or susceptible chimpanzee. The flow is assumed to occur through immigration at a rate π_c a fraction δ_c is assumed to be infected chimpanzee while the remaining $(1-\delta_c)$ is susceptible chimpanzee. The susceptible chimpanzee move to the infected class by acquiring SIV through sexual contact with the infected chimpanzee at a rate $\beta_3 c_3 I_c$, where β_3 is the probability of SIV infection and c_3 is the number of sexual partners of the chimpanzees. The chimpanzee population suffers a natural death at a rate μ_c. These assumptions lead to the following combined system of ordinary differential equations which define the progress of the disease as follows:

$$\left.\begin{aligned}
\frac{dS_h}{dt} &= (1-\delta_h)\pi_h - \beta_1 c_1 I_h S_h - \beta_2 c_2 I_c S_h - \mu_h S_h \\
\frac{dI_h}{dt} &= \delta_h \pi_h + \beta_1 c_1 I_h S_h + \beta_2 c_2 I_c S_h - (\mu_h + \alpha) I_h \\
\frac{dA_h}{dt} &= \alpha I_h - (\mu_h + \mu_0) A_h \\
\frac{dS_c}{dt} &= (1-\delta_c)\pi_c - \beta_3 c_3 I_c S_c - \mu_c S_c \\
\frac{dI_c}{dt} &= \delta_c \pi_c + \beta_3 c_3 I_c S_c - \mu_c I_c
\end{aligned}\right\} \quad (20.1)$$

20.3 Basic Properties of the Model

Definition A set M is invariant if and only if for all $x \in M$, $\varphi(x,t) \in M$ for all t. A set is positively (negatively) invariant if for all $x \in M, \varphi(x,t) \in M$ for all $t > 0, (t < 0)$ [15].

Lemma 1.1 *The model* Eq. (20.1) *above has a solution which are contained in the feasible region* $\Omega = \Omega_h + \Omega_c$

$$\frac{dN_h}{dt} = \pi_h - \mu_h N_h$$

Hence, by standard comparison theorem as found in [16] *it appears that* $0 \leq N_h \leq \frac{\pi_h}{\mu_h}$ *so that* $(\pi_h - \mu_h N_h) \geq Ae^{-\mu_h t}$, *as A is constant.*

Thus, for model Eq. (20.1) *all the possible solutions of the human and chimpanzee populations are in the region,*

$$\left.\begin{array}{l} \Omega_h = \left\{(S_h, I_h, A_h) \in R_+^3 \ : N_h \leq \frac{\pi_h}{\mu_h}\right\} \\ \Omega_c = \left\{(S_c, I_c) \in R_+^2 \ : N_c \leq \frac{\pi_c}{\mu_c}\right\} \end{array}\right\}$$

Therefore, the feasible set for model (20.1) *is given by*

$$(S_h, I_h, A_h, S_c, I_c) \in R_+^5 \ : S_h, I_h, A_h, S_c, I_c \geq 0, \quad N_h \leq \frac{\pi_h}{\mu_h}, N_c \leq \frac{\pi_c}{\mu_c}$$

20.3.1 Disease Free Equilibrium (DFE)

The model system (20.1) has a DFE, obtained by setting the right-hand sides of the Eq. (20.1) to zero.

$$E_0 = \left(S_h^*, I_h^*, A_h^*, S_h^*, I_h^*\right) = \left(\frac{\pi_h}{\mu_h}, 0, 0, \frac{\pi_c}{\mu_c}, 0\right) \tag{20.2}$$

The linear stability of an HIV/AIDS model can be established using the next generation operator method [17]. Thus; the basic reproduction number is given by $\Re_0 = \rho(FV^{-1})$, where ρ the spectral radius is given as

$$\Re_0 = \frac{\beta_1 c_1 \pi_h \beta_3 c_3 \pi_c}{\mu_h(\mu_h + \alpha)\mu_c^2} \tag{20.3}$$

Theorem 1.1 *The disease free equilibrium (DFE) of the model system* (20.1), *given by* (20.3), *is locally asymptotically stable (LAS) if* $\Re_0 < 1$ *and unstable if* $\Re_0 > 1$.

20.4 Optimal Control Analysis of HIV/AIDS Model

In order to investigate the optimal control measures of the virus, some control parameters are observed. Two control strategies were considered; treatment for the infected humans and culling for the pan troglodytes chimpanzees. Culling is a control strategy for chimpanzees by removing them to a reserved area as found in [18, 19]. The model Eq. (20.1) was extended by incorporating treatment parameter into the human population as $(1-u_1)$, where $u_1 \in [0,1]$ is the control on the use of treatment for infected humans and $(1-u_2)$, where $u_2 \in [0,1]$, is the control on the use of culling on the p.t troglodytes' chimpanzees. Removing this chimpanzee species to a reserved area will reduce the contact between chimpanzees and human, which in turn reduce the spread of HIV/AIDS. Therefore the model system (20.1), above becomes

$$\left.\begin{aligned}
\frac{\mathrm{d}S_\mathrm{h}}{\mathrm{d}t} &= (1-\delta_\mathrm{h})\pi_\mathrm{h} - \beta_1 c_1 I_\mathrm{h}(1-\mu_1)S_\mathrm{h} - \beta_2 c_2 I_\mathrm{c} S_\mathrm{h} - \mu_\mathrm{h} S_\mathrm{h} \\
\frac{\mathrm{d}I_\mathrm{h}}{\mathrm{d}t} &= \delta_\mathrm{h}\pi_\mathrm{h} + \beta_1 c_1 I_\mathrm{h}(1-\mu_1)S_\mathrm{h} + \beta_2 c_2 I_\mathrm{c} S_\mathrm{h} - (\mu_\mathrm{h}+\alpha) I_\mathrm{h} \\
\frac{\mathrm{d}A_\mathrm{h}}{\mathrm{d}t} &= \alpha I_\mathrm{h} - (\mu_\mathrm{h}+\mu_0)A_\mathrm{h} \\
\frac{\mathrm{d}S_\mathrm{c}}{\mathrm{d}t} &= (1-\delta_\mathrm{c})\pi_\mathrm{c} - \beta_3 c_3 I_\mathrm{c}(1-\mu_2)S_\mathrm{c} - \mu_\mathrm{c} S_\mathrm{c} \\
\frac{\mathrm{d}I_\mathrm{c}}{\mathrm{d}t} &= \delta_\mathrm{c}\pi_\mathrm{c} + \beta_3 c_3 I_\mathrm{c}(1-\mu_2)S_\mathrm{c} - \mu_\mathrm{c} I_\mathrm{c}
\end{aligned}\right\} \quad (20.4)$$

Consider the objective function as

$$J(u) = \min_{u_1,u_2} \int_0^T \left[B_1 I_\mathrm{h} + B_2 I_\mathrm{c} + B_3 u_1^2 + B_4 u_2^2\right] \mathrm{d}t \quad (20.5)$$

The aim is to minimize the number of infected humans and chimpanzees, $I_\mathrm{h}(t), I_\mathrm{c}(t)$, while minimizing the cost of control $u_1(t), u_2(t)$ and the coefficients (B_1, B_2, B_3, B_4) are positive weights to balance the factors. A quadratic cost on the controls is chosen that is similar to other literatures, see for instance [10, 20]. T is the final time, thus, we seek an optimal control u_1^*, u_2^* such that

$$J(u_1^*, u_2^*) = \min_{u_1,u_2}\{J(u_1,u_2) | u_1, u_2 \in \upsilon\} \quad (20.6)$$

As the control set

$$U = \left\{ (u_1^*, u_2^*) | u_i : [0,T] \to [0,1], \text{lebesgue measurable}, i = 1,2 \right\} \quad (20.7)$$

20.4.1 Existence of Control Problem

The terms B_1I_h and B_2I_c are the cost of infection while $B_3u_1^2$ and $B_4u_2^2$ are the costs of treatment and culling respectively. The necessary conditions that an optimal control must satisfy come from pontryagins maximum principle [20]. The principle converts (20.5) and (20.6) into a problem of minimizing point wise a Hamiltonian H, with respect to U.

$$\left.\begin{aligned} H = B_1I_h + B_2I_c + B_3u_1^2 + B_4u_2^2 + \lambda_1\{(1-\delta_h)\pi_h - \beta_1c_1I_h(1-u_1)S_h - \beta_2c_2I_cS_h - \mu_hS_h\} \\ + \lambda_2\{\delta_h\pi_h + \beta_1c_1I_h(1-u)S_h + \beta_2c_2I_cS_h - (\mu_h+\alpha)I_h\} \\ + \lambda_3\{\alpha I_h - (\mu_h+\mu_0)A_h\} \\ + \lambda_4\{(1-\delta_c)\pi_c - \beta_3c_3I_c(1-u_2)S_c - \mu_cS_c\} \\ + \lambda_5\{\delta_c\pi_c + \beta_3c_3I_c(1-u_2)S_c - \mu_cS_c\} \end{aligned}\right\} \tag{20.8}$$

where λ_i for $i = 1, 2 \ldots 5$ are ad-joint variables or co-state variables.

Theorem *For the optimal control u_1^*, u_2^* that minimizes $J(u_1, u_2)$ over U, then there exist ad-joint variables λ_i* for $i = 1, 2 \ldots 5$ satisfying

$$\left.\begin{aligned} -\frac{d\lambda_1}{d_t} &= -\beta_1c_1I_h(u_1-1)(\lambda_1-\lambda_2) + \beta_2c_2I_c(\lambda_1-\lambda_2) + \mu_h\lambda_1 \\ -\frac{d\lambda_2}{d_t} &= -B_1 - \beta_1c_1(u_1-1)S_h(\lambda_1-\lambda_2) - \alpha(\lambda_2-\lambda_3) + \mu_h\lambda_2 \\ -\frac{d\lambda_3}{d_t} &= \lambda_3(\mu_h+\mu_0) \\ -\frac{d\lambda_4}{d_t} &= -\beta_3c_3I_c(u_2-1)(\lambda_4-\lambda_5) + \mu_c\lambda_4 + \mu_c\lambda_5 \\ -\frac{d\lambda_5}{d_t} &= -B_2 + \beta_2c_2S_h(\lambda_1-\lambda_2) - \beta_3c_3(u_2-1)(\lambda_4-\lambda_5) \end{aligned}\right\} \tag{20.9}$$

With transversality condition,$\lambda_i = 0$, for $i = 1, 2 \ldots 5$ and the controls u_1^*, u_2^* satisfying optimality conditions.

$$\left.\begin{aligned} u_1^* &= \max\left\{0, \min\left(1, -\frac{\beta_1c_1I_h(\lambda_1-\lambda_2)S_h}{2B_3}\right)\right\} \\ u_2^* &= \max\left\{0, \min\left(1 - \frac{\beta_3c_3I_c(\lambda_4-\lambda_5)S_c}{2B_4}\right)\right\} \end{aligned}\right\} \tag{20.10}$$

Proof Hamiltonian function is differentiated in order to find the governing equations of the ad-joints variables, evaluated at the optimal control. Then the ad-joints system can be written as

$$\left.\begin{aligned} -\frac{d\lambda_1}{d_t} &= \frac{\partial H}{\partial S_h} = -\beta_1c_1I_h(u_1-1)(\lambda_1-\lambda_2) + \beta_2c_2I_c(\lambda_1+\lambda_2) + \mu_h\lambda_1 \\ -\frac{d\lambda_2}{d_t} &= \frac{\partial H}{\partial I_h} = -B_1 - \beta_1c_1(u_1-1)S_h(\lambda_1-\lambda_2) - \alpha(\lambda_2-\lambda_3) + \mu_h\lambda_2 \\ -\frac{d\lambda_3}{d_t} &= \frac{\partial H}{\partial A_h} = \lambda_3(\mu_h+\mu_0) \\ -\frac{d\lambda_4}{d_t} &= \frac{\partial H}{\partial S_c} = -\beta_3c_3I_c(u_2-1)(\lambda_4-\lambda_5) + \mu_c\lambda_4 + \mu_c\lambda_5 \\ -\frac{d\lambda_5}{d_t} &= \frac{\partial H}{\partial I_c} = -B_2 + \beta_2c_2S_h(\lambda_1-\lambda_2) - \beta_3c_3(u_2-1)(\lambda_4-\lambda_5) \end{aligned}\right\} \tag{20.11}$$

With transversality condition, $\lambda_i = 0, \text{for}, i = 1, 2\ldots5$ on the interior of the control set, where $0 < u_1 < 1, 0 < u_2 < 1$ we have

$$\left.\begin{aligned} \frac{\partial H}{\partial u_1} &= 2B_3u_1 + \lambda_1\beta_1c_1I_hS_h - \lambda_2\beta_1c_1I_hS_h \\ \frac{\partial H}{\partial u_2} &= 2B_4u_2 + \lambda_4\beta_3c_3I_cS_c - \lambda_5\beta_3c_3I_cS_c \end{aligned}\right\}$$

Thus, it is obtained that

$$\left.\begin{aligned} u_1^* &= -\frac{\beta_1c_1I_h(\lambda_1-\lambda_2)S_h}{2B_3} \\ u_2^* &= -\frac{\beta_3c_3I_c(\lambda_4-\lambda_5)}{2B_4} \end{aligned}\right\} \tag{20.12}$$

Therefore by standard control arguments involving the bounds on the controls, it can be concluded that (Table 20.1)

$$u_1^* = \begin{cases} 0 & \text{if } \xi_1^* \le 0 \\ \xi_1^* & \text{if } 0 < \xi_1^* < 1 \\ 1 & \text{if } \xi_1^* \ge 1 \end{cases}, \quad u_2^* = \begin{cases} 0 & \text{if } \xi_2^* \le 0 \\ \xi_2^* & \text{if } 0 < \xi_2^* < 1 \\ 1 & \text{if } \xi_2^* \ge 1 \end{cases}$$

$$\xi_1^* = \frac{-\beta_1c_2I_h(\lambda_1-\lambda_2)S_h}{2B_3}, \quad \xi_2^* = \frac{-\beta_3c_3I_c(\lambda_4-\lambda_5)S_c}{2B_4}$$

Table 20.1 Parameter description

Parameter	Description	Est. value	Reference
β_1	Probability of HIV infection	0.15	Previous study
β_2	Probability of SIV infection by humans	0.012	Assumed
β_3	Probability of SIV infection among chimpanzees	0.015	Assumed
δ_h	Fraction of infected humans	0.0065	Assumed
δ_c	Fraction of infected chimpanzees	0.0008	Assumed
c_1	Number of sexual partners for humans	3.0	Previous study
c_2	Number of infected chimpanzees	2.0	Assumed
c_3	Number of sexual partners for chimpanzees	2.0	Assumed
π_h	Recruitment rate of human	0.029	Previous study
π_c	Recruitment rate of chimpanzees	0.0012	Assumed
μ_h	Natural death rate of human	0.02	Previous study
μ_c	Natural death rate of chimpanzees	0.033	Assumed
μ_o	Disease-induced death of human	0.33	Previous study
α	Rate of developing AIDS	0.045	Previous study

20.5 Numerical Simulation

In this section, the extended deterministic model with control strategies was examined, to study the effect of treatment and culling, on the dynamics of HIV/AIDS in a community. The numerical simulation is carried out while the optimal set is found by solving the optimality system consisting of the state and ad-joint systems. Control (u_1) is used to optimize the objective functional. In Fig. 20.1, the results shows a significant difference in the I_h and A_h with optimal control strategy compared to I_h and A_h without control. It was observed in Fig. 20.1a that the infected humans I_h decrease as a result of control strategies against the increase in the uncontrolled case. In Fig. 20.1b, similar situation has also been observed in the case of human AIDS. Similarly control (u_2) is used to optimize the objective functional. In Fig. 20.1c, the results show a significant difference in (I_c) with optimal control strategy compared to I_c without control. It was observed in Fig. 20.1c, that the infected chimpanzees decrease as a result of control strategy against the increase in the uncontrolled case.

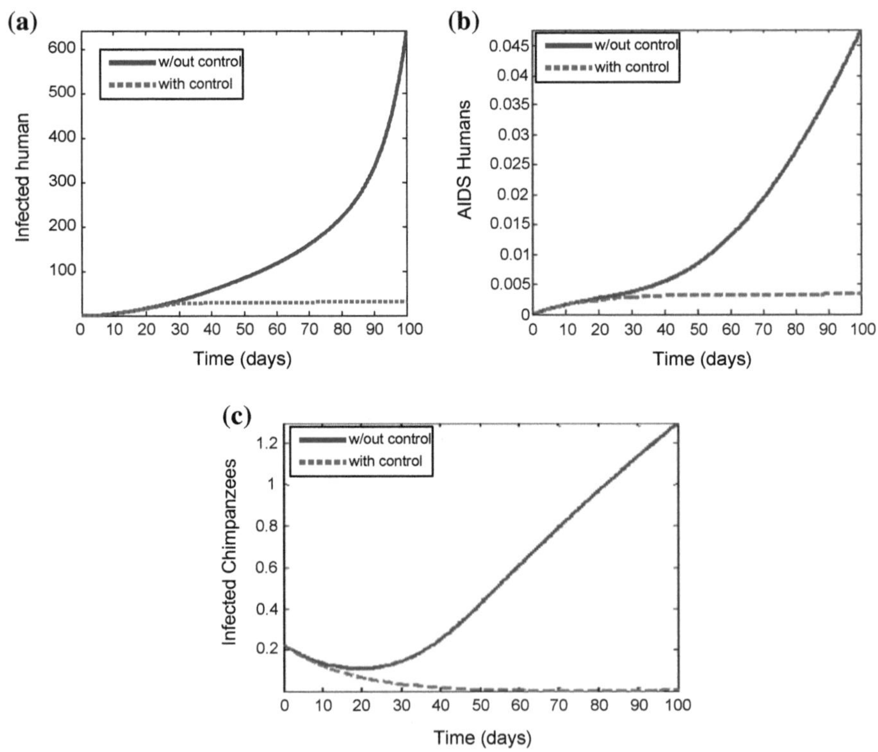

Fig. 20.1 **a**, **b** Simulations of HIV/AIDS model showing the effect of optimal use of treatment on the infected human and AIDS human respectively. **c** Simulations of HIV/AIDS model showing effect of optimal use of Culling on the spread of SIV

20.6 Conclusion

In this paper, a mathematical model on the spread of HIV/AIDS was developed. The formulation incorporates enrolment of human and chimpanzees populations through a constant immigration, with a fraction of infective immigrants with time-dependent control strategies. The impact of the control strategies, treatment for the infected and AIDS individuals with culling for the infected chimpanzees are considered. The condition for optimal control of HIV/AIDS and SIV were derived and analyzed with time-dependent preventive measures. The optimal control has a very desirable effect for reducing the HIV/AIDS and SIV. However, the simulation result indicate that an effective optimal use of the control programs that follows this two strategies, treatment and culling can effectively reduce the spread of the disease in a community.

References

1. Gayle, H.: An overview of the global HIV/AIDS epidemic, with a focus on the United States. AIDS (London, England) **14**, S8-17 (2000)
2. Karim, Q.A.: Prevention of HIV by male circumcision. Br. Med. J. **335**(7609), 4 (2007)
3. Deblonde, J.: HIV testing in Europe: mapping policies and exploring practices in the era of increased treatment availability. Discussion Faculty of Medicine and Health Sciences, Ghent University (2014)
4. The foundation for AIDS, www.amfar.org/worldwide-aids-stats/
5. HIV transmission Question and Answers, www.avert.org/hiv-transmission-questions-answers.htm#sthash.IfVu2W5U.dpuf
6. Gamkrelidze, A.: World health organization's HIV/AIDS policy and Georgia. Georgian Medical (2008)
7. Sharp, P.M., Shaw, G.M., Hahn, B.H.: Simian immunodeficiency virus infection of chimpanzees. J. Virol. **79**(7), 3891–3902 (2005)
8. Hahn, B.H., et al.: AIDS as a zoonosis: scientific and public health implications. Science **287** (5453), 607–614 (2000)
9. Sharp, P.M., Hahn, B.H.: The evolution of HIV-1 and the origin of AIDS. Philos. Trans. R. Soc. London B Biol. Sci. **365**(1552), 2487–2494 (2010)
10. Seifried, J.: Mechanisms of down-regulation of immune activation and B-cell responses in the natural hosts of simian immunodeficiency virus. Discussion Freie Universität Berlin (2012)
11. Kaadan, A.N., Zaiat, K.A.-D.: The Story of AIDS Discovery. Institute for the history of Arabic science, Aleppo University (2009)
12. Artois, M., Bengis R., Delahay R.J., Duchêne M.-J., Duff J.P., Ferroglio E., Gortazar C.: Wildlife disease surveillance and monitoring. In: Management of Disease in Wild Mammals, Springer, Japan (2009)
13. Delahay, R.J., Smith, G.C., Hutchings M.R.: Management of disease in wild mammals, Springer, USA (2009)
14. Anderson, R.M.: A preliminary study of the transmission dynamics of the human immunodeficiency virus (HIV), the causative agent of AIDS. Math. Med. Biol. **3**(4), 229–263 (1986)
15. Wiggins, S.: Introduction to applied nonlinear dynamical systems and chaos, vol. 2. Springer Science & Business Media, Berlin (2003)

16. Lakshmikantham, V., Leela, S., Martynyuk, A.A.: Stability Analysis of Nonlinear Systems. CRC Press, Boca Raton (1988)
17. Diekmann, O, Heesterbeek, J.A.P., Metz J.A.J.: On the definition and the computation of the basic reproduction ratio R 0 in models for infectious diseases in heterogeneous populations. J. Math. Biol. **28**(4), 365–382 (1990)
18. Gourley, S.A., Liu, R., Wu J.: Eradicating vector-borne diseases via age-structured culling. J. Math. Biol. (2007)
19. Hu, X., Liu, Y., Wu, J.: Culling structured hosts to eradicate vector-borne diseases. Math. Biosci. Eng. MBE **6**(2), 301–319 (2009)
20. Pontryagin, L.S., Boltyanskii, V.G, Gamkrelidze, R.V., Mishchenko, E.: The mathematical theory of optimal processes (Translated by D.E. Brown) A Pergamon Press Book. The Macmillan Company New York (1967)

Chapter 21
Analysis of Non-Newtonian Fluid Boundary Layer Flows Due to Surface Tension Gradient

Azhani Mohd Razali and Seripah Awang Kechil

Abstract The paper considers non-Newtonian power-law fluid flow driven by the convection of surface tension gradient in the presence of transverse magnetic field along a free-surface. The aim is to examine the effects of temperature and solute-dependent surface tension and magnetic field on the boundary layer flow of non-Newtonian fluids. The governing partial differential equations are reduced to a set of ordinary differential equations using similarity transformations. The coupled nonlinear system of equations is solved using a numerical approach known as the fourth-order finite difference scheme with shooting method. The numerical calculations were carried out for various values of fluid physical properties: power-law index, thermosolutal Marangoni number, and magnetic parameter. The results indicate that the free-surface velocity decreases with the increasing number of power-law index. As the thermosolutal Marangoni number increases, the thickness of thermal boundary layer decreases, while the surface temperature and concentration gradients increases. The magnetic field modifies the flow patterns of the non-Newtonian power-law fluid by reducing the fluid velocity at the surface.

Keywords Non-Newtonian fluid · Power-law model · Surface tension · Marangoni convection · Finite difference · Shooting method

Nomenclature

A	Surface temperature variation
B	Surface concentration variation
σ_{T}	Temperature coefficient of surface tension
σ_{C}	Concentration coefficient of surface tension

A.M. Razali (✉)
Plant Assessment Technology Group, Industrial Technology Division Malaysian Nuclear Agency, Bangi, 43000 Kajang, Malaysia
e-mail: azhani@nuclearmalaysia.gov.my

S.A. Kechil
Faculty of Computer and Mathematical Sciences, Universiti Teknologi MARA, 40450 Shah Alam, Selangor, Malaysia
e-mail: seripah@tmsk.uitm.edu.my

A.-R. Ahmad et al. (eds.), *Proceedings of the International Conference on Computing, Mathematics and Statistics (iCMS 2015)*,
DOI 10.1007/978-981-10-2772-7_21

σ_0	Surface tension at origin
u	Velocity components along x direction
v	Velocity components along y direction
K	Consistency coefficient
$f(\eta)$	Dimensionless stream function
$\theta(\eta)$	Dimensionless temperature function
$\phi(\eta)$	Dimensionless concentration function
T	Temperature
T_0	Surface temperature
ρ	Density of fluid
n	Power-law index
γ_m	Electrical conductivity
α	Thermal diffusivity of fluid
D	Mass diffusivity of fluid
η	Similarity variable
ψ	Stream function
C	Concentration
C_0	Surface concentration

21.1 Introduction

With the increasingly fast growing technology, many industrial processes, especially in manufacturing, involve the flows of non-Newtonian fluids. Many new products also exhibit non-Newtonian behaviour. Examples of industrial non-Newtonian fluids are multiphase mixtures, pharmaceutical formulations, molten plastics, cosmetics, toiletries, synthetic lubricants, biological fluids, polymer solutions and foodstuffs [1, 2]. It is also interesting to note that many salt solutions are also non-Newtonian. Study on the flow behaviour of salt solutions would be helpful for new generation of Molten-Salt Reactors in the world. However, the study of non-Newtonian fluids is relatively new compared to Newtonian fluid. Few studies were done on thermocapillary Marangoni convection flow of non-Newtonian power-law fluid. Zhang et al. [3] studied the flow of non-Newtonian power-law fluid along a horizontal surface with variable surface temperature using analytical approximation techniques. Lin et al. [4] investigated the heat transfer of an electrically conducting power-law fluid driven by temperature gradient at the surface using the method of Runge Kutta. The present study will investigate the boundary layer flow of non-Newtonian power-law fluid induced by thermosolutal surface tension gradient in the presence of transverse magnetic field with the application of a fourth-order finite difference scheme. The non-Newtonian fluid behaviour, the surface tension gradient effect and the magnetic effect are observed based on the power-law index, n, Marangoni number, M_{T} and M_{C}, and the magnetic parameter, M_{g}, respectively.

21.2 Mathematical Formulation of Problem

Consider a laminar and steady flow of an incompressible non-Newtonian fluid layer along a free-upper surface as shown in Fig. 21.1. The horizontal axis x is taken along the free-surface and y is taken normal to it. The non-Newtonian fluid is assumed to be an electrically conductive power-law fluid, fully driven by the surface tension variation in the presence of a uniform magnetic field, $B = B_{\mathrm{m}}^2$. The surface tension, σ changes linearly with both temperature and concentration given by the relation, $\sigma = \sigma_0 - \sigma_{\mathrm{T}}(T - T_0) - \sigma_{\mathrm{C}}(C - C_0)$. At the free stream, the velocity $U_\infty = 0$ with temperature T_∞ and concentration C_∞.

Following Zhang et al. [3], a set of two-dimensional partial differential equations which model the boundary layer flows is extended with a concentration conservation equation and a transverse uniform magnetic field of strength, $B = B_{\mathrm{m}}^2$. Given that $\tau_{yx} = K\left|\frac{\partial u}{\partial y}\right|^{n-1}\frac{\partial u}{\partial y}$ is the sheer stress of power-law non-Newtonian fluid, a new system of equations is written as follows:

$$\frac{\partial u}{\partial x} + \frac{\partial v}{\partial y} = 0, \tag{21.1}$$

$$u\frac{\partial u}{\partial x} + v\frac{\partial u}{\partial y} = \frac{K}{\rho}\frac{\partial}{\partial x}\left(\left|\frac{\partial u}{\partial y}\right|^{n-1}\frac{\partial u}{\partial y}\right) - \frac{\gamma_{\mathrm{m}} B_{\mathrm{m}}^2}{\rho}u, \tag{21.2}$$

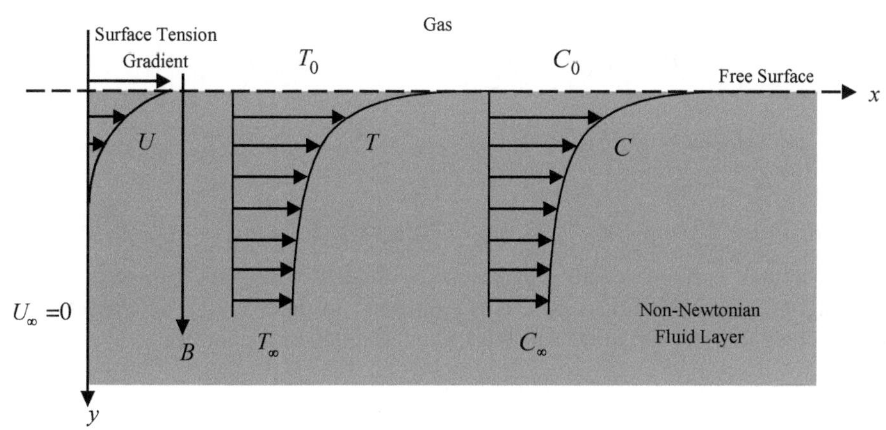

Fig. 21.1 Physical model for the boundary layer flow along a free-surface

$$u\frac{\partial T}{\partial x}+v\frac{\partial T}{\partial y}=\alpha\frac{\partial}{\partial y}\left(\left|\frac{\partial u}{\partial y}\right|^{n-1}\frac{\partial T}{\partial y}\right), \tag{21.3}$$

$$u\frac{\partial C}{\partial x}+v\frac{\partial C}{\partial y}=D\frac{\partial}{\partial y}\left(\left|\frac{\partial u}{\partial y}\right|^{n-1}\frac{\partial C}{\partial y}\right). \tag{21.4}$$

The boundary conditions are based on the surface geometry and surrounding conditions:

$$K\left|\frac{\partial u}{\partial y}\right|^{n-1}\frac{\partial u}{\partial y}=\frac{\partial \sigma}{\partial T}\cdot\frac{\partial T}{\partial x}+\frac{\partial \sigma}{\partial C}\cdot\frac{\partial C}{\partial x},\ v=0,$$
$$T=T_0+Ax,\text{ and } C=C_0+Bx \text{ at } y=0, \tag{21.5}$$

$$u\to 0,\ T\to T_0,\text{ and } C\to C_0 \text{ as } y\to\infty. \tag{21.6}$$

The system is simplified to a set of ordinary differential equations by the introduction of the non-dimensional similarity variables. Adapted from Zhang et al. [3], the similarity variables are given as

$$\psi(x,y)=\left(\frac{K(A\sigma_0\sigma_{\mathrm{T}})^{2n-1}}{\rho}\right)^{\frac{1}{3n}}x^{\frac{2}{3}}f(\eta),\ \eta=\left(\frac{\rho^n}{K^2(A\sigma_0\sigma_{\mathrm{T}})^{n-2}}\right)^{\frac{1}{3n}}x^{-\frac{1}{3}}y,$$
$$T=T_0+Ax\theta(\eta)\text{ and } C=C_0+Bx\phi(\eta). \tag{21.7}$$

By applying the similarity transformation (21.7), the governing Eqs. (21.1)–(21.4) are transformed into

$$f'^2-2ff''+M_gf'=3\left\lfloor(n-1)|f''|^{n-2}f'''f''+f'''|f''|^{n-1}\right\rfloor \tag{21.8}$$

$$3M_{\mathrm{T}}f'\theta-2M_{\mathrm{T}}f\theta'=3\left\lfloor(n-1)|f''|^{n-2}f'''\theta'+\theta''|f''|^{n-1}\right\rfloor, \tag{21.9}$$

$$\text{and } 3M_{\mathrm{C}}f'\phi-2M_{\mathrm{C}}f\phi'=3\left\lfloor(n-1)|f''|^{n-2}f'''\phi'+\phi''|f''|^{n-1}\right\rfloor, \tag{21.10}$$

given that $\varepsilon=\frac{\sigma_{\mathrm{C}}\cdot B}{\sigma_{\mathrm{T}}\cdot A}=\frac{M_{\mathrm{C}}}{M_{\mathrm{T}}}$, $M_{\mathrm{C}}=\frac{K}{D\rho}$, $M_{\mathrm{T}}=\frac{K}{\alpha\rho}$ and $M_{\mathrm{g}}=3\frac{\gamma_{\mathrm{m}}B_{\mathrm{m}}^2}{\rho}\left(\frac{K\rho^n}{(A\sigma_0\sigma_{\mathrm{T}})^{n+1}}\right)^{\frac{1}{3n}}x^{\frac{2}{3}}$ are the thermosolutal surface tension parameter, solutal Marangoni number, thermal Marangoni number and magnetic parameter, respectively. The transformed Eqs. (21.8)–(21.10) are subject to the following boundary conditions:

$$f(0) = 0,\ f'(\infty) = 0,\ |f''(0)|^{n-1}(f''(0)) = -1 - \varepsilon, \tag{21.11}$$

$$\theta(0) = 1,\ \theta(\infty) = 0,\ \phi(0) = 1,\ \text{and}\ \phi(\infty) = 0.$$

21.3 Numerical Approach

A numerical method of fourth-order finite difference scheme is applied as it is one of the effective methods in solving the nonlinear cases of boundary value problems. MATLAB bvp4c solver is chosen as the solving tool due to its high order of accuracy, better error tolerance and due to its ability to solve problems at multipoint conditions and of unknown parameters. However, MATLAB bvp4c solver requires a system of first-order linear or nonlinear differential equations. Therefore, new dependent variable, $F_i(\eta)$ for $i = 1, 2, \ldots, N$ is introduced to reduce the high order ordinary differential equations to first-order ordinary differential equations. i must start from 1 since MATLAB index can only process real positive integers or logical. By assuming $F_1 = f(\eta)$, $F_4 = \theta(\eta)$, $F_6 = \phi(\eta)$ and $N = 7$, Eqs. (21.8)–(21.10) are reduced to

$$\frac{\mathrm{d}F_1}{\mathrm{d}\eta} = F_2, \tag{21.12}$$

$$\frac{\mathrm{d}F_2}{\mathrm{d}\eta} = F_3, \tag{21.13}$$

$$\frac{\mathrm{d}F_3}{\mathrm{d}\eta} = \frac{(F_2)^2 - 2F_1F_3 + M_gF_2}{3\left[(n-1)|F_3|^{n-2}F_3 + |F_3|^{n-1}\right]}, \tag{21.14}$$

$$\frac{\mathrm{d}F_4}{\mathrm{d}\eta} = F_5, \tag{21.15}$$

$$\frac{\mathrm{d}F_5}{\mathrm{d}\eta} = \frac{3MF_2F_4 - 2MF_1F_5 - 3(n-1)|F_3|^{n-2}\frac{\mathrm{d}F_3}{\mathrm{d}\eta}F_5}{3|F_3|^{n-1}}, \tag{21.16}$$

$$\frac{\mathrm{d}F_6}{\mathrm{d}\eta} = F_7, \tag{21.17}$$

$$\frac{\mathrm{d}F_7}{\mathrm{d}\eta} = \frac{3MF_2F_6 - 2MF_1F_7 - 3(n-1)|F_3|^{n-2}\frac{\mathrm{d}F_3}{\mathrm{d}\eta}F_7}{3|F_3|^{n-1}}, \tag{21.18}$$

Accordingly, the boundary conditions (21.11) are also reduced to the first-order boundary conditions as follows:

$$F_1(0) = 0,\ F_2(\eta_\infty) = 0,\ F_3(0) = -1 - \varepsilon,\ F_4(0) = 1, \qquad (21.19)$$

$$F_4(\eta_\infty) = 0,\ F_6(0) = 1,\ F_6(\eta_\infty) = 0.$$

It is assumed that $M_C = M_T$ so that the value of ε will be assumed constant ε = 1.0. Since $M_C = M_T$, therefore the effects of solutal and thermal Marangoni numbers are similar. η_∞ is referred as a 'dimensionless infinity', which is the shooting value that needs to be guessed. The convergence criterion of boundary layer profiles will greatly depend on fairly good guesses of η_∞ and the mesh points.

21.4 Results and Discussions

Table 21.1 presents the values of fluid velocity at the surface, $f'(0)$ for $n = 0.8, 0.9, 1.0, 1.1$ and 1.2. It is observed in Table 4.1, the free-surface velocity decreases with the increasing value of n, resulting to a slow motion of the fluid flow along the surface.

Thermosolutal Marangoni number is related to the formation of thermal gradient at the interface. In Table 21.2, it is shown that the surface temperature gradient is an increasing function of Marangoni numbers. The incremental values indicate that the heat and mass transfer rates at the surface are increasing, leading to the possibilities of instabilities of the fluid flow. In contrary, the temperature profiles of non-Newtonian power-law fluid decreased with the increasing value of thermosolutal Marangoni number as illustrated in Fig. 21.2.

Magnetic field is often used to suppress Marangoni instability at the wall or interface. Figure 21.3 shows the reduction of fluid velocity at surface with increasing number of magnetic parameter. The reduction in velocity is due to the presence of a drag force that acts normal to the flow direction. The force from the applied transverse magnetic field has a tendency to resist the fluid flow, increasing friction between fluid layers and hence reduces the movement of the fluid flow at the surface.

Table 21.1 Free-surface velocity, $f'(0)$ for various values of n given that $M_T = M_C = 1.0$, $M_g = 0.0$ and $\eta_\infty = 2.8$

n	$f'(0)$
0.8	2.200641
0.9	2.094145
1.0	2.034548
1.1	2.007980
1.2	2.006955

Table 21.2 Free-surface temperature gradient, $-\theta'(0)$ for various values of n, M_T and M_C given that $M_g = 0.0$

n	$M_T = M_C = 1.0$ $\eta_\infty = 3.8$	$M_T = M_C = 2.0$ $\eta_\infty = 3.9$	$M_T = M_C = 3.0$ $\eta_\infty = 3.9$	$M_T = M_C = 4.0$ $\eta_\infty = 3.9$
0.8	1.223867	1.939439	2.456880	2.889285
0.9	1.329666	2.019876	2.545722	2.987302
1.0	1.409448	2.112596	2.650484	3.102747
1.1	1.494779	2.217374	2.769740	3.234214
1.2	1.587239	2.333647	2.902723	3.381002

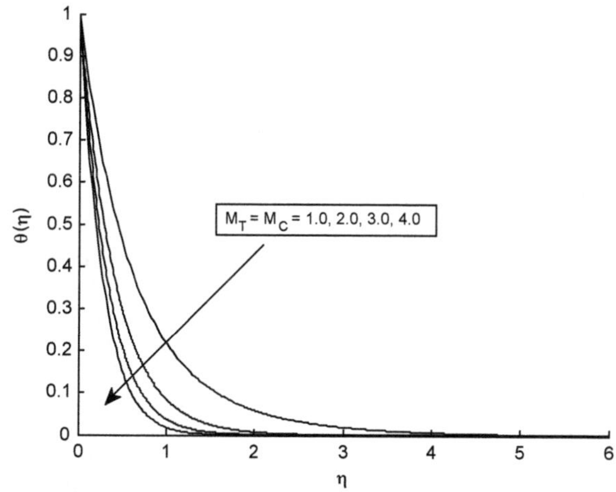

Fig. 21.2 Temperature profiles for various values of M_T and M_C with $n = 1.2$, $M_g = 0.0$ and $\eta_\infty = 6.0$

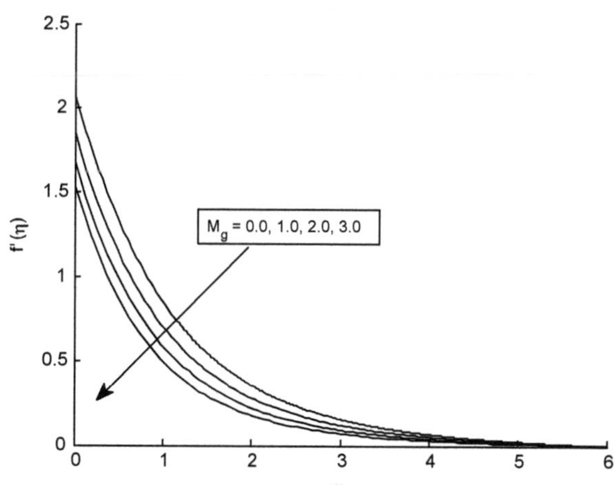

Fig. 21.3 Velocity profiles for various values of M_g with $n = 1.2$, $M_T = M_C = 1.0$ and $\eta_\infty = 6.0$

21.5 Conclusions

This paper presents a numerical study of an electrically conducting non-Newtonian power-law fluid flow induced by thermosolutal surface tension gradients in the presence of a transverse magnetic field. Some of the findings in this study are: (i) the free-surface velocity decreases with the increasing number of power-law index, (ii) as the thermosolutal Marangoni number increases, the thickness of thermal boundary layer decreases, while the surface temperature and concentration gradients increases, and, (iii) magnetic field modifies the flow patterns of the non-Newtonian power-law fluid by reducing the fluid velocity at the surface.

Acknowledgements This work was supported by Public Service Department (JPA) and Ministry of Science Technology and Innovation (MOSTI).

References

1. Xu, H., Liao, S.J., Pop, I.: Series solution of unsteady boundary layer flows of non-Newtonian fluids near a forward stagnation point. J. Non-Newton. Fluid Mech. **139**, 31–43 (2006)
2. Reddy, B.S., Kishan, N., Rajasekhar, M.N.: MHD boundary layer flow of a non-Newtonian power-law fluid on a moving flat plate. Adv. Appl. Sci. Res. **3**, 1472–1481 (2012)
3. Zhang, Y., Zheng, L., Wang, X., Song, G.: Analysis of marangoni convection of non-Newtonian power-law fluids with linear temperature distribution. Therm. Sci. **15**, S45–S52 (2011)
4. Lin, Y., Zheng, L., Zhang, X.: Magnetohydrodynamics thermocapillary marangoni convection heat transfer of power-law fluids driven by temperature gradient. ASME. J. Heat Transf. **135** (5), 051702-1–051706-6 (2013)

Chapter 22
Chebyshev Wavelet Operational Matrix of Fractional Derivative Through Wavelet-Polynomial Transformation and Its Applications on Fractional Order Differential Equations

Abdulnasir Isah and Phang Chang

Abstract In this paper, new numerical method based on Chebyshev wavelet operational matrix of fractional derivative is presented. The known Chebyshev Wavelets is presented first. Then, we derived the operational matrix of fractional order derivative (OMOFOD), through wavelet transformation matrix which was utilized together with spectral and collocation methods to reduce the linear and nonlinear fractional differential equations (FDEs) to a system of algebraic equations respectively. Our results in solving different linear and nonlinear FDEs confirm the applicability and accuracy of the proposed method.

Keywords Shifted Chebyshev polynomials · Chebyshev wavelets · Transformation matrix · Operational matrix of fractional derivative · Fractional differential equation

22.1 Introduction

Fractional calculus is a generalization of ordinary differentiation and integration to an arbitrary order. It is a valuable tool for modeling of many phenomena in the fields of physics, chemistry, engineering, aerodynamics, etc. [1, 2]. Modeling and simulation of systems and processes, based on the description of their properties in terms of fractional derivatives, naturally lead to the formation of fractional differential equations (FDEs), and hence the necessity of solving them. Few analytic and

A. Isah (✉) · P. Chang
Department of Mathematics and Statistics, University Tun Hussein Onn Malaysia, 86400 Parit Raja, Batu Pahat, Johor, Malaysia
e-mail: abdulnasir.isah@gmail.com

P. Chang
e-mail: pchang@uthm.edu.my

A.-R. Ahmad et al. (eds.), *Proceedings of the International Conference on Computing, Mathematics and Statistics (iCMS 2015)*,
DOI 10.1007/978-981-10-2772-7_22

numerical methods for solving FDEs are available in the literature, such as Adomian decomposition method [3], variational iteration method [4], and homotopy perturbation method [5]. Recently, the operational matrices of fractional order integration for Haar wavelets, Legendre wavelets, Chebyshev wavelets, Bernoulli wavelets have been developed in [6–9] respectively, to solve FDEs. The motivation of this research is to propose a more simple technique of obtaining the operational matrix with straight forward applicability to the FDEs in comparison to the existing operational matrices, which most of them are obtained through block pulse functions [8]. In this paper, we introduced a new method based on shifted Chebyshev polynomials to derive the operational matrix of fractional order derivative (OMOFOD), through wavelet-polynomial transformation matrix, which is used to solve FDEs. The method reduces the FDEs to a system of algebraic equations. The paper is organized as follows. Section 22.2 introduces some basic definitions and mathematical preliminaries of fractional calculus. In Sect. 22.3, we first define shifted Chebyshev polynomial, and then describe the basics of wavelets and Chebyshev wavelets. In Sect. 22.4, we introduce the Chebyshev operational matrix of fractional order derivative and Chebyshev wavelet. In Sect. 22.5, application of the Chebyshev wavelets operational matrix of fractional order derivative together with collocation methods are shown. In Sect. 22.6, the method is used on numerical examples to demonstrate the accuracy of the scheme.

22.2 Preliminaries

22.2.1 *Fractional Derivative and Integration*

Here, we recollect some basic definitions and properties of fractional calculus that are used in this article. We use Caputo's definition of fractional derivative in this research due to its advantage and reliability in applications over other definitions.

Definition 2.1 The Caputo fractional derivative of order α (D^α) of a function $f(t)$ is defined as

$$D^\alpha f(t) = \frac{1}{\Gamma(n-\alpha)} \int_0^t \frac{f^{(n)}(\tau)}{(t-\tau)^{\alpha-n+1}} \mathrm{d}\tau,\ n-1<\alpha\leq n, \quad n\in \mathrm{N}. \tag{22.1}$$

The following are some properties of Caputo fractional derivatives,

$$D^\alpha C = 0.\ (C \,\text{is a constant}) \tag{22.2}$$

$$D^\alpha t^\beta = \begin{cases} 0, & \beta\in \mathrm{N}\cup\{0\} \text{ and } \beta<\lceil\alpha\rceil \\ \frac{\Gamma(\beta+1)}{\Gamma(\beta+1-\alpha)} t^{\beta-\alpha}, & \beta\in \mathrm{N}\cup\{0\} \text{ and } \beta\geq\lceil\alpha\rceil \text{ or } \beta\notin \mathrm{N} \text{ and } \beta>\lfloor\alpha\rfloor \end{cases}. \tag{22.3}$$

Where $\lceil \alpha \rceil$ denotes the smallest integer greater than or equal to α and $\lfloor \alpha \rfloor$ denotes the largest integer less than or equal to α.

22.3 Chebyshev Polynomials and Chebyshev Wavelets

22.3.1 Shifted Chebyshev Polynomial of First Kind

The Chebyshev polynomials $T_i(t), i = 0, 1, \ldots$ are defined on the interval [−1, 1]. In order to use these polynomials on the interval [0, 1], we defined the shifted Chebyshev polynomials of first kind by introducing the change of variable, $t = 2x - 1$. Let the shifted Chebyshev polynomials $T_i(2x-1)$ be denoted by $T_i(x)$ which satisfies the orthogonality condition

$$\int_0^1 T_i(x)T_j(x)w(x)\mathrm{d}x = \begin{cases} \varsigma_k \pi/2, & \text{for } i = j \\ 0, & \text{for } i \neq j \end{cases}, \tag{22.4}$$

where $w(x) = (\sqrt{1-x^2})^{-1}$, $\varsigma_0 = 2$ and $\varsigma_k = 1,\ k \geq 1$. The analytic form of $T_i(x)$ is

$$T_i(x) = i\sum_{k=0}^{i} (-1)^{i-k} \frac{(i+k-1)!2^{2k}}{(i-k)!(2k)!} x^k, \tag{22.5}$$

where $T_i(0) = (-1)^i$ and $T_i(1) = 1$. $T_i(x)$ may be generated by the recurrence formula, $T_{i+1}(x) = 2(2x-1)T_i(x) - T_{i-1}(x),\ i = 1, 2, \ldots,$ given that $T_0(x) = 1$ and $T_1(x) = 2x - 1$.

22.3.2 Wavelets and Chebyshev Wavelets

Wavelets constitute a family of functions constructed from dilation and translation of a single function called mother wavelet. When the dilation parameter, a, and the translation parameter, b, varies continuously, we have the following family of continuous wavelets as [10], i.e., $\psi_{a,b}(t) = |a|^{1/2}\psi((t-b)/a)\quad a, b \in R$. If we restrict the parameters a and b to discrete values such that $a = a_0^{-k}, b = nb_0 a_0^{-k}, a_0 > 1, b_0 > 1,$ where n and k are positive integers, the family of discrete wavelets are defined as $\psi_{n,k}(t) = |a_0|^{1/2}\psi\left(a_0^k t - nb_0\right)$, where $\psi_{n,k}$ forms a wavelet basis for $L^2(R)$. Chebyshev wavelets $\psi_{n,m}(t) = \psi(k, n, m, t)$

have four arguments: k can assume any positive integer, m is the order for Chebyshev polynomials, and t is the normalized time. They are defined on the interval [0, 1) by

$$\psi_{n,m}(t) = \begin{cases} 2^{\frac{k+1}{2}}\tilde{T}_m(2^{k+1}t - 2n + 1), & \text{for } \frac{n}{2^k} \leq t < \frac{n+1}{2^k} \\ 0, & \text{otherwise} \end{cases}, \tag{22.6}$$

where

$$\tilde{T}_m(x) = \begin{cases} 1/\sqrt{\pi}, & m = 0 \\ \sqrt{2/\pi}T_m(t), & m \geq 1. \end{cases}$$

and $m = 0, 1, \ldots, M$ and $n = 0, 1, \ldots, 2^k - 1$. We should note that when dealing with Chebyshev wavelets the weight function, $w(x)$, have to be dilated and translated as $w_n(x) = w(2^k t - 2n + 1)$. Using shifted Chebyshev polynomial, $T_m(x)$, as defined in Sect. 3.1, one can write the Chebyshev Wavelets as

$$\psi_{n,m}(t) = \begin{cases} 2^{\frac{k+1}{2}}\sqrt{2/(\pi\varsigma_k)}\tilde{T}_m(2^k t - n), & \text{for } n/2^k \leq t < (n+1)/2^k \\ 0, & \text{otherwise} \end{cases}, \tag{22.7}$$

with the same range of m and n as in (22.6).

22.3.3 Function Approximations

Any function $f(t)$ which is square integrable in the interval [0, 1) can be expanded into Chebyshev wavelet series as

$$f(t) = \sum_{n=0}^{\infty}\sum_{m=0}^{\infty} C_{n,m}\psi_{n,m}, \tag{22.8}$$

where the coefficient $C_{n,m}$ is given by $C_{n,m} = \langle f(t),\ \psi_{n,m}(t)\rangle$. If the infinite series in (22.8) is truncated, then it can be written as

$$f(t) \approx \sum_{n=0}^{2^k-1}\sum_{m=0}^{M} C_{n,m}\psi_{n,m} = C^{\mathrm{T}}\Psi(t) \tag{22.9}$$

where C and Ψ are $2^k(M+1) \times 1$ matrices given by

$$C = \left[c_{0,0}, c_{0,1}, \ldots, c_{0,M}, c_{1,0}, c_{1,1} \ldots c_{1,M}, \ldots, c_{(2^k-1),0}, c_{(2^k-1),1}, \ldots, c_{(2^k-1),M}\right]^{\mathrm{T}} \tag{22.10}$$

$$\Psi(t) = \left[\psi_{0,0}, \psi_{0,1}, \ldots, \psi_{0,M}, \psi_{1,0}, \ldots, \psi_{1,M}, \ldots, \psi_{(2^k-1),0}, \psi_{(2^k-1),1}, \ldots, \psi_{(2^k-1),M}\right]^{\mathrm{T}} \tag{22.11}$$

22.4 Chebyshev Wavelet Operational Matrix of Fractional Order Derivative

In this section, we derive the Chebyshev wavelet operational matrix of the fractional derivative by first transforming the wavelets to shifted Chebyshev polynomials. Then, we make use of the shifted Chebyshev operational matrix of the fractional derivative derived in [11], and finally we derive the Chebyshev wavelet operational matrix of the fractional derivative.

22.4.1 Transformation Matrix of the Chebyshev Wavelets to Chebyshev Polynomials

An arbitrary function $y(x) \in L^2[0,1]$ can be expanded into shifted Chebyshev polynomials as $y(x) \approx \sum_{m=0}^{M} r_m T_m(x) = R\Psi'(x)$ where the shifted Chebyshev coefficient vector R and the shifted Chebyshev vector $\Psi'(x)$ are given by

$$R = [r_0, r_1, \ldots, r_M], \Psi'(x) = [T_0(x), T_1(x), \ldots, T_M(x)]^{\mathrm{T}} \tag{22.12}$$

The Chebyshev wavelet may be expanded into $(M + 1)$-terms shifted Chebyshev polynomials as

$$\Psi_{2^k(M+1)\times 1}(t) = \Phi_{2^k(M+1)\times(M+1)} \Psi'_{(M+1)\times 1}, \tag{22.13}$$

where Φ is transformation matrix of Chebyshev wavelet to Chebyshev polynomial.

22.4.2 Shifted Chebyshev Operational Matrix of Fractional Order Derivative

The fractional derivative of order α of the vector $\Psi'(t)$ as shown in [11] can be expressed as

$$D^{\alpha}\Psi'(t) = F^{(\alpha)}\Psi'(t), \tag{22.14}$$

where $F^{(\alpha)}$ is the $(m + 1) \times (m + 1)$ operational matrix of fractional derivative of order α defined as

$$F^{(\alpha)} = \begin{bmatrix} 0 & 0 & \cdots & 0 \\ \vdots & \vdots & \cdots & \vdots \\ 0 & 0 & \cdots & 0 \\ s_{\alpha}(\lceil\alpha\rceil, 0) & s_{\alpha}(\lceil\alpha\rceil, 1) & \cdots & s_{\alpha}(\lceil\alpha\rceil, m) \\ \vdots & \vdots & \cdots & \vdots \\ s_{\alpha}(i, 0) & s_{\alpha}(i, 1) & \cdots & s_{\alpha}(i, m) \\ \vdots & \vdots & \cdots & \vdots \\ s_{\alpha}(m, 0) & s_{\alpha}(m, 1) & \cdots & s_{\alpha}(m, m) \end{bmatrix},$$

where, $s(i,j)$ is given by

$$s(i,j) = \sum_{k=0}^{i} \frac{(-1)^{i-k}(2i)(i+k-1)!\Gamma(k-\alpha+1/2)}{\varsigma_j \Gamma(k+1/2)(i-k)!\Gamma(k-\alpha-j+1)\Gamma(k+j-\alpha+1)} \tag{22.15}$$

See [11] for more details on $F^{(\alpha)}$.

22.4.3 Chebyshev Wavelet Operational Matrix of Fractional Order Derivative

We now derive Chebyshev wavelet operational matrix of fractional order derivative. Let

$$D^{\alpha}\Psi(x) = H^{(\alpha)}\Psi(x), \tag{22.16}$$

where, $H^{(\alpha)}$ is the Chebyshev wavelet operational matrix of fractional derivative. Using (22.13) and (22.14), we get

$$D^{\alpha}\Psi(x) = D^{\alpha}\Phi\Psi'(x) = \Phi D^{\alpha}\Psi'(x) = \Phi F^{(\alpha)}\Psi'(x). \tag{22.17}$$

From (22.16) and (22.17), we have

$$H^{(\alpha)}\Psi(x) = H^{(\alpha)}\Phi\Psi'(x) = \Phi F^{(\alpha)}\Psi'(x). \tag{22.18}$$

Thus, Chebyshev wavelet operational matrix of fractional derivative $H^{(\alpha)}$ is given by

$$H^{(\alpha)} = \Phi F^{(\alpha)} \Phi^{-1} \tag{22.19}$$

22.5 Applications of Chebyshev Wavelet Operational Matrix of Fractional Order Derivative on FDEs

Here, we are going to apply the Chebyshev wavelet operational matrix of fractional derivative to solve linear and non-linear multi-order FDEs.

22.5.1 *Linear Multi-order Fractional Differential Equation*

Consider the linear FDE

$$D^{\alpha}y(x) = a_1 D^{\mu_1}y(x) + \cdots + a_s D^{\mu_n}y(x) + a_{s+1}y(x) + a_{s+2}g(x), \tag{22.20}$$

with initial conditions $y^{(i)}(0) = d_i, \quad i = 0, 1, \cdots r$, where a_j, for $j = 1, \ldots s+2$ are real constant coefficients. We have $r < \alpha \leq r+1$, $0 < \mu_1 < \mu_2 < \cdots < \mu_s < \alpha$ and D^{α} denotes the Caputo fractional derivative of order α. To solve problem (22.20), we first approximate $y(x)$ and $g(x)$ by the Chebyshev wavelets as

$$y(x) \approx \sum_{n=0}^{2^k-1} \sum_{m=0}^{M} c_{n,m}\psi_{n,m} = C^{\mathrm{T}}\Psi(x) \quad g(x) \approx \sum_{n=0}^{2^k-1} \sum_{m=0}^{M} z_{n,m}\psi_{n,m} = G^{\mathrm{T}}\Psi(x) \tag{22.21}$$

Where $G = \left[z_{0,0}, z_{0,1}, \ldots z_{0,M}, z_{1,0}, z_{1,1} \ldots, z_{1,M}, \ldots, z_{(2^k-1),0}, z_{(2^k-1),1}, \ldots, z_{(2^k-1),M}\right]^{\mathrm{T}}$ is known but C as defined in (22.10) is the unknown vector. Now, using (22.16) we get

$$D^{\alpha}y(x) \approx C^{\mathrm{T}} D^{\alpha}\Psi(x) \approx C^{\mathrm{T}} H^{(\alpha)}\Psi(x) \tag{22.22}$$

$$D^{\mu_j}y(x) \approx C^{\mathrm{T}} D^{\mu_j}\Psi(x) \approx C^{\mathrm{T}} H^{(\mu_j)}\Psi(x) \tag{22.23}$$

Employing (22.21–22.23), the residual $R(x)$ for Eq. (22.20) can be written as

$$\begin{aligned} R(x) \approx (&C^{\mathrm{T}} H^{(\alpha)}\Psi(x) - a_1 C^{\mathrm{T}} H^{(\mu_1)}\Psi(x) - \cdots - a_k C^{\mathrm{T}} H^{(\mu_k)}\Psi(x) \\ &- a_{k+1} C^{\mathrm{T}}\Psi(x) - a_{k+2} G^{T}\Psi(x)) \end{aligned} \tag{22.24}$$

As in a typical tau method, we generate $2^k(M+1) - r$ linear equations by applying

$$\langle R(x), \Psi_j(x)\rangle = \int_0^1 R(x)\Psi_j(x)\mathrm{d}x = 0 \quad j = 1, \ldots 2^k(M+1) - r. \tag{22.25}$$

Also, by substituting initial conditions into (22.21–22.23), we have

$$\begin{aligned} y(0) &\approx C^{\mathrm{T}}\Psi(0) = d_0 \\ y^{(1)}(0) &\approx C^{\mathrm{T}}H^{(1)}\Psi(0) = d_1 \\ &\vdots \\ y^{(r)}(0) &\approx C^{\mathrm{T}}H^{(r)}\Psi(0) = d_r \end{aligned} \tag{22.26}$$

Equations (22.25–22.26) generate $2^k(M+1)$ set of linear equations. These linear equations can be solved for unknown coefficients of the vector C. Consequently, $y(x)$ given in Eq. (22.21) can be calculated.

22.5.2 Nonlinear Multi-order Fractional Differential Equation

Consider the nonlinear multi-order fractional differential equation

$$D^{\alpha}y(x) = F(x, y(x), D^{\mu_1}y(x), \ldots, D^{\mu_s}y(x)) \tag{22.27}$$

with initial conditions $y^{(i)}(0) = d_i$, $i = 0, 1, \ldots r$ where $r < \alpha \leq r+1$, $0 < \mu_1 < \mu_2 < \cdots < \mu_s < \alpha$. D^{α} Denotes the Caputo fractional derivative of order α and F can be nonlinear in general. To solve this problem, we first approximate $y(x)$, $D^{\alpha}y(x)$, $D^{\mu_j}y(x)$ for $j = 1, \cdots, s$ as (22.21), (22.22), and (22.23) respectively, and by substituting these equations in (22.27), we get

$$C^{\mathrm{T}}H^{(\alpha)}\Psi(x) \approx F(x, C^{\mathrm{T}}\Psi(x), C^{\mathrm{T}}H^{(\mu_1)}\Psi(x), \ldots, C^{\mathrm{T}}H^{(\mu_s)}\Psi(x)), \tag{22.28}$$

Also for the initial conditions, we approximate as (22.26). Now, to find the solution $y(x)$, we collocate (22.28) at $2^k(M+1) - r$ points. For suitable collocation points, we use the first $2^k(M+1) - r$ shifted Chebyshev polynomial roots $T_{2^k(M+1)-r}(x)$. The equations obtained together with (22.26) generate $2^k(M+1)$ nonlinear equations which can be solved using Newton's iterative method. Consequently $y(x)$ given in (22.21) can be calculated.

22.6 Illustrative Examples

Example 6.1 Consider the equation in [12], $D^2y(x)+3Dy(x)+2D^{\mu_1}y(x)+D^{\mu_2}y(x)+5y(x)=g(x)$, $y(0)=1, y'(0)=0$ with $g(x)=1+3x+(\Gamma(3-\mu_1))^{-1}x^{2-\mu_1}+(\Gamma(3-\mu_2))^{-1}x^{2-\mu_2}+5(1+x^2/2)$ where $\mu_1=0.1379$ and $\mu_2=0.0159$. The exact solution of this problem as in [12] is given by $1+x^2/2$. By applying the technique described in Sect. 5.1 with $M=2$ and $k=0$, we approximate the following as

$$y(x) \approx C^{\mathrm{T}}\Psi(x), D^2y(x) \approx C^{\mathrm{T}}H^{(2)}\Psi(x), D^{\mu_1}y(x) \approx C^{\mathrm{T}}H^{(\mu_1)}\Psi(x)$$
$$D^{\mu_2}y(x) \approx C^{\mathrm{T}}H^{(\mu_2)}\Psi(x), g(x) = G^{\mathrm{T}}\Psi(x),$$

Thus, we have

$$H^{(1)} = \begin{bmatrix} 0 & 0 & 0 \\ 8 & 0 & 0 \\ 0 & 16 & 0 \end{bmatrix}, H^{(2)} = \begin{bmatrix} 0 & 0 & 0 \\ 0 & 0 & 0 \\ 128 & 0 & 0 \end{bmatrix}, G = \begin{bmatrix} 15.8980 \\ 1.4242 \\ 0.1395 \end{bmatrix}$$

$$H^{\mu_1} = \frac{1}{\sqrt{\pi}}\begin{bmatrix} 0 & 0 & 0 \\ 4.3307 & 2.0050 & -0.0441 \\ -3.5795 & 0.9215 & 1.9790 \end{bmatrix},$$
$$H^{\mu_2} = \frac{1}{\sqrt{\pi}}\begin{bmatrix} 0 & 0 & 0 \\ 3.6289 & 1.7999 & -0.0047 \\ -3.5567 & 0.0868 & 1.7975 \end{bmatrix}$$

Therefore, using (22.25) we obtain

$$128c_{0,2}+24c_{0,1}+12.2902c_{0,1}/\sqrt{\pi}-10.7157c_{0,2}/\sqrt{\pi}+5c_{0,0}=15.8980. \tag{22.29}$$

Then, applying (22.26), we get

$$1/\pi\left[\sqrt{2}c_{0,0}-2\sqrt{2}c_{0,1}+2\sqrt{2}c_{0,0}\right]=1,\ 1/\pi\left[8\sqrt{2}c_{0,1}-32\sqrt{2}c_{0,2}\right]=0 \tag{22.30}$$

Solving (22.29) and (22.30), we have $c_{0,0}=2.3257$, $c_{0,1}=0.0695, c_{0,2}=0.0173$. Thus,

$$y(x) = \left[c_{0,0}, c_{0,1}, c_{0,2}\right]\left(\frac{\sqrt{2}}{\pi}\begin{bmatrix} 1 \\ 2(4x-1) \\ 2(32x^2-16x+1) \end{bmatrix}\right) = 1+0.5x^2,$$

this is the exact solution.

Example 6.2 We consider the following nonlinear initial value problem (see [11])

$$D^{(3)}y(x) + D^{(5/2)}y(x) + y^2(x) = x^4$$
$$y(0) = 0, y'(0) = 0, y''(0) = 2$$

The exact solution is known to be $y(x) = x^2$ [11]. We used the technique described in Sect. 5.2 to solve this problem with $M = 3$ and $k = 0$. First, we approximate the equation with Chebyshev wavelet as follows:

$$C^{\mathrm{T}}H^{(3)}\Psi(x) + C^{\mathrm{T}}H^{(5/2)}\Psi(x) + (C^{\mathrm{T}}\Psi(x))^2 - x^4 = 0, \tag{22.31}$$

which is collocated at the first roots of fourth-order shifted Chebyshev polynomial, i.e., 0.03806023374 and we obtain the following nonlinear algebraic equation,

$$\begin{aligned} &3072\sqrt{2}c_{0,3}/\pi + 3292.345747\sqrt{2}c_{0,3}/\pi^{5/2} - 0.0000002098388050 \\ &\quad + \left(1/\pi\left[\sqrt{2}c_{0,0} - 1.695518130\sqrt{2}c_{0,1} + 0.8747817294\sqrt{2}c_{0,2} + 0.212309848\sqrt{2}c_{0,3}\right]\right)^2 = 0 \end{aligned} \tag{22.32}$$

Also from the initial conditions, the following equations are generated.

$$\begin{aligned} &1/\pi\left[\sqrt{2}c_{0,0} - 2\sqrt{2}c_{0,1} + 2\sqrt{2}c_{0,2} - 2\sqrt{2}c_{0,3}\right] = 0 \\ &1/\pi\left[\sqrt{2}(8c_{0,1} + 24c_{0,3}) - 32\sqrt{2}c_{0,2} + 48\sqrt{2}c_{0,3}\right] = 0 \\ &1/\pi\left[128\sqrt{2}c_{0,2} - 768\sqrt{2}c_{0,3}\right] = 2 \end{aligned} \tag{22.33}$$

Thus, solving (22.32) and (22.33), we get $c_{0,0} = 0.2082601378, c_{0,1} = 0.1388400919, c_{0,2} = 0.03471002297$, and $c_{0,3} = 4.120027869 \times 10^{-16}$. Consequently,

$$y(x) = \left[c_{0,0}, c_{0,1}, c_{0,2}, c_{0,3}\right]\left(\sqrt{\frac{2}{\pi}}\begin{bmatrix} 1 \\ 2(4x-1) \\ 2(32x^2 - 16x + 1) \\ 2(256x^3 - 192x^2 + 36x - 1) \end{bmatrix}\right) = x^2$$

This is the exact solution.

22.7 Conclusion

A general formulation for the Chebyshev wavelet operational matrix of fractional order derivative has been derived. This matrix is used to approximate numerical solution of FDEs. Our numerical findings are compared with exact solutions. The solutions obtained using the present method shows that this approach can solve problems effectively.

Acknowledgments This work was supported in part by FRGS Grant Vot 1433. We also thank UTHM for financial support through GIPS U060.

References

1. Kilbas, A., Aleksandrovich, A., Srivastava, H.M., Trujillo, J.J.: Theory and Applications of Fractional Differential Equations, vol. 204. Elsevier Science Limited (2006)
2. Podlubny, I.: Fractional Differential Equations: An Introduction to Fractional Derivatives, Fractional Differential Equations, to Methods of their Solution and Some of their Applications, vol. 198. Academic press (1998)
3. Ray, S.S., Chaudhuri, K.S., Bera, R. K.: Analytical approximate solution of nonlinear dynamic system containing fractional derivative by modified decomposition method. Appl. Math. Comput. **182**(1) (2006)
4. Yang, S., Xiao, A., Su, H.: Convergence of the variational iteration method for solving multi-order fractional differential equations. Comput. Math. Appl. **60**(10) (2010)
5. Odibat, Z.: On Legendre polynomial approximation with the VIM or HAM for numerical treatment of nonlinear fractional differential equations. J. Comput. Appl. Math. **235**(9) (2011)
6. Li, Y., Zhao, W.: Haar wavelet operational matrix of fractional order integration and its applications in solving the fractional order differential equations. Appl. Math. Comput. **216**(8) (2010)
7. Jafari, H., et al.: Application of Legendre wavelets for solving fractional differential equations. Comput. Math. Appl. **62**(3) (2011)
8. Yuanlu, L.: Solving a nonlinear fractional differential equation using Chebyshev wavelets. Commun. Nonlinear Sci. Numer. Simul. **15**(9) (2010)
9. Keshavarz, E., Ordokhani, Y., Razzaghi, M.: Bernoulli wavelet operational matrix of fractional order integration and its applications in solving the fractional order differential equations. Appl. Math. Model. **38**(24) (2014)
10. Guf, J.-S., Jiang, W.-S.: The Haar wavelets operational matrix of integration. Int. J. Syst. Sci. **27**(7) (1996)
11. Doha, E.H., Bhrawy, A.H., Ezz-Eldien, S.S.: A Chebyshev spectral method based on operational matrix for initial and boundary value problems of fractional order. Comput. Math. Appl. **62**(5) (2011)
12. El-Sayed, A.M.A., El-Mesiry, A.E.M., El-Saka, H.A.A.: Numerical solution for multi-term fractional (arbitrary) orders differential equations. Comput. Appl. Math. **23**(1) (2004)

Chapter 23
Developing New Block Method for Direct Solution of Second-Order Ordinary Differential Equations

J.O. Kuboye and Z. Omar

Abstract This paper focuses on the derivation of a new block method for solving second-order initial value problems. The method is developed via interpolation and collocation approach. Power series approximate solution is considered as an interpolation equation while its second derivative is assigned as the collocation equation. The new method was then applied to solve some second-order initial value problems of ordinary differential equations and the results generated are better in terms of accuracy when compared with existing methods.

Keywords Power series · Interpolation · Collocation · Block method

23.1 Introduction

In most application, second-order ordinary differential equations of the form

$$y'' = f(x, y, y'),\ y(a) = y_0,\ y'(a) = y'_0 \tag{23.1}$$

are solved by reducing to its equivalent system of first-order ordinary differential equations. Then, any suitable numerical methods would be used to solve the resulting system of equations. This approach is extensively discussed by scholars such as [1–6]. It was observed that this reduction process has a lot of setbacks such as difficulties in writing computer program for the method and computational burden which may jeopardize the accuracy of the method in terms of error.

Therefore, it is appropriate if direct methods of solving (23.1) are employed instead as suggested by [7–9]. The development of Linear Multistep Method

J.O. Kuboye (✉) · Z. Omar
School of Quantitative Sciences, College of Art and Sciences,
Universiti Utara Malaysia, Sintok, Malaysia
e-mail: Kubbysholly2007@yahoo.com

Z. Omar
e-mail: zurni@uum.edu.my

A.-R. Ahmad et al. (eds.), *Proceedings of the International Conference on Computing, Mathematics and Statistics (iCMS 2015)*,
DOI 10.1007/978-981-10-2772-7_23

(LMM) through the predictor–corrector mode has been considered by scholars such as [6, 10–12] among others. These researchers independently suggested a multi-derivative Linear Multistep Method (LMM) which was implemented in a predictor–corrector mode where Taylor series algorithm is used to supply the starting values. The method is associated with computational burden due to the evaluation of many functions per iteration.

In order to overcome the difficulties mentioned in predictor–corrector method, block method was developed [5]. This computes numerical solution at more than one grid point simultaneously [13] proposed a uniform order six block method, whereby the points of interpolation were made at the beginning of the interval. However, the numerical results generated when the method was applied to second-order problems are not encouraging in terms of error. Furthermore, a family of implicit integrators was also developed by [14] using a step-length $k = 4(1)6$, whereby collocation was only made at one points. However, the accuracy of the methods was very low.

In order to increase the accuracy of the existing methods, this paper proposes a new block method with step-length $k = 4$ for solving (23.1) directly. The interpolation and collocation technique was used in deriving the method where interpolation was made at the second point before the last point and collocation is at all the grid points.

23.2 Derivation of the Method

We consider power series of the form

$$y(x) = \sum_{j=0}^{k+2} a_j x^j \tag{23.2}$$

as an approximate solution to (23.1) where the step-length $k = 4$. The first and second derivatives of (23.2) are

$$y'(x) = \sum_{j=1}^{k+2} j a_j x^{j-1} \tag{23.3}$$

$$y''(x) = \sum_{j=2}^{k+2} j(j-1) a_j x^{j-2} = f(x, y, y') \tag{23.4}$$

Interpolating (23.2) at the points $x = x_{n+i},\ i = 1, 2$ and collocating (23.4) at the points $x = x_{n+i},\ i = 0(1)4$ produces

$$\begin{pmatrix} 1 & x_{n+!} & x_{n+1}^2 & x_{n+1}^3 & x_{n+1}^4 & x_{n+!}^5 & x_{n+!}^6 \\ 1 & x_{n+2} & x_{n+2}^2 & x_{n+2}^3 & x_{n+2}^4 & x_{n+2}^5 & x_{n+2}^6 \\ 0 & 0 & 2 & 6x_n & 12x_n^2 & 20x_n^3 & 30x_n^4 \\ 0 & 0 & 2 & 6x_{n+1} & 12x_{n+1}^2 & 20x_{n+1}^3 & 30x_{n+1}^4 \\ 0 & 0 & 2 & 6x_{n+2} & 12x_{n+2}^2 & 20x_{n+2}^3 & 30x_{n+2}^4 \\ 0 & 0 & 2 & 6x_{n+3} & 12x_{n+3}^2 & 20x_{n+3}^3 & 20x_{n+3}^4 \\ 0 & 0 & 2 & 6x_{n+4} & 12x_{n+4}^2 & 20x_{n+4}^3 & 20x_{n+4}^4 \end{pmatrix} \begin{pmatrix} a_0 \\ a_1 \\ a_2 \\ a_3 \\ a_4 \\ a_5 \\ a_6 \end{pmatrix} = \begin{pmatrix} y_{n+1} \\ y_{n+2} \\ f_n \\ f_{n+1} \\ f_{n+2} \\ f_{n+3} \\ f_{n+4} \end{pmatrix} \tag{23.5}$$

In finding the values of *as* in (23.5), Gaussian elimination method is employed. Substituting the values of *as* into Eq. (23.2) and on simplifying gives a continuous linear multistep method of the form:

$$y(x) = \sum_{j=1}^{k-2} \alpha_j(x) y_{n+j} + h^2 \sum_{j=0}^{k} \beta_j(x) f_{n+j} \tag{23.6}$$

Using the transformation $z = \frac{x - x_{n+k-1}}{h}$ implies

$$x = zh + x_n + 3h \tag{23.7}$$

Substitute (23.7) into (23.6) gives

$$y(z) = \sum_{j=1}^{k-2} \alpha_j(z) y_{n+j} + h^2 \sum_{j=0}^{k} \beta_j(z) f_{n+j} \tag{23.8}$$

where

$$\begin{aligned}
\alpha_1(z) &= -1 - z \\
\alpha_2(z) &= 2 + z \\
\beta_0(z) &= \frac{1}{4320}\left(-18 + 15z - 60z^3 - 15z^4 + 18z^5 + 6z^6\right) \\
\beta_1(z) &= \frac{1}{360}\left(36 + 18z + 30z^3 + 5z^4 - 9z^5 - 2z^6\right) \\
\beta_2(z) &= \frac{1}{720}\left(582 + 747z - 180z^3 + 15z^4 + 36z^5 + 6z^6\right) \\
\beta_3(z) &= \frac{1}{320}\left(36 + 154z + 180z^2 + 50z^3 - 25z^4 - 15z^5 - 2z^6\right) \\
\beta_4(z) &= \frac{1}{1440}\left(-6 - 27z + 60z^3 + 55z^4 + 18z^5 + 2z^6\right)
\end{aligned}$$

Evaluating (23.8) at the non-interpolating points, that is, at $z = -3$, 0 and 1 produces the following schemes:

$$240y_{n+2} - 480y_{n+1} + 240y_n = h^2(-f_{n+4} + 4f_{n+3} + 14f_{n+2} + 204f_{n+1} + 19f_n) \tag{23.9}$$

$$240y_{n+3} - 480y_{n+3} + 240y_{n+1} = h^2(-f_{n+4} + 24f_{n+3} + 194f_{n+2} + 24f_{n+1} - f_n) \tag{23.10}$$

$$\begin{aligned} 240y_{n+4} - 720y_{n+2} + 4800y_{n+1} = h^2(&17f_{n+4} + 252f_{n+3} + 402f_{n+2} + 52f_{n+1} \\ &- 3f_n) \end{aligned} \tag{23.11}$$

Differentiate (23.8) once gives

$$y'(z) = \sum_{j=1}^{k-2} \alpha'_j(z) y_{n+j} + h^2 \sum_{j=0}^{k} \beta'_j(z) f_{n+j} \tag{23.12}$$

where

$$\begin{aligned}
\alpha'_1(z) &= -1 \\
\alpha'_2(z) &= 1 \\
\beta'_0(z) &= \frac{1}{1440}(5 - 60z^2 - 20z^3 + 30z^4 + 12z^5) \\
\beta_1(z) &= \frac{1}{360}(18 + 90z^2 + 20z^3 - 45z^4 - 12z^5) \\
\beta_2(z) &= \frac{1}{720}(249 - 180z^2 + 20z^3 + 60z^4 + 12z^5) \\
\beta_3(z) &= \frac{1}{320}(154 + 360z + 150z^2 - 100z^3 - 75z^4 - 12z^5) \\
\beta_4(z) &= \frac{1}{1440}(-27 + 180z^2 + 220z^3 + 90z^4 + 12z^5)
\end{aligned}$$

Equation (23.12) is evaluated at all the grid points. That is, at $z = -3, -2, -1, 0$ and 1 yields

$$\begin{aligned} 1440hy'_n - 1440y_{n+2} + 1440y_{n+1} = h^2(&27f_{n+4} - 140f_{n+3} + 198f_{n+2} \\ &- 1764f_{n+1} - 481f_n). \end{aligned} \tag{23.13}$$

$$\begin{aligned} 1440hy'_{n+1} - 1440y_{n+2} + 1440y_{n+1} = h^2(&-11f_{n+4} + 72f_{n+3} - 330f_{n+2} \\ &- 472f_{n+1} + 21f_n). \end{aligned} \tag{23.14}$$

$$1440hy'_{n+2} - 1440y_{n+2} + 1440y_{n+1} = h^2(11f_{n+4} - 76f_{n+3} + 582f_{n+2} + 220f_{n+1} - 17f_n)\,. \quad (23.15)$$

$$1440hy'_{n+3} - 1440y_{n+2} + 1440y_{n+1} = h^2(-27f_{n+4} + 616f_{n+3} + 1494f_{n+2} + 72f_{n+1} + 5f_n). \quad (23.16)$$

$$1440hy'_{n+4} - 1440y_{n+2} + 1440y_{n+1} = h^2(475f_{n+4} + 1908f_{n+3} + 966f_{n+2} + 284f_{n+1} - 33f_n). \quad (23.17)$$

Combining Eqs. (23.9)–(23.11) and (23.13), we get

$$\begin{pmatrix} -480 & 240 & 0 & 0 \\ 240 & -480 & 240 & 0 \\ 480 & -720 & 0 & 240 \\ 1440 & -1440 & 0 & 0 \end{pmatrix}\begin{pmatrix} y_{n+1} \\ y_{n+2} \\ y_{n+3} \\ y_{n+4} \end{pmatrix} = \begin{pmatrix} 0 & 0 & 0 & -240 \\ 0 & 0 & 0 & 0 \\ 0 & 0 & 0 & 0 \\ 0 & 0 & 0 & 0 \end{pmatrix}\begin{pmatrix} y_{n-3} \\ y_{n-2} \\ y_{n-1} \\ y_n \end{pmatrix}$$
$$+ h\begin{pmatrix} 0 & 0 & 0 & 0 \\ 0 & 0 & 0 & 0 \\ 0 & 0 & 0 & 0 \\ 0 & 0 & 0 & -1440 \end{pmatrix}\begin{pmatrix} y'_{n-3} \\ y'_{n-2} \\ y'_{n-1} \\ y'_n \end{pmatrix}$$
$$+ h^2\begin{pmatrix} 204 & 14 & 4 & -1 \\ 24 & 194 & 24 & -1 \\ 52 & 402 & 252 & 17 \\ -1764 & 198 & -140 & 27 \end{pmatrix}\begin{pmatrix} f_{n+1} \\ f_{n+2} \\ f_{n+3} \\ f_{n+4} \end{pmatrix}$$
$$+ h^2\begin{pmatrix} 0 & 0 & 0 & 19 \\ 0 & 0 & 0 & -1 \\ 0 & 0 & 0 & -3 \\ 0 & 0 & 0 & -481 \end{pmatrix}\begin{pmatrix} f_{n-3} \\ f_{n-2} \\ f_{n-1} \\ f_n \end{pmatrix}.$$

Multiplying the above equation with the inverse of the matrix

$$\begin{pmatrix} -480 & 240 & 0 & 0 \\ 240 & -480 & 240 & 0 \\ 480 & -720 & 0 & 240 \\ 1440 & -1440 & 0 & 0 \end{pmatrix}$$

gives

$$\begin{pmatrix} 1 & 0 & 0 & 0 \\ 0 & 1 & 0 & 0 \\ 0 & 0 & 1 & 0 \\ 0 & 0 & 0 & 1 \end{pmatrix}\begin{pmatrix} y_{n+1} \\ y_{n+2} \\ y_{n+3} \\ y_{n+4} \end{pmatrix} = \begin{pmatrix} 0 & 0 & 0 & 1 \\ 0 & 0 & 0 & 1 \\ 0 & 0 & 0 & 1 \\ 0 & 0 & 0 & 1 \end{pmatrix}\begin{pmatrix} y_{n-3} \\ y_{n-2} \\ y_{n-1} \\ y_n \end{pmatrix} + h\begin{pmatrix} 0 & 0 & 0 & 1 \\ 0 & 0 & 0 & 2 \\ 0 & 0 & 0 & 3 \\ 0 & 0 & 0 & 4 \end{pmatrix}\begin{pmatrix} y'_{n-3} \\ y'_{n-2} \\ y'_{n-1} \\ y'_n \end{pmatrix}$$

$$+h^2\begin{pmatrix} \frac{162}{432} & \frac{-141}{720} & \frac{87}{1080} & \frac{-63}{4320} \\ \frac{24}{15} & \frac{-3}{9} & \frac{24}{135} & \frac{-3}{90} \\ \frac{351}{120} & \frac{81}{240} & \frac{8}{24} & \frac{-27}{480} \\ \frac{192}{45} & \frac{48}{45} & \frac{192}{135} & 0 \end{pmatrix}\begin{pmatrix} f_{n+1} \\ f_{n+2} \\ f_{n+3} \\ f_{n+4} \end{pmatrix}$$

$$+h^2\begin{pmatrix} 0 & 0 & 0 & \frac{1101}{4320} \\ 0 & 0 & 0 & \frac{159}{270} \\ 0 & 0 & 0 & \frac{441}{480} \\ 0 & 0 & 0 & \frac{168}{135} \end{pmatrix}\begin{pmatrix} f_{n-3} \\ f_{n-2} \\ f_{n-1} \\ f_n \end{pmatrix}$$

which leads to

$$y_{n+1} = y_n + hy'_n + \frac{h^2}{4320}(-63f_{n+4} + 348f_{n+3} - 846f_{n+2} + 1620f_{n+1} + 1101f_n) \tag{23.18}$$

$$y_{n+2} = y_n + 2hy'_n + \frac{h^2}{270}(-9f_{n+4} + 48f_{n+3} - 90f_{n+2} + 432f_{n+1} + 159f_n) \tag{23.19}$$

$$y_{n+3} = y_n + 3hy'_n + \frac{h^2}{480}(-27f_{n+4} + 180f_{n+3} + 162f_{n+2} + 1404f_{n+1} + 441f_n) \tag{23.20}$$

$$y_{n+4} = y_n + 4hy'_n + \frac{h^2}{135}(192f_{n+3} + 144f_{n+2} + 576f_{n+1} + 168f_n) \tag{23.21}$$

Substituting (23.18) and (23.19) into (23.14)–(23.17) gives the derivative of the block method

$$y'_{n+1} = y'_n + \frac{h}{720}(-19f_{n+4} + 106f_{n+3} - 264f_{n+2} + 646f_{n+1} + 251f_n) \tag{23.22}$$

$$y'_{n+2} = y'_n + \frac{h}{90}(-f_{n+4} + 4f_{n+3} + 24f_{n+2} + 124f_{n+1} + 29f_n) \tag{23.23}$$

$$y'_{n+3} = y'_n + \frac{h}{80}(-3f_{n+4} + 42f_{n+3} + 72f_{n+2} + 102f_{n+1} + 27f_n) \tag{23.24}$$

$$y'_{n+4} = y'_n + \frac{h}{45}(14f_{n+4} + 64f_{n+3} + 24f_{n+2} + 64f_{n+1} + 14f_n) \tag{23.25}$$

23.3 Test Problems

In order to examine the accuracy of the newly developed block method, the following second-order initial value problems are considered.

Problem 1: $y'' - 100y = 0,\ y(0) = 1,\ y'(0) = -10,\ h = 0.01$
Exact Solution: $y(x) = e^{-10x}$.

Problem 2: $y'' - x(y')^2 = 0,\ y(0) = 1,\ y'(0) = \frac{1}{2},\ h = 0.003125$
Exact Solution: $y(x) = 1 + \frac{1}{2} In\left(\frac{2+x}{2-x}\right)$

Problem 3: $2yy'' - (y')^2 + 4y^2 = 0,\ y\left(\frac{\pi}{6}\right) = \frac{1}{4},\ y'\left(\frac{\pi}{6}\right) = \frac{\sqrt{3}}{2},\ h = \frac{1}{320}$
Exact Solution: $y(x) = \sin^2 x$

23.4 Numerical Results

The tables displayed below show the numerical results when the new block method with step-length $k = 4$ was applied to differential equations above. The generated numerical results are compared with the existing methods of the same step-length or higher step-length (Tables 23.1, 23.2, 23.3).

Table 23.1 Comparison of the new method with block method (23.14) for solving Problem 1

x	Exact solution	Computed solution	Error in new method, $k = 4$	Error in (23.14), $k = 4$	Error in (23.14), $k = 5$
0.01	0.904837418035959520	0.904837418933206020	8.972465E−10	1.1067E−05	1.2413E−06
0.02	0.818730753077981820	0.818730755238518570	2.160537E−09	3.1403E−05	3.4226E−06
0.03	0.740818220681717880	0.740818224125312510	3.443595E−09	5.2700E−05	5.7008E−06
0.04	0.670320046035639330	0.670320050447407390	4.411768E−09	7.4521E−05	8.0308E−06
0.05	0.606530659712633420	0.606530633744317260	2.596832E−08	8.2312E−05	1.0439E−05
0.06	0.548811636094026500	0.548811579725073550	5.636895E−08	9.7067E−05	1.1244E−05
0.07	0.496585303791409470	0.496585216456416680	8.733499E−08	1.1323E−04	1.2725E−05
0.08	0.449328964117221560	0.449328844707934880	1.194093E−07	1.3052E−04	1.4369E−05
0.09	0.406569659740599170	0.406569486018967530	1.737216E−07	1.3614E−04	1.6156E−05
0.10	0.367879441171442330	0.367879211559240750	2.296122E−07	1.4725E−04	1.8102E−05
0.11	0.332871083698079500	0.332870795896438030	2.878016E−07	1.6012E−04	1.8649E−05
0.12	0.301194211912202080	0.301193862883732230	3.490285E−07	1.7459E−04	1.9725E−05

Table 23.2 Comparison of the new method with block method (23.15) and predictor–corrector method (23.16) for solving Problem 2

x	Exact solution	Computed solution	Error in new method, $k = 4$	Error in (23.15), $k = 4$	Error in (23.16), $k = 4$
0.1	1.050041729278491400	1.050041729278491600	2.220446E−16	9.992E−15	6.550E−11
0.2	1.100335347731075600	1.100335347731075800	2.220446E−16	5.456E−14	5.480E−10
0.3	1.151140435936467000	1.151140435936465600	1.332268E−15	4.700E−13	1.925E−09
0.4	1.202732554054082300	1.202732554054076800	5.551115E−15	1.637E−12	4.802E−09
0.5	1.255412811882995500	1.255412811882979900	1.554312E−14	4.664E−12	1.000E−08
0.6	1.309519604203111600	1.309519604203074800	3.685940E−14	1.116E−11	1.872E−08
0.7	1.365443754271396000	1.365443754271314000	8.193446E−14	2.501E−11	3.274E−08
0.8	1.423648930193601300	1.423648930193427600	1.736389E−13	5.215E−11	5.396E−08
0.9	1.484700278594051100	1.484700278593693100	3.579359E−13	1.076E−11	8.800E−08
1.0	1.549306144334053700	1.549306144333322500	7.311929E−13	2.170E−10	1.435E−07

Table 23.3 Comparison of the new method with predictor–corrector methods (23.16) and (23.17) solving Problem 3

x	Exact solution	Computed solution	Error in new method, $k = 4$	Error in (23.16), $k = 4$	Error in (23.17), $k = 4$
1.1	0.795642473567278130	0.795642473550267400	1.701073E−11	0.416327E −06	0.469215E−06
1.2	0.869859011321088560	0.869859011298550480	2.253808E−11	0.458667E −06	0.408029E−06
1.3	0.929330437451130750	0.929330437423069090	2.806166E−11	0.409282E −06	0.228974E−06
1.4	0.971685813870537320	0.971685813837236290	3.330103E−11	0.262955E −06	0.812872E−07
1.5	0.995236565381586220	0.995236565343565420	3.802081E−11	0.455387E −07	0.524472E−06
1.6	0.999043797840225900	0.999043797798192860	4.203304E−11	0.480549E −06	0.108974E−05
1.7	0.982955728902747180	0.982955728857558770	4.518841E−11	0.103225E −05	0.175373E−05
1.8	0.947613739108849830	0.947613739061481160	4.736866E−11	0.167850E −05	0.248148E−05
1.9	0.894426802063518540	0.894426802015042990	4.847556E−11	0.238575E −05	0.322842E−05
2.0	0.825515313105665170	0.825515313057240800	4.842438E−11	0.311084E −05	0.394301E−05

23.5 Conclusion

A block method with step-length $k = 4$ for solving second-order initial value problems of ODEs has been successfully developed in this paper. The results obtained when the new block method was applied to solve second-order initial value problems are compared with some existing methods and found to be better in terms of accuracy.

References

1. Spiegel, R.M.: Theory and Problems of Advance Mathematics for Engineers and Scientist. McGraw Hill, Inc. New York (1971)
2. Lambert, J.D.: Computational Methods in Ordinary Differential Equations. Introductory Mathematics for Scientists and Engineers. Wiley, New Jersey (1973)
3. Goult, R.J., Hoskins, R.F., Milner, J.A., Pratt, M.J.: Applicable Mathematics: Course for Scientists and Engineers. Macmillan press Ltd., London (1973)
4. Lambert, J.D., Watson, I.A.: Symmetric multistep methods for periodic initial value problems. IMA J. Appl. Math. **18**(2), 189–202 (1976)
5. Fatunla, S.O.: Numerical Methods for Initial Value Problems in Ordinary Differential Equations. Academic Press Inc. Harcourt Brace Jovanovich Publishers, New York (1988)
6. Awoyemi, D.O.: A class of continuous methods for general second order initial value problems in ordinary differential equations. Int. J. Comput. Math. **72**(1), 29–37 (1999)
7. Dahlquist, G.: Convergence and stability in the numerical integration of ordinary differential equations. Mathematica Scandinavia **4**, 33–53 (1959)
8. Henrici, P.: Discrete Variable Methods in Ordinary Differential Equations (1962)
9. Jeltsh, R.: Note on a—stability of multi-step multi derivative methods. BIT **16**, 7–78 (1976)
10. Brown, R.L.: Some characteristics of implicit multistep multi-derivative integration formulas. SIAM J Numer. Anal. **14**(6), 982–993 (1977)
11. Awoyemi, D.O.: A new sixth-order algorithm for general second order ordinary differential equations. Int. J. Comput. Math. **77**(1), 117–124 (2001)
12. Kayode, S.J.: An efficient zero-stable numerical method for fourth-order differential equations. Int. J. Math. Math. Sci. **2008**, 1–10 (2008)
13. Badmus, A.M., Yahaya, Y.A.: An accurate uniform order 6 block method for direct solution of general second order ordinary differential equations. Pac. J. Sci. Technol. **10**(2), 248–254 (2009)
14. Awari, Y.S., Abada, A.A.: A class of seven point zero stable continuous block method for solution of second order ordinary differential equation. Int. J. Math. Stat. Invent. (IJMSI) **2**(1), 47–54 (2014)

Chapter 24
Eccentric Connectivity Index of Certain Classes of Cycloalkenes

R.S. Haoer, K.A. Atan, A.M. Khalaf, M.R. Md. Said and R. Hasni

Abstract Let G be a simple connected molecular graph. The eccentric connectivity index $\xi(G)$ is defined as $\xi(G) = \sum_{v \epsilon V(G)} \deg(v)\, \mathrm{ec}(v)$, where $\deg(v)$ denotes the degree of vertex v and $\mathrm{ec}(v)$ is the largest distance between v and any other vertex $u \epsilon G$. In this paper, we establish the general formulas for the eccentric connectivity index of molecular graphs of cycloalkenes.

Keywords Eccentric connectivity index · Molecular graphs · Eccentricity · Cycloalkenes

24.1 Introduction

The study of the quantitative structure–activity relationship (QSAR) aims to rapidly and effectively predict the physic-chemical, pharmacological, and toxicological properties of a compound directly from its molecular structure. In chemistry, a molecular graph represents the topology of a molecule. One can model a molecule by a graph using the points to represent the atoms, and the edges to symbolize the

R.S. Haoer (✉) · A.M. Khalaf
Department of Mathematics College of Computer Sciences and Mathematics, University of Kufa, Najaf, Iraq
e-mail: raadsehen@gmail.com

A.M. Khalaf
e-mail: abduljaleel.khalaf@uokufa.edu.iq

R.S. Haoer · K.A. Atan · M.R.Md.Said
Institute for Mathematical Research, Universiti Putra Malaysia, 43400 Serdang, Selangor, Malaysia
e-mail: kamel@upm.edu.my

R. Hasni
Department of Mathematics Faculty of Science and Technology, University Malaysia Terengganu, 21030 Kuala Terengganu, Terengganu, Malaysia
e-mail: hroslan@umt.edu.my

A.-R. Ahmad et al. (eds.), *Proceedings of the International Conference on Computing, Mathematics and Statistics (iCMS 2015)*,
DOI 10.1007/978-981-10-2772-7_24

covalent bonds. So, it is natural for scholars to study relevant properties of these graph models. During this process, a number of graph invariants are proposed. During the past several decades, many such graph invariant 'topological indices' have been raised and extensively studied, such as, Wiener index, Randi '*c* index, Hosoya index, and so on.

More recently, Sharma et al. [1] introduced a new topological index, Eccentric Connectivity Index, is defined as $\xi(G) = \sum_{v \epsilon V(G)} \deg(v)\, \mathrm{ec}(v)$, where $\deg(v)$ denotes the degree of vertex v and $\mathrm{ec}(v)$ is the largest distance between v and any other vertex $u \epsilon G$. This topological model has been shown to give a high degree of predictability of pharmaceutical properties, and may provide leads for the development of safe and potent anti-HIV compounds. We encourage the reader to consult papers [2, 3] for some applications and papers [4, 5] for the mathematical properties of this topological index. More recently, Morgan et al. [6] gave a sharp upper bound for eccentric connectivity index of trees of given order, and sharp upper and lower bounds for eccentric connectivity index of trees with given order and diameter. Jianguang Yang and Fangli Xia [7], obtained exact formulas for calculating the eccentric connectivity index of dendrimers. Libing Zhang and Hongbo Hua [8] investigated the eccentric connectivity index of unicyclic graphs. Dureja et al. [9] studied the super augmented eccentric connectivity index and have been investigated to be beneficial indicators in chemistry study.

In this paper, we establish the general formulas for the eccentric connectivity index of molecular graphs of cycloalkenes.

24.2 Eccentric Connectivity Index of Cycloalkenes

In this section, we construct the general formulas for the eccentric connectivity index of cycloalkenes and the chemical graphs that are constructed by attaching alkyl R_i instead of each hydrogen atoms in the cycloalkenes.

We denote for group of alkyl by R_i; $i = 1, 2, 3, \ldots, r$ and $R_1, R_2, R_3, \ldots, R_r$ are denote for methyl, ethyl, propyl ..., respectively, as shown in Fig. 24.1.

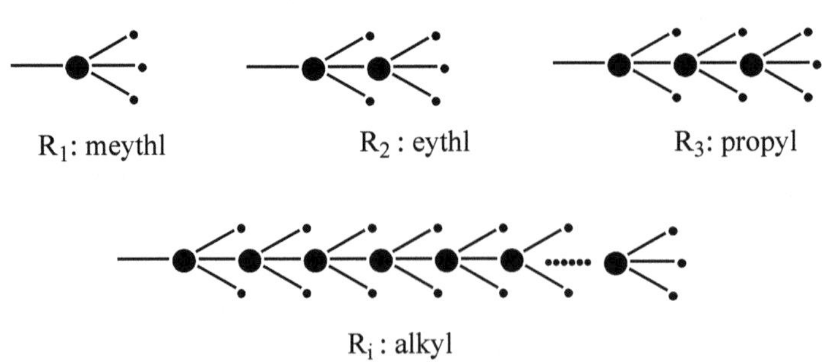

Fig. 24.1 Group of alkyl

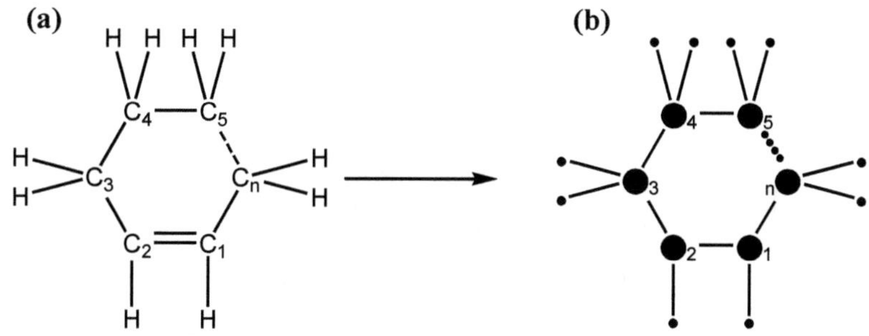

Fig. 24.2 **a** is the molecular structure of cycloalkenes; **b** is the molecular graph representing the chemical compound of cycloalkenes

We denote for cycloalkenes by C_n^{2n-2} [cycloalkenes obtained by attaching $2n-2$ pendant vertices (Hydrogen atoms) to vertices of a cycle (Carbon atoms)] as shown in Fig. 24.2. When we put alkyl instead of each hydrogen atoms in the cycloalkenes we get new classes of cycloalkenes. We denote for chemical graphs that formed by attaching R_i instead of each hydrogen atom in cycloalkenes by $C_n^{R_i}$ as shown in Fig. 24.3.

In Fig. 24.2 above (a) displays the molecular structures of cycloalkenes and (b) represents the molecular graph of cycloalkenes as a chemical compound. Also Fig. 24.3 displays the molecular structures of $C_n^{R_i}$ and Fig. 24.4 represents the molecular graph of $C_n^{R_i}$. The total number of vertices in molecular structure and molecular graph is equal. We calculate the eccentric connectivity index of the graph in Fig. 24.2(b) and Fig. 24.4 as follows:

Theorem 2.1 *Let n be a positive integer, then the eccentric connectivity index of a graph* $G = C_n^{2n-2}$ (Fig. 24.2b) *is*

$$\xi(G) = \begin{cases} 3n^2 + 6n - 6; & \text{for } n \text{ even}, n \geq 4 \\ 3n^2 + 3n - 4; & \text{for } n \text{ odd}, n \geq 3 \end{cases}.$$

Proof Let C_i be the vertices of the positive Carbon atoms; $i = 1, 2, 3, \ldots, n$, and H_i vertices of the positive Hydrogen atoms; $j = 1, 2, 3, \ldots, 2n-2$.

Case 1. If n is even:

We have n of the carbon atoms that have the same eccentricity $\frac{1}{2}n + 1$, but two of them have degree 3 and the remaining $n-2$ have degree 4. We also have $2n-2$ of the hydrogen atoms have the same degree 1, also the same eccentricity $\frac{1}{2}n + 2$.

Hence by the definition of eccentric connectivity index, we get

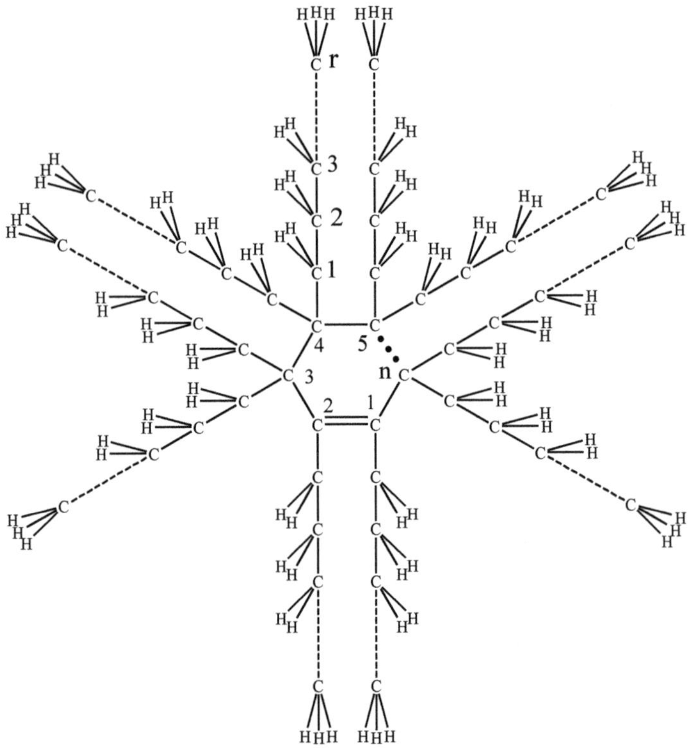

Fig. 24.3 Molecular structure of $C_n^{R_i}$

$$
\begin{aligned}
\xi(G) &= \sum_{i=1}^{2} d_{v_i} e(v_i) + \sum_{i=3}^{n} d_{v_i} e(v_i) + \sum_{j=1}^{2n-2} d_{v_j} e(v_j) \\
&= (2)(3)\left(\frac{1}{2}n+1\right) + (n-2)(4)\left(\frac{1}{2}n+1\right) + (2n-2)(1)\left(\frac{1}{2}n+2\right) \\
&= (3n+6) + \left(2n^2-8\right) + \left(n^2+3n-4\right) \\
&= 3n^2+6n-6.
\end{aligned}
$$

Case 2. If n is odd:

We have n of the carbon atoms that have the same eccentricity $\frac{1}{2}(n+1)$, but two of them have degree 3 and the remaining $n-2$ have degree 4. We also have $2n-2$ of the hydrogen atoms have the same degree 1 and also have the same eccentricity $\frac{1}{2}(n+3)$.

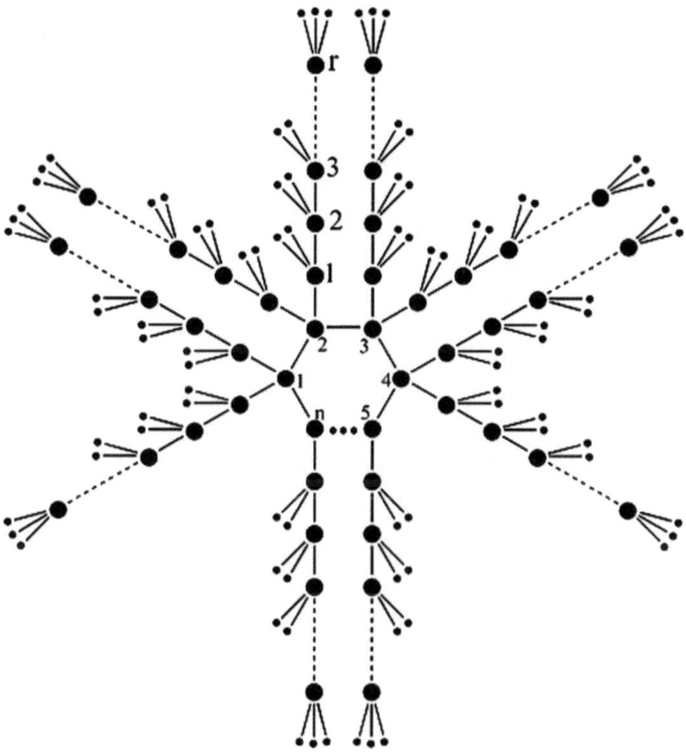

Fig. 24.4 Molecular graph representing the chemical compound of $C_n^{R_i}$

Hence by the definition of eccentric connectivity index, we get

$$\begin{aligned}\xi(G) &= \sum_{i=1}^{2} d_{v_i} e(v_i) + \sum_{i=3}^{n} d_{v_i} e(v_i) + \sum_{j=1}^{2n-2} d_{v_j} e(v_j) \\ &= (2)(3)\left(\frac{1}{2}(n+1)\right) + (n-2)(4)\left(\frac{1}{2}(n+1)\right) + (2n-2)(1)\left(\frac{1}{2}(n+3)\right) \\ &= (3n+3) + \left(2n^2 - 2n - 4\right) + \left(n^2 + 3n - n - 3\right) \\ &= 3n^2 + 3n - 4.\end{aligned}$$

∎

Theorem 2.2 *Let n and r be a positive integer, then the eccentric connectivity index of a graph* $G = C_n^{R_i}$ (Fig. 24.4) *is*

$$\xi(G) = \begin{cases} 3n^2 + 6n^2r + 18nr^2 + 24nr + 6n - 18r^2 - 28r - 6; & \textit{for n even}, n \geq 4 \\ 3n^2 + 6n^2r + 18nr^2 + 18nr + 3n - 18r^2 - 22r - 4; & \textit{for n odd}, n \geq 3 \end{cases}$$

Proof **Case 1**. If n is even:

We have $n + 2nr - 2r$ carbon atoms. n of them represent vertices of the cycle that have the same eccentricity $\frac{1}{2}n + r + 1$but two of them have degree 3 and the remaining $n - 2$ have the same degree 4. The remaining $2nr - 2r$ that have the same degree 4 and eccentricity $\frac{1}{2}n + r + i + 1;\ i = 1, 2, 3, \ldots, r$.

As for the hydrogen atoms, we have $4nr + 2n - 4r - 2$ located in the terminal and thus the degree of each of them is 1. Also $6n - 6$ of them that are pendent in the terminal carbon atoms have the same eccentricity $\frac{1}{2}n + 2r + 2$. The remaining $4nr - 4n - 4r + 4$ of the hydrogen atoms that have the same eccentricity $\frac{1}{2}n + r + i + 1;\ i = 2, 3, \ldots, r$.

Hence, by the definition of eccentric connectivity index, we get

$$\begin{aligned}
\xi(G) &= (2)(3)\left(\frac{1}{2}n + r + 1\right) + (n-2)(4)\left(\frac{1}{2}n + r + 1\right) \\
&\quad + \sum_{i=1}^{r}(2n-2)(4) + \left(\frac{1}{2}n + r + i + 1\right) + (6n-6)\left(\frac{1}{2}n + 2r + 2\right) \\
&\quad + \sum_{i=2}^{r}(4n-4) + \left(\frac{1}{2}n + r + i + 1\right) \\
&= (3n + 6r + 6) + \left(2n^2 + 4nr - 8r - 8\right) + (8n-8)(r)\left(\frac{1}{2}n + r + 1\right) \\
&\quad + (8n-8)\left(\sum_{i=1}^{r} i\right) + \left(3n^2 + 12nr + 9n - 12r - 12\right) \\
&\quad + (4n-4)(r-1)\left(\frac{1}{2}n + r + 1\right) + (4n-4)\left(\sum_{i=2}^{r} i\right) \\
&= (3n + 6r + 6) + \left(2n^2 + 4nr - 8r - 8\right) \\
&\quad + \left(4n^2r + 12nr^2 + 8nr - 12r^2 - 12r\right) \\
&\quad + \left(3n^2 + 12nr + 9n - 12r - 12\right) \\
&\quad + \left(2n^2r + 6nr^2 - 6n - 2n^2 - 6r^2 - 2r + 8\right) \\
&= 3n^2 + 6n^2r + 18nr^2 + 24nr + 6n - 18r^2 - 28r - 6.
\end{aligned}$$

Case 2. If n is odd

We have $n + 2nr - 2r$ carbon atoms. Where n of them represent vertices of the cycle that have the same eccentricity $\frac{1}{2}(n + 2r + 1)$ but two of them have degree 3 and the remaining $n - 2$ have the same degree 4. The remaining $2nr - 2r$ have the same degree 4 and eccentricity $\frac{1}{2}(n + 2r + 1) + i;\ i = 1, 2, 3, \ldots, r$. Also for the hydrogen atoms, we have $4nr+2n-4r-2$ located in the terminal and thus the degree of each of them is 1. Where $6n-6$ of them are pendent in the terminal carbon atoms have the same eccentricity $\frac{1}{2}n+4r+3$. The remaining $4nr-4n-4r+4$ of the hydrogen atoms that have the same eccentricity equal $\frac{1}{2}(n+2r+1)+i;\ i=2,3,\ldots,r$.

Hence, by the definition of eccentric connectivity index, we get

$$\begin{aligned}
\xi(G) &= 2\times 3\times\frac{1}{2}(n+2r+1)+(n-2)(4)\times\frac{1}{2}(n+2r+1)\\
&\quad+\left(\sum_{i=1}^{r}(2n-2)(4)+\left(\frac{1}{2}(n+2r+1)+i\right)\right)\\
&\quad+(6n-6)\times\frac{1}{2}(n+4r+3)+\left(\sum_{i=2}^{r}(4n-4)+\left(\frac{1}{2}(n+2r+1)+i\right)\right)\\
&= (3n+6r+3)+\left(2n^2+4nr-2n-8r-4\right)\\
&\quad+(8n-8)(r)\times\frac{1}{2}(n+2r+1)\\
&\quad+\left((8n-8)\sum_{i=1}^{r}i\right)+\left(3n^2+12nr+6n-12n-9\right)\\
&\quad+(2n-2)(r-1)(n+2r+1)+\left((2n-2)\sum_{i=2}^{r}i\right)\\
&= (3n+6r+3)+\left(2n^2+4nr-2r-8r-4\right)\\
&\quad+\left(4n^2r+12nr^2+4nr-12r^2-8r\right)\\
&\quad+\left(3n^2+12nr+6n-12r-9\right)\\
&\quad+\left(2n^2r+6nr^2-2nr-4n-2n^2-6r^2+6\right)\\
&= 3n^2+6n^2r+18nr^2+18nr+3n-18r^2-22r-4
\end{aligned}$$

■

24.3 Conclusion

In this paper, we computed the eccentric connectivity index of molecular graphs of cycloalkenes. It is interesting to investigate this index for other chemical structures via their molecular graphs. The characterization of graphs with extremum eccentric connectivity index has been an active area of research and we hope to consider this problem in future. We close this paper with following question: Which graphs will attain the maximum or minimum eccentric connectivity index?

References

1. Sharma, V., Goswami, R., Madan, A.K.: Eccentric connectivity index: a novel highly discriminating topological descriptor for structure-property and structure-activity studies. J. Chem. Inf. Model. **37**, 273–282 (1997)
2. Dureja, H., Madan, A.K.: Topochemical models for prediction of cyclindependent kinase 2 inhibitory activity of indole-2-ones. J. Mol. Model. **11**, 525–531 (2005)
3. Dureja, H., Madan, A.K.: Topochemical models for the prediction of permeability through blood-brain barrier. Int. J. Pharm. **323**, 27–33 (2006)
4. Ashrafi, A.R., Saheli, M., Ghorbani, M.: The eccentric connectivity index of nanotubes and nanotori. J. Comput. Appl. Math. (2010)
5. Morgan, M.J., Mukwembi, S., Swart, H.C.: On the eccentric connectivity index of a graph. Disc. Math. (2010)
6. Ilic, A., Gutman, I.: Eccentric connectivity index of chemical trees. MATCH Commun. Math. Comput. Chem. **65**, 731–744 (2011)
7. Yang, J., Xia, F.: The eccentric connectivity index of dendrimers. Int. J. Contemp. Math. **5**(45), 2231–2236 (2010)
8. Zhang, L., Hua, H.: The eccentric connectivity index of unicyclic graphs. Int. J. Contemp. Math. **5**(46), 2257–2262 (2010)
9. Dureja, H., Gupta, S., Madan, A.K.: Predicting anti-HIV-1 activity of 6-arylbenzonitriles: computational approach using superaugmented eccentric connectivity topochemical indices. J. Mol. Graph. Model. **26**, 1020–1029 (2008)

Chapter 25
Numerical Solution of the Gardner Equation

W.K. Tiong, K.G. Tay, C.T. Ong and S.N. Sze

Abstract The Gardner equation is commonly used to describe wave propagation in weakly nonlinear dispersive medium. The Gardner equation has a higher order nonlinear term, which could make the numerical calculation inaccurate. In this paper, the Gardner equation is solved using two numerical methods, i.e., the method of lincs and pscudospcctral mcthod. Thc cfficicncy and accuracy of both mcthods were studied. Our results show that both methods are accurate and efficient methods to solve the Gardner equation. By comparing the accuracy of both the methods, the method of lines performs better than pseudospectral method most of the time.

Keywords Gardner equation · The method of lines · Pseudospectral method

25.1 Introduction

The Gardner equation, more commonly known as the extended Korteweg-de Vries (eKdV) equation

W.K. Tiong (✉) · S.N. Sze
Department of Computational Science and Mathematics, Universiti Malaysia Sarawak, 94300 Kota Samarahan, Sarawak, Malaysia
e-mail: wktiong@fit.unimas.my

S.N. Sze
e-mail: snsze@fit.unimas.my

K.G. Tay
Department of Communication Engineering, Universiti Tun Hussein Onn Malaysia, 86400 Parit Raja, Batu Pahat, Johor, Malaysia
e-mail: tay@uthm.edu.my

C.T. Ong
Department of Mathematical Sciences, Universiti Teknologi Malaysia, 81300 Skudai, Johor, Malaysia
e-mail: octiong@utm.my

A.-R. Ahmad et al. (eds.), *Proceedings of the International Conference on Computing, Mathematics and Statistics (iCMS 2015)*,
DOI 10.1007/978-981-10-2772-7_25

$$u_t + 6\alpha u u_x + 6\beta u^2 u_x + \gamma u_{xxx} = 0, \tag{25.1}$$

is an universal model to describe waves propagation in weakly nonlinear dispersive medium where the higher order nonlinearity effects is important. One of the most common applications of the Gardner equation is to describe the propagation of the large-amplitude internal waves in stratified fluid flow (see [1, 2] and references therein). Because internal waves are often of large amplitudes, it is sometimes useful to include a cubic nonlinear term in (25.1). The coefficient β in (25.1) can be both positive and negative depending on the physical problems. In the context of internal waves, it depends on the stratification of the fluid. When $\beta = 0$, then (25.1) becomes the well-known KdV equation

$$u_t + 6\alpha u u_x + \gamma u_{xxx} = 0. \tag{25.2}$$

When the coefficients α, β, γ are constant, then (25.1) has a steady-state solitary wave solution

$$u(x,t) = \frac{H}{1 + B\cos hK(x - Vt)}$$

where

$$V = \alpha H = \gamma K^2, \quad B^2 = 1 + \frac{\beta\gamma K^2}{\alpha^2}.$$

The solitary wave solution (25.1) is characterised by a single parameter B. The wave amplitude is

$$a = \frac{H}{1 + B}.$$

When $\beta\gamma < 0$, then (25.1) has a single family of solitary waves such that $0 < B < 1$. When the amplitude is small, (25.1) transforms into the KdV-type solitary waves ($B \to 1$). When $B \to 0$, (25.1) describes the so-called "thick" solitary wave (also known as "table-top" soliton) (refer to Fig. 25.1). On the other hand, (25.1) has two families of solitary waves, both with $B^2 \to 1$ when $\beta\gamma > 0$. The first has $1 < B < \infty$ and ranges from small KdV-type solitary waves when $B \to 1$, to large wave with a "sech" profile as $B \to \infty$. The other family of solitary waves has the opposite polarity with $-\infty < B < -1$ and ranges from large waves with a "sech" profile to a limiting algebraic solitary wave of amplitude $-2\alpha/\beta$ (refer to Fig. 25.1) [2].

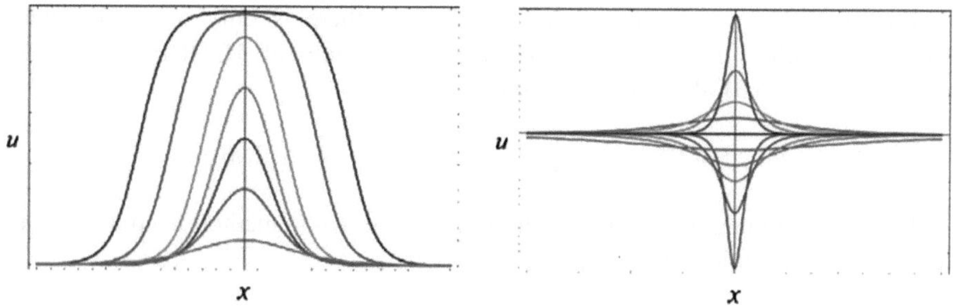

Fig. 25.1 The shape of solitary wave when *left* $\alpha > 0$, $\beta < 0$, $\gamma > 0$. *Right* $\alpha > 0$, $\beta > 0$, $\gamma > 0$

25.2 Numerical Methods

In this paper, two numerical methods, i.e. the method of lines (MOL) and pseudospectral are considered to solve the Gardner equation (25.1).

25.2.1 The Method of Lines

MOL is a powerful method to solve partial differential equations. It involves making an approximation to the space derivatives and reducing the problem into a system of ordinary differential equations (ODEs). Then this system of ODEs can be solved using a time integrator [3, 4]. In this paper, the spatial derivatives in (25.1) are first discretised using central finite-difference formulae as follows:

$$u \approx \frac{u_{j+1}+u_j+u_{j-1}}{3},\; u_x \approx \frac{u_{j+1}-u_{j-1}}{2\Delta x},\; u_{xxx} \approx \frac{u_{j+2}-2u_{j+1}+2u_{j-1}-u_{j-2}}{2(\Delta x)^3}, \tag{25.2}$$

where j is the index denoting the position along x-axis and Δx is the spacing along the axis. x-interval is divided into M points with $j = 0, 1, 2, \ldots, M-2, M-1$. Therefore, MOL approximation of (25.1) is given by

$$\begin{aligned}\frac{\mathrm{d}u_j}{\mathrm{d}t} &= -\frac{\alpha}{\Delta x}\left(u_{j+1}+u_j+u_{j-1}\right)\left(u_{j+1}-u_{j-1}\right) - \frac{\beta}{3\Delta x}\left(u_{j+1}+u_j+u_{j-1}\right)^2\left(u_{j+1}-u_{j-1}\right)\\ &\quad -\frac{\gamma}{2(\Delta x)^3}\left(u_{j+2}-2u_{j+1}+2u_{j-1}-u_{j-2}\right) \equiv f(u_j).\end{aligned} \tag{25.3}$$

Equation (25.3) is written as an ODE since there is only one independent variable, which is t. Also, (25.3) represents a system of M ODEs. The initial condition for (25.3) after discretisation is given by

$$u(x_j, t = 0) = u_0(x_j), \quad j = 0, 1, 2, \ldots, M - 2, M - 1.$$

For the time integration, the fourth-order Runge–Kutta method is applied. Thus, the numerical solution at time t_{i+1} is

$$u_{i+1,j} = u_{i,j} + \frac{1}{6}(a_{i,j} + 2b_{i,j} + 2c_{i,j} + d_{i,j}),$$

where

$$a_{i,j} = \Delta t f(u_{i,j}),\ b_{i,j} = \Delta t f\left(u_{i,j} + \frac{1}{2}a_{i,j}\right),\ c_{i,j} = \Delta t f\left(u_{i,j} + \frac{1}{2}b_{i,j}\right),$$
$$d_{i,j} = \Delta t f(u_{i,j} + c_{i,j}).$$

Here, Δt is the step size of the temporal coordinate.

25.2.2 Pseudospectral Method

For pseudospectral method, we adopted the Chan and Kerkhoven [5] scheme to solve the Gardner equation (25.1). The Gardner equation (25.1) is integrated in time t by the leapfrog finite-difference scheme in the spectral space x. The infinite interval is replaced by $-L < x < L$ with L sufficiently large such that the periodicity assumptions hold.

By introducing $X = sx + \pi$, where $s = \pi/L$, $U(x, t)$ will be transformed into $V(x, t)$.

$$V_t + 6\alpha s V V_X + 6\beta s V^2 V_X + \gamma s^3 V_{XXX} = 0. \tag{25.4}$$

By letting $W(x,t) = \frac{1}{2}sV^2$, and $F(x,t) = \frac{1}{3}sV^3$ the nonlinear terms $6\alpha s V V_X$ and $6\beta s V^2 V_X$ can be written as $6\alpha W_X$ and $6\beta V^2 V_X$, respectively, so that (25.4) becomes

$$V_t + 6\alpha s W_X + 6\beta F_X + \gamma s^3 V_{XXX} = 0. \tag{25.5}$$

In order to obtain the numerical solution of (25.5), the interval $[0, 2\pi]$ is discretised by $N + 1$ equidistant points. Let $X_0 = 0, X_1, X_2, \rightleftharpoons X_N = 2\pi$, so that Discrete Fourier Transform (DFT) of $V(X_j, t)$ for $j = 0, 1, 2, \ldots, N - 1$, denoted by $\hat{V}(p, t)$ is given by

$$\widehat{V}(p,t) = \frac{1}{\sqrt{N}} \sum_{j=0}^{N-1} V(x_j,t) \mathrm{e}^{-\left(\frac{2\pi jp}{N}\right)i},$$

where $j = 0, 1, 2, \ldots, N-1$. The DFT of (25.5) with respect to X gives

$$\widehat{V}_t(p,t) + 6i\alpha p \widehat{W}(p,t) + 6i\beta p \widehat{F}(p,t) - i\gamma s^3 p^3 \widehat{V}(p,t) = 0. \qquad (25.6)$$

Using the following approximations

$$\widehat{V}_t(p,t) \approx \frac{\widehat{V}(p,t+\Delta t) - \widehat{V}(p,t-\Delta t)}{2\Delta t}, \quad \widehat{V}(p,t) \approx \frac{\widehat{V}(p,t+\Delta t) + \widehat{V}(p,t-\Delta t)}{2},$$

and denoting $\widehat{V}(p,t+\Delta t)$ by $\widehat{V}_{pt}$, $\widehat{V}(p,t-\Delta t)$ by $\widehat{V}_{mt}$ and $\widehat{V}(p,t)$ by $\hat{V}_t$, (25.6) becomes

$$\frac{\widehat{V}_{pt} - \widehat{V}_{mt}}{2\Delta t} + 6i\alpha p \widehat{W}(p,t) + 6i\beta p \widehat{F}(p,t) - i\gamma s^3 p^3 \left[\frac{\widehat{V}_{pt} + \widehat{V}_{mt}}{2}\right] = 0. \qquad (25.7)$$

By multiplying (25.7) with $2\Delta t$ yields

$$\widehat{V}_{pt} - \widehat{V}_{mt} + 12i\alpha p \Delta t \widehat{W}(p,t) + 12i\beta p \Delta t \widehat{F}(p,t) - i\gamma s^3 p^3 \Delta t (\widehat{V}_{pt} + \widehat{V}_{mt}) = 0. \qquad (25.8)$$

Collecting the terms in (25.8) gives

$$[1 - i\gamma s^3 p^3 \Delta t] \hat{V}_{pt} = [1 + i\gamma s^3 p^3 \Delta t] \hat{V}_{mt} - 12i\alpha p \Delta t \hat{W}_{pt} - 12i\beta p \Delta t \hat{F}_{pt}.$$

This reduces to the forward scheme given by

$$\widehat{V}_{pt} = \frac{[1 + i\gamma s^3 p^3 \Delta t] \widehat{V}_{mt} - 12i\alpha p \Delta t \widehat{W}_{pt} - 12i\beta p \Delta t \widehat{F}_{pt}}{1 - i\gamma s^3 p^3 \Delta t}. \qquad (25.9)$$

This is a three-level scheme, in which one needs to know the first level, initial condition that is $\widehat{V}_{mt}$ and subsequent second level, $\hat{V}_t$, then one can get the third level, $\hat{V}_{pt}$. The process is repeated till the desired $\hat{V}_{pt}$ is obtained. To get the second level, $\hat{V}_t$, the interval between $\hat{V}_{mt}$ and $\hat{V}_t$ is divided by ten sub intervals. Later, we substitute Δt in (25.9) by $\Delta t/10$ in order to get the equation for $\hat{V}_t$ as

$$\hat{V}_t = \frac{\left[1 + i\gamma s^3 p^3 \frac{\Delta t}{10}\right] \hat{V}_{mt} - 12i\alpha p \frac{\Delta t}{10} \hat{W}_{pt} - 12i\beta p \frac{\Delta t}{10} \hat{F}_{pt}}{1 - i\gamma s^3 p^3 \frac{\Delta t}{10}}. \qquad (25.10)$$

Equation (25.10) is evaluated for ten times to get $\hat{V}_t$ since the interval between $\hat{V}_{mt}$ and $\hat{V}_t$ is divided by ten sub intervals.

25.3 Results and Discussions

Numerical results for both cases, i.e. $\beta\gamma < 0$ and $\beta\gamma > 0$ are presented in this section. To check the stability and quality of our numerical results between two methods, we will examine to which extent the mass and energy are conserved numerically using our scheme. This is because these quantities are not numerically conserved due to the unavoidable numerical errors during numerical time integration. The conservation of mass and energy of the Gardner equation (25.1) are defined as

$$M = \int_{-\infty}^{\infty} u\mathrm{d}x \quad \text{and} \quad E = \int_{-\infty}^{\infty} u^2\mathrm{d}x.$$

25.3.1 *Case 1:* $\beta\gamma < 0$

First, we consider the case where $\beta\gamma < 0$. Let $\alpha = 1$, $\beta = -1$ and $\gamma = 1$. Thus, the eKdV Eq. (25.1) becomes

$$u_t + 6uu_x - 6u^2u_x + u_{xxx} = 0.$$

The initial condition for this case is given by

$$u(x,0) = \frac{1 - B^2}{1 + B\,\cos h\sqrt{1 - B^2}(x)}.$$

In this case, the solitary wave solution becomes KdV-type solution when $B \to 1$. When $B \to 0$, then the solitary wave solution becomes table-top wave. In our work, we set $B = 0.5$ and $B = 0.0001$ in order to get both KdV-type soliton and table-top solutions and the corresponding numerical results are shown in Figs. 25.2 and 25.3 respectively. Both sets of plots looks the same.

The conservations of mass and energy of the solitary wave before and after numerical integration for $B = 0.5$ and $B = 0.0001$ are shown in Tables 25.1 and 25.2. The values for mass and energy at $t = 0$ are obtained analytically using Maple. However, the mass and energy conservations at $t = 100$ and $t = 200$ are obtained by performing numerical integration using MATLAB. From the Table 25.1, it is clear that MOL performs better than pseudospectral method for $B = 0.5$ as MOL produces smaller errors compared to pseudospectral method. However, when

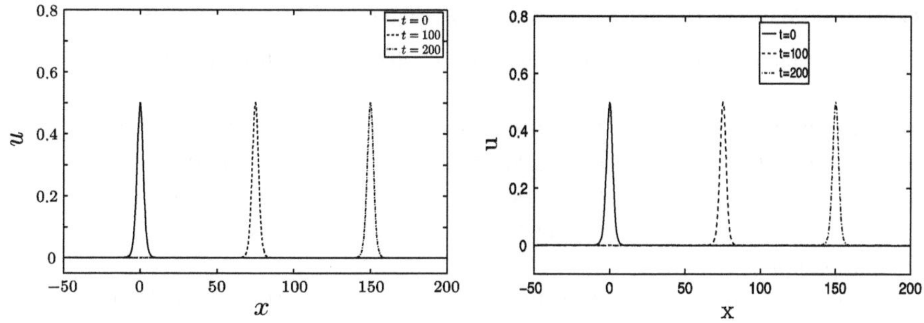

Fig. 25.2 $B = 0.5$. *Left plot* MOL. *Right plot* Pseudospectral method

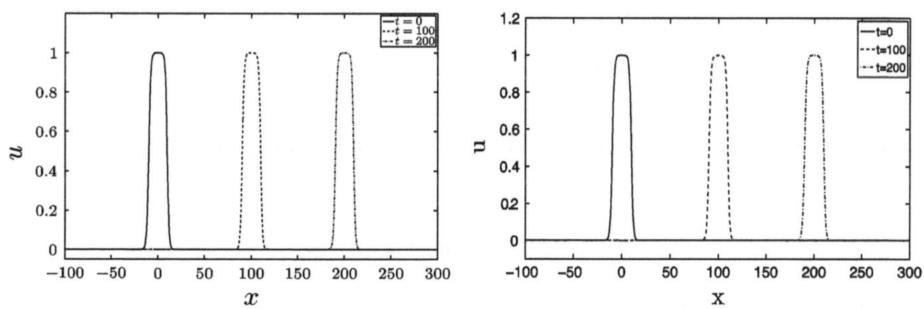

Fig. 25.3 $B = 0.0001$. *Left plot* MOL. *Right plot* Pseudospectral method

Table 25.1 Mass and energy conservations for $B = 0.5$

Before integration $t = 0$	After integration $t = 100$	After integration $t = 200$
MOL		
$M \approx 2.633915793$	$M \approx 2.633918300$	$M \approx 2.633923100$
$E \approx 0.901864986$	$E \approx 0.901865010$	$E \approx 0.901865064$
Error	$M \approx 2.51 \times 10^{-6}$	$M \approx 7.31 \times 10^{-6}$
	$E \approx 2.4 \times 10^{-8}$	$E \approx 7.8 \times 10^{-8}$
Pseudospectral		
$M \approx 2.633915793$	$M \approx 2.633930925$	$M \approx 2.633950228$
$E \approx 0.901864986$	$E \approx 0.901907508$	$E \approx 0.901709747$
Error	$M \approx 1.51 \times 10^{-5}$	$M \approx 3.44 \times 10^{-5}$
	$E \approx 4.25 \times 10^{-5}$	$E \approx 1.55 \times 10^{-4}$

$B = 0.0001$, pseudospectral performs slightly better than MOL (refer to Table 25.2). Nonetheless, both methods are efficient in solving the Gardner equation as the errors produced by both methods are small.

Table 25.2 Mass and energy conservations for B = 0.0001

Before integration t = 0	After integration t = 100	After integration t = 200
MOL		
$M \approx 19.806975100$	$M \approx 19.806986900$	$M \approx 19.806968450$
$E \approx 17.806975137$	$E \approx 17.806974068$	$E \approx 17.806975105$
Error	$M \approx 1.18 \times 10^{-5}$	$M \approx 6.65 \times 10^{-6}$
	$E \approx 1.07 \times 10^{-6}$	$E \approx 3.2 \times 10^{-8}$
Pseudospectral		
$M \approx 19.806975100$	$M \approx 19.806969384$	$M \approx 19.806970219$
$E \approx 17.806975137$	$E \approx 17.806954554$	$E \approx 17.806889750$
Error	$M \approx 5.71 \times 10^{-6}$	$M \approx 4.88 \times 10^{-6}$
	$E \approx 2.06 \times 10^{-5}$	$E \approx 8.54 \times 10^{-5}$

25.3.2 Case 2: $\beta\gamma > 0$

Now, we let $\alpha = 1$, $\beta = 1$ and $\gamma = 1$ so that $\beta\gamma > 0$. Thus, the Gardner equation becomes

$$u_t + 6uu_x + 6u^2u_x + u_{xxx} = 0,$$

with the initial condition

$$u(x,0) = \frac{B^2 - 1}{1 + B\,\cos h\sqrt{B^2 - 1}(x)}.$$

In this case, we expect to have a wave with a "sech" profile when $B \to \infty$. Here, we let B = 2 and the numerical results for both methods are shown in Fig. 25.4. Again, both plots look the same. The error comparison of the mass and energy conservation are shown in Table 25.3. It is clear that both methods are suitable methods for solving the Gardner equation as the error produced by both methods are small. Comparing both methods, MOL performs better than pseudospectral method.

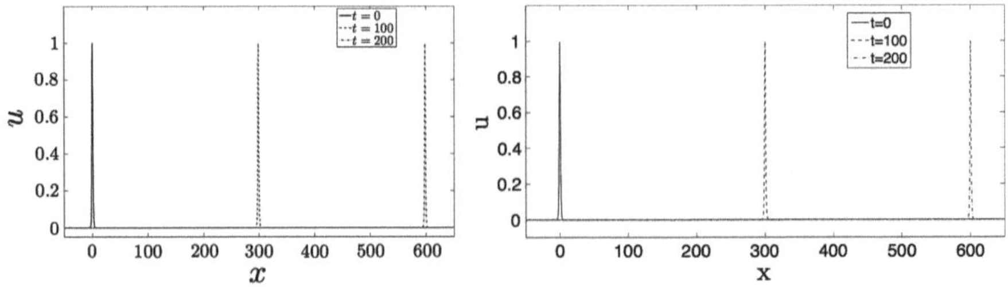

Fig. 25.4 B = 2. *Left plot* MOL. *Right plot* Pseudospectral method

Table 25.3 Mass and energy conservations for $B = 2$

Before integration $t = 0$	After integration $t = 100$	After integration $t = 200$
MOL		
$M \approx 2.094395103$	$M \approx 2.094383200$	$M \approx 2.094381550$
$E \approx 1.369706513$	$E \approx 1.369705905$	$E \approx 1.369705449$
Error	$M \approx 1.19 \times 10^{-5}$	$M \approx 1.36 \times 10^{-5}$
	$E \approx 6.08 \times 10^{-7}$	$E \approx 1.06 \times 10^{-6}$
Pseudospectral		
$M \approx 2.094395103$	$M \approx 2.094255849$	$M \approx 2.094402511$
$E \approx 1.369706513$	$E \approx 1.369414439$	$E \approx 1.368102345$
Error	$M \approx 1.39 \times 10^{-4}$	$M \approx 7.41 \times 10^{-6}$
	$E \approx 2.92 \times 10^{-4}$	$E \approx 1.6 \times 10^{-3}$

25.4 Conclusion

In this paper, we have solved the Gardner equation (25.1) using two different numerical methods, i.e. the method of lines (MOL) and pseudospectral method. In general, both methods are good and efficient method to solve the Gardner equation. The mass and energy of the solitary wave of the Gardner equation are both conserved during numerical integration. However, MOL performs better compared to pseudospectral most of the time as the differences between mass and energy conservations obtained numerically and analytically are smaller.

Acknowledgments This work was supported by Universiti Malaysia Sarawak through Small Grant Scheme—01(S128)/1022/2013(12) and RAGS R026 Universiti Tun Hussein Onn. The authors would also like to thank Universiti Teknologi Malaysia for providing the resources used in the conduct of this study.

References

1. Grimshaw, R., Pelinovsky, E., Talipova, T.: Damping of large-amplitude solitary waves. Wave Motion **37**, 351–364 (2013)
2. Grimshaw, R., Pelinovsky, E., Talipova, T., Kurkina, O.: Internal solitary waves: propagation, deformation and disintegration. Nonlinear Proc. Geophys. **17**, 633–649 (2010)
3. Schiesser, W.: Method of lines solution of the Korteweg-de Vries equation. Comput. Math. Appl. **28**, 147–154 (1994)
4. Schiesser, W.: The Numerical Method of Lines: Integration of Partial Differential Equations. Academic Press, San Diego (1991)
5. Chan, T.F., Kerhoven, T.: Fourier methods with extended stability intervals for the Korteweg-de Vries equation. SIAM J. Numer. Anal. **22**, 441–454 (1985)

Chapter 26
Solving Nonlinear Schrodinger Equation with Variable Coefficient Using Homotopy Perturbation Method

Nazatulsyima Mohd Yazid, Kim Gaik Tay, Yaan Yee Choy and Azila Md Sudin

Abstract In this paper, the application of the homotopy perturbation method (HPM) to the nonlinear Schrodinger equation with variable coefficient (NLSV) is presented to obtain approximate analytical solution. The procedure of the method is systematically illustrated. The result derived by this method is then compared with the progressive wave solution to verify the accuracy of the HPM solution. The solution obtained by the HPM is an infinite series for appropriate initial condition that can be expressed in a closed form to the exact solution. The absolute errors of the HPM solution of the NLSV equation with the progressive wave solution will later be carried out using the MAPLE program. The results of the HPM solution are of high accuracy, verifying that the method is indeed effective and promising. The HPM is found to be a powerful mathematical tool which can be used to solve nonlinear partial differential equations.

Keywords Homotopy perturbation method · NLSV equation · Progressive wave · Nonlinear partial differential equations

---

N.M. Yazid (✉)
Faculty of Science, Technology and Human Development, Universiti Tun Hussein Onn Malaysia, 86400 Parit Raja, Batu Pahat, Johor, Malaysia
e-mail: nazatulsyima91@yahoo.com

K.G. Tay
Department of Communication Engineering, Universiti Tun Hussein Onn Malaysia, Parit Raja, Malaysia
e-mail: tay@uthm.edu.my

Y.Y. Choy · A.M. Sudin
Department of Mathematics and Statistics, Universiti Tun Hussein Onn Malaysia, Parit Raja, Malaysia
e-mail: yychoy@uthm.edu.my

A.M. Sudin
e-mail: azzila@uthm.edu.my

A.-R. Ahmad et al. (eds.), *Proceedings of the International Conference on Computing, Mathematics and Statistics (iCMS 2015)*,
DOI 10.1007/978-981-10-2772-7_26

26.1 Introduction

Rapid improvement of nonlinear science in the last two decades has spurred interest among researchers, physicists, mathematicians, and architects in analytical techniques for nonlinear problems. One of the techniques proposed to solve these nonlinear problems is by homotopy perturbation method (HPM). HPM was first proposed by He [1] and was developed and improved by He [2–4]. This method deforms a difficult problem into a simple problem. The HPM is a coupling of the traditional perturbation method and homotopy in topology constructed with an embedding parameter $p \in [0, 1]$ which is considered as a "small parameter". The HPM yields a very rapid convergence of the solution series to the exact solution if such a solution exists. It offers certain advantages over routine numerical methods. The HPM does not involve discretization of variables as most traditional numerical methods do. Recently, many researchers applied this method to various problems and demonstrated its efficiency in solving nonlinear problems [5–7].

Various physical phenomena in engineering and physics can be described by some nonlinear differential equations. The nonlinear Schrodinger (NLS) occurs in many areas of physics such as optics, hydrodynamics, plasma physics, superconductivity, and quantum mechanics. The NLS equation with cubic nonlinearity in the form

$$i\frac{\partial U}{\partial \tau} + \mu_1 \frac{\partial^2 U}{\partial \xi^2} + \mu_2 |U|^2 U = 0, \quad i^2 = -1, \tag{26.1}$$

where function U is a complex-valued function of the spatial coordinate ξ and time τ. μ_1 and μ_2 are real parameters corresponding to focusing $(\mu_1\mu_2 > 0)$ or defocusing $(\mu_1\mu_2 < 0)$ effects of the nonlinearity. The NLS equation is the simplest representative equation describing the self-modulation of one-dimensional monochromatic plane waves in dispersive media. It exhibits a balance between nonlinearity and dispersion.

Due to its application on arterial mechanism, Ravindran and Prasad [8] showed that for a linear elastic tube wall model, the nonlinear self-modulation of pressure waves is governed by the NLS Eq. (26.1). A study on the modulation of nonlinear wave in fluid-filled elastic or tapered tubes filled with an inviscid fluid was studied by Antar and Demiray [9] and Bakirtas and Demiray [10]. They obtained the NLS Eq. (26.1) and nonlinear Schrodinger equation with variable coefficient (NLSV). The NLSV is given as follows:

$$i\frac{\partial U}{\partial \tau} + \mu_1 \frac{\partial^2 U}{\partial \xi^2} + \mu_2 |U|^2 U - \mu_3 h_1(\tau) U = 0, \tag{26.2}$$

where $\mu_3 h_1(\tau) U$ denotes the variable coefficient term. Choy [11] studied the nonlinear wave modulation of a thin elastic tube with a symmetrical stenosis. Using an approximate equation of an inviscid fluid, she showed that the governing equations can be reduced to the NLSV Eq. (26.2) too.

The NLS Eq. (26.1) is integrable when the particles move in one-dimensional space. It can be solved with the inverse scattering method [12] for limited initial conditions. Numerical solutions of the NLS Eq. (26.1) had been carried out by researchers using various methods such as Crank–Nicolson finite-difference method [13], split-step finite-difference method [14], compact split-step finite-difference method [15], Adomian decomposition method (ADM) [16, 17], and HPM [18].

Many studies have been devoted to the numerical solution of the NLS Eq. (26.1) and the HPM solution of the NLS Eq. (26.1). However, none of the literature has dealt with the HPM solution of the NLSV Eqs. (26.2) yet. Motivated by numerical works as well as the HPM solution of the NLS Eq. (26.1) by previous researchers, we are going to approximate the NLSV Eq. (26.2) using the HPM method. Using initial conditions from a previous study, we then compare the solution with progressive wave obtained by [11].

26.2 Basic Idea of Homotopy Perturbation Method

To demonstrate the fundamental thought of HPM, we consider the following nonlinear differential equation:

$$A(u) - f(r) = 0, \qquad r \in \Omega \tag{26.3}$$

with the following boundary conditions:

$$B\left(u, \frac{\partial u}{\partial n}\right) = 0, \qquad r \in \Gamma, \tag{26.4}$$

where A is a general operator, B is a boundary operator, $f(r)$ is known as an analytical function and Γ is the boundary of the domain Ω. The operator A can be divided into two parts namely linear, L and nonlinear N. Therefore, Eq. (26.3) can be rewritten as follows:

$$L(u) + N(u) - f(r) = 0. \tag{26.5}$$

Using the homotopy technique, we construct a homotopy $v(r,p) : \Omega \times [0,1] \rightarrow R$ which satisfies

$$H(v,p) = (1-p)[L(v) - L(u_0)] + p[A(v) - f(r)] = 0, \quad p \in [0,1], \quad r \in \Omega \tag{26.6a}$$

or

$$H(v,p) = L(v) - L(u_0) + pL(u_0) + p[N(v) - f(r)] = 0, \tag{26.6b}$$

where $p \in [0, 1]$ is an embedding parameter, u_0 is an initial approximation of Eq. (26.3) which satisfies the boundary condition. Obviously, from Eq. (26.6a, 26.6b) we have

$$H(v,0) = L(v) - L(u_0) = 0, \tag{26.7}$$

$$H(v,1) = A(v) - f(r) = 0, \tag{26.8}$$

and the changing process of p from zero to unity is just that of $v(r,p)$ changing from $u_0(r)$ to $u(r)$. In topology, this is called deformation, $L(v) - L(u_0)$ and $A(v) - f(r)$ are called homotopic.

According to the HPM, first we use the embedding parameter p as a small parameter, and assume that the solution of Eq. (6) can be written as a power series in p:

$$v = v_o + pv_1 + p^2 v_2 + p^3 v_3 + \cdots. \tag{26.9}$$

Setting $p = 1$, results in approximate solution of Eq. (26.3):

$$u = \lim_{p \to 1} v = v_0 + v_1 + v_2 + v_3 + \cdots. \tag{26.10}$$

The series (26.10) is convergent for most cases. However, convergent rate depends on the nonlinear operator $A(v)$. The following opinion is suggested by He [1–4].

a. The second derivative on $N(v)$ with respect to v must be small, because the parameter p may be relatively large, i.e., $p \to 1$.
b. The norm of $L^{-1}\partial N/\partial v$ must be smaller than one in order for the series to converge.

26.3 An Application of HPM Solution for the NLSV Equation

Consider the following NLSV Eq. (26.2) with initial condition given by [11] as follows:

$$U(\xi, 0) = a \tanh\left[\left(-\frac{\mu_2}{2\mu_1}\right)^{\frac{1}{2}} a\xi\right] e^{iK\xi}. \tag{26.11}$$

Using HPM, we construct a homotopy in the following form:

$$H(v,p) = (1-p)\left[i\frac{\partial V}{\partial \tau} - i\frac{\partial U_0}{\partial \tau}\right] + p\left[i\frac{\partial V}{\partial \tau} + \mu_1\frac{\partial^2 V}{\partial \xi^2} + \mu_2|V|^2 V - \mu_3 h_1(\tau)V\right] \\ = 0, \tag{26.12}$$

or

$$H(v,p) = (1-p)\left[i\frac{\partial V}{\partial \tau} - i\frac{\partial U_0}{\partial \tau}\right] + p\left[i\frac{\partial V}{\partial \tau} + \mu_1\frac{\partial^2 V}{\partial \xi^2} + \mu_2 V^2\bar{V} - \mu_3 h_1(\tau)V\right] = 0, \tag{26.13}$$

where $U_0(\xi,\tau) = V_0(\xi,0) = U(\xi,0)$, $|V|^2 = V\bar{V}$ and $\bar{V}$ is the conjugate of V. Suppose that the series solution of Eq. (26.12) and its conjugate have the following forms:

$$V = V_0(\xi,\tau) + pV_1(\xi,\tau) + p^2V_2(\xi,\tau) + p^3V_3(\xi,\tau) + \cdots, \tag{26.14}$$

$$\bar{V} = \bar{V}_0(\xi,\tau) + p\bar{V}_1(\xi,\tau) + p^2\bar{V}_2(\xi,\tau) + p^3\bar{V}_3(\xi,\tau) + \cdots. \tag{26.15}$$

Substituting Eqs. (26.14) and (26.15) into Eq. (26.13), and collecting the terms with identical powers of p, leads to

$$p_0 = i\frac{\partial V_0}{\partial \tau} - i\frac{\partial U_0}{\partial \tau} = 0, \tag{26.16}$$

$$p_1 = i\frac{\partial V_1}{\partial \tau} + i\frac{\partial U_0}{\partial \tau} + \mu_1\frac{\partial^2 V_0}{\partial \xi^2} + \mu_2 V_0^2\bar{V}_0 - \mu_3 h_1(\tau)V_0 = 0, \tag{26.17}$$

$$p_2 - i\frac{\partial V_2}{\partial \tau} + \mu_1\frac{\partial^2 V_1}{\partial \xi^2} + \mu_2\left(V_0^2\bar{V}_1 + 2V_0V_1\bar{V}_0\right) - \mu_3 h_1(\tau)V_1 = 0, \tag{26.18}$$

$$\vdots$$

The given initial value admits the use of

$$V_i(\xi,0) = \begin{cases} a\ \tanh\left[\sqrt{-\frac{\mu_2}{2\mu_1}}a\xi\right]e^{iK\xi}, & i = 0, \\ 0, & i = 1,\ 2,\ 3,\cdots \end{cases}. \tag{26.19}$$

The solutions read

$$V_0(\xi,\tau) = a\ \tanh\left[\sqrt{-\frac{\mu_2}{2\mu_1}}a\xi\right] e^{iK\xi}, \tag{26.20}$$

$$\begin{aligned} V_1(\xi,\tau) = {} & ia\ \tanh\left[\sqrt{-\frac{\mu_2}{2\mu_1}}a\xi\right] (a^2\mu_2 - \mu_1 K^2)\tau e^{iK\xi} \\ & - \frac{10}{3} ia\mu_3 \tanh\left[\sqrt{-\frac{\mu_2}{2\mu_1}}a\xi\right] \text{arc}\ \tan(\sinh[0.3\tau]) e^{iK\xi} \\ & - 2a^2\mu_1\sqrt{-\frac{\mu_2}{2\mu_1}}a\left(\text{sech}^2\left[\sqrt{-\frac{\mu_2}{2\mu_1}}a\xi\right]\right)\tau K e^{iK\xi} \end{aligned}, \tag{26.21}$$

$$\vdots$$

and so on. The solution of the NLSV Eq. (26.2) can be obtained by setting $p = 1$ in Eq. (26.9). Thus, we have

$$\begin{aligned} U(\xi,\tau) = {} & V(\xi,\tau) = V_0(\xi,\tau) + V_1(\xi,\tau) + \cdots \\ = {} & a\ \tanh\left[\sqrt{-\frac{\mu_2}{2\mu_1}}a\xi\right] e^{iK\xi} + ia\ \tanh\left[\sqrt{-\frac{\mu_2}{2\mu_1}}a\xi\right] (a^2\mu_2 - \mu_1 K^2)\tau e^{iK\xi} \\ & - \frac{10}{3} ia\mu_3 \tanh\left[\sqrt{-\frac{\mu_2}{2\mu_1}}a\xi\right] \text{arc}\ \tan(\sinh[0.3\tau]) e^{iK\xi} \\ & - 2a^2\mu_1\sqrt{-\frac{\mu_2}{2\mu_1}}a\left(\text{sech}^2\left[\sqrt{-\frac{\mu_2}{2\mu_1}}a\xi\right]\right)\tau K e^{iK\xi} + \cdots \end{aligned} \tag{26.22}$$

The progressive wave solution (quite close to exact solution) of the NLSV Eq. (26.2) is given by [11] as follows:

$$U(\xi,\tau) = a\ \tanh\left[\sqrt{-\frac{\mu_2}{2\mu_1}}a(\xi - 2\mu_1 K\tau)\right] e^{i\left(K\xi - \Omega\tau - \frac{\mu_3}{\delta}\arctan(\sinh(\delta\tau))\right)}. \tag{26.23}$$

where $\Omega = \mu_1 K^2 - \mu_2 a^2$. Given that these numerical values of the coefficient by [11] $h_1(\tau) = \text{sech}(0.3\tau)$, $\mu_1 = -0.1548$, $\mu_2 = 26.4295$, $\mu_3 = 7.3572$ provided $a = 1$, $\delta = 0.3$, and $K = 2$.

Figures 26.1 and 26.3 show the 3D-plot of the HPM solution and progressive wave solution of the NLSV Eq. (26.2) with spatial parameter, ξ and time, τ, respectively, while Figs. 26.2 and 26.4 show the 2D-plot of HPM solution and the progressive wave solution of the NLSV Eq. (26.2) with spatial parameter, ξ at time

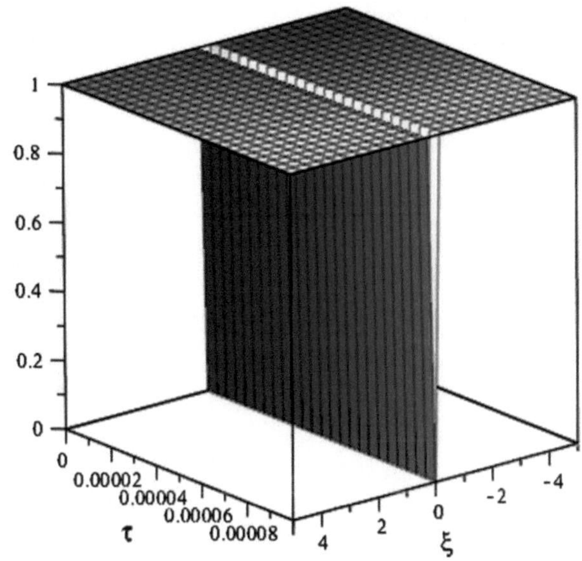

Fig. 26.1 3D-plot of the HPM solution of the NLSV Eq. (26.2) under initial condition (26.11)

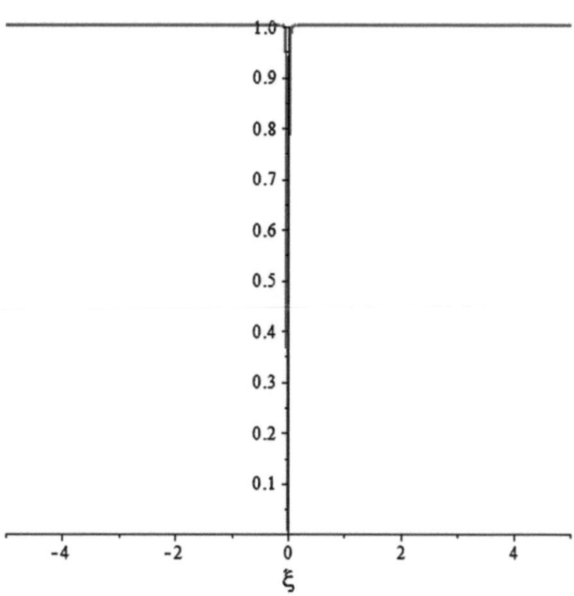

Fig. 26.2 2D-plot of the HPM solution of the NLSV Eq. (26.2) under initial condition (26.11)

$\tau = 0.0001$, respectively. Table 26.1 shows the absolute error between the progressive wave and HPM solutions for certain spatial points, ξ and at time, $\tau = 0.0001$.

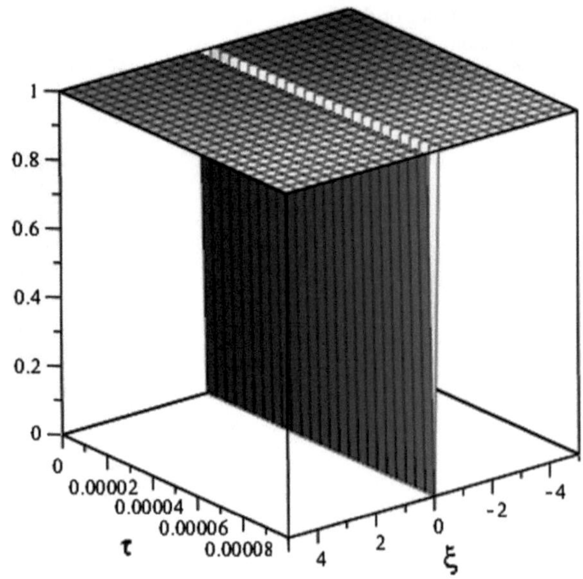

Fig. 26.3 3D-plot of the exact solution of the NLSV Eq. (26.2)

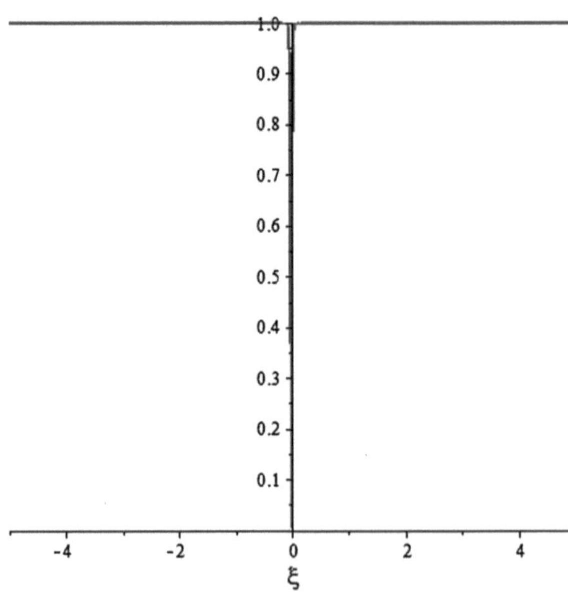

Fig. 26.4 2D-plot of the exact solution of the NLSV Eq. (26.2)

Table 26.1 The absolute error of the NLSV equation for different spatial values, ξ at $\tau = 0.0001$

Spatial Parameter, ξ	-5	-3	0	3	5
L_∞	1.2×10^{-3}	1.2×10^{-3}	0	1.2×10^{-3}	1.2×10^{-3}

26.4 Conclusion

In this work, HPM is successfully applied to solve the NLSV Eq. (26.2). The solution obtained by HPM is an infinite series for appropriate initial condition that can be expressed in a closed form to the exact solution. The solution obtained by HPM is found to be a powerful mathematical tool which can be used to solve nonlinear partial differential equations.

Acknowledgments The authors wish to express appreciation to the Ministry of Education Malaysia for financial support and Universiti Tun Hussein Onn Malaysia for the RAGS's Grant Vote R026.

References

1. He, J.H.: Homotopy perturbation technique. Comput. Methods Appl. Mech. Eng. **178**, 257–262 (1999)
2. He, J.H.: A coupling method of homotopy technique and perturbation technique for nonlinear problems. Int. J. Non-Linear Mech. **35**, 37–43 (2000)
3. He, J.H.: Homotopy perturbation method: a new nonlinear analytical technique. Appl. Math. Comput. **135**, 73–79 (2003)
4. He, J.H.: Comparison of homotopy perturbation method and homotopy analysis method. Appl. Math. Comput. **156**, 528–539 (2004)
5. Li, J.L.: Adomian's decomposition method and homotopy perturbation method in solving nonlinear equations. J. Comput. Appl. Math. **228**(1), 168–173 (2009)
6. Noor, M.A., Khan, W.A.: New iterative methods for solving nonlinear equation by using homotopy perturbation method. Appl. Math. Comput. **219**(8), 3365–3374 (2012)
7. Ghasemi, M., Tavassoli, K.M., Davari, A.: Numerical solution of two dimensional nonlinear differential equation by homotopy perturbation method. Appl. Math. Comput. **189**(1), 341–345 (2007)
8. Ravindran, R., Prasad, P.: A mathematical analysis of nonlinear waves in a fluid-filled viscoelastic tube. Acta Mech. **31**(3–4), 253–280 (1979)
9. Antar, N., Demiray, H.: Nonlinear wave modulation in a prestressed fluid-filled thin elastic tube. Int. J. Nonlinear Mech. **34**, 123–138 (1999)
10. Bakirtaş, İ., Demiray, H.: Amplitude modulation of nonlinear waves in a fluid-filled tapered elastic tube. Appl. Math. Comput. **154**(3), 747–767 (2004)
11. Choy, Y.Y.: Nonlinear Wave Modulation in a Fluid-Filled Thin Elastic Stenosed Artery. Doctorial dissertation, UTM, Skudai (2014)
12. Ablowitz, M.A., Clarkson, P.A.: Solitons, Nonlinear Evolution Equations and Inverse Scattering. Cambridge University Press, Cambridge (1992)
13. Taha, T.B., Ablowitz, M.J.: Analytical and numerical aspects of certain nonlinear evolution equations. II. Numerical, nonlinear Schrödinger equation. J. Comput. Phys. **55**(2), 203–209 (1984)
14. Wang, H.: Numerical studies on the split-step finite difference method for nonlinear Schrödinger equations. Appl. Math. Comput. **170**(1), 17–35 (2005)
15. Dehghan, M., Taleei, A.: A compact split-step finite difference method for solving the nonlinear Schrödinger equations with constant and variable coefficients. Comput. Phys. Commun. **181**(1), 43–51 (2010)
16. El-Sayed, S.M., Kaya, D.: A numerical solution and an exact explicit solution of the NLS equation. Appl. Math. Comput. **172**(2), 1315–1322 (2006)

17. Bratsos, A., Ehrhardt, M., Famelis, I.T.: A discrete Adomian decomposition method for discrete nonlinear Schrödinger equations. Appl. Math. Comput. **172**(1), 190–205 (2008)
18. Mousa, M.M., Ragab, S.F.: Application of the homotopy perturbation method to linear and nonlinear Schrodinger equations. Verlag der Zeitschrift Fur Natueforschung **63**(3–4), 140–144 (2008)

Chapter 27
The Atom Bond Connectivity Index of Some Trees and Bicyclic Graphs

Mohanad A. Mohammed, K.A. Atan, A.M. Khalaf, M.R. Md. Said and R. Hasni

Abstract The atom bond connectivity (*ABC*) index is one of the recently most investigated degree-based molecular structure descriptors that have applications in chemistry. For a graph G, the ABC index is defined as $ABC(G) = \sum_{uv \in E(G)} \sqrt{d_v + d_u - 2/d_v \cdot d_u}$, where d_u denotes the degree of a vertex u in G. In this paper, we obtain the general formula for ABC index of some special, chemical trees, and bicyclic graphs.

Keywords The atom bond connectivity (*ABC*) index · Bicyclic graphs · Molecular graphs · Chemical tree

27.1 Introduction

Molecular descriptors have found a wide application in quantitative structure–activity relationship analysis (QSAR) and in quantitative structure–property relationship (QSPR) (targets the estimation of specific characteristics based on the structures of the compounds under study) studies [10]. Among them, topological

M.A. Mohammed (✉) · A.M. Khalaf
Department of Mathematics College of Computer Sciences and Mathematics, University of Kufa, Najaf, Iraq
e-mail: mohanadalim@gmail.com

A.M. Khalaf
e-mail: abduljaleel.khalaf@uokufa.edu.iq

M.A. Mohammed · K.A. Atan · M.R.Md.Said
Institute for Mathematical Research, Universiti Putra Malaysia, 43400 Serdang, Selangor, Malaysia
e-mail: kamel@upm.edu.my

R. Hasni
Department of Mathematics Faculty of Science and Technology, University Malaysia Terengganu, 21030 Kuala Terengganu, Terengganu, Malaysia
e-mail: hroslan@umt.edu.my

A.-R. Ahmad et al. (eds.), *Proceedings of the International Conference on Computing, Mathematics and Statistics (iCMS 2015)*,
DOI 10.1007/978-981-10-2772-7_27

indices have a prominent place. One of the best known and widely used is the connectivity index, χ, introduced in Randic, M., On Characterization of Molecular Branching, J. Am. Chem. Soc. 97 1975 by Milan Randic [5, 9], who has shown this index to reflect molecular branching. To keep the spirit of the Randic index, Ernesto Estrada et al. proposed an index, nowadays known as the Atom Bond Connectivity (*ABC*) index [3]. The introduction of graph theoretic concepts in chemistry is well known and the reader is referred to the following references for definitions and notations [2, 7].

The atom bond connectivity (*ABC*) index of a graph $G = (V, E)$ is defined as $ABC(G) = \sum_{uv\in E(G)} \sqrt{d_v + d_u - 2/d_v \cdot d_u}$. This index was shown [3, 6] to be well correlated with the heats of formation of alkanes, and that it thus can serve for predicting their thermodynamic properties. In addition to this, Estrada [4] elaborated a novel quantum theory-like justification for this topological index, showing that it provides a model for taking into account 1, 2-, 1, 3-, and 1, 4- interactions in the carbon atom skeleton of saturated hydrocarbons, and that it can be used for rationalizing steric effects in such compounds. These results triggered a number of mathematical investigations of the *ABC* index [5]. The (chemical) trees, unicyclic, bicyclic graph(s) with maximal *ABC* index were determined [5]. However, the problem of characterizing general connected graph(s) with minimal-*ABC* index appears to be difficult, even in the case of trees. Recently, Chen and Liu [1] gave the sharp upper and lower bounds on *ABC* index of chemical bicyclic graphs that are obtained. In this paper we have two sections, in the first section, we present the general formula of atom bond connectivity index of some special trees graphs. In second section, we provide a general formula of *ABC* index of bicyclic graphs.

27.2 Preliminaries

A vertex of a graph is said to be a pendant if its neighborhood contains exactly one vertex. An edge of a graph is said to be pendant if one of its vertices is a pendant vertex.

Let G be a graph and $u, v \in V(G)$. A $(u, v) -$ walk is a finite sequence of vertices $a_0, a_1, a_2, \ldots, a_k$, with $k \geq 0$, $u = a_0$ and $v = a_k$, such that $a_{i-1}a_i \in E(G)$ for each $i = 1, 2, 3, \ldots, k$. The length of the walk is k that is the number of edges in it. A $(u, v) -$ walk is called a path if no vertices are repeated.

Let B_i be a branch of a tree T formed by attaching i pendant path of length 2 to the vertex v such that the degree of v in T is i + 1. Let B_i^* be a branch of a tree T formed by attaching i − 1 pendant path of length 2 and a pendant path of length 3 to the vertex v such that the degree of v in T is i + 1 (see Fig. 27.1 for an illustration), the minimal-*ABC* trees were fully characterized by Hosseini et al. [8].

The Curtain graph is a tree graph obtained by attaching m branches of B_k to each vertex of path P_n. We denote such graph by $T(n, B_k^m)$ (see Fig. 27.2).

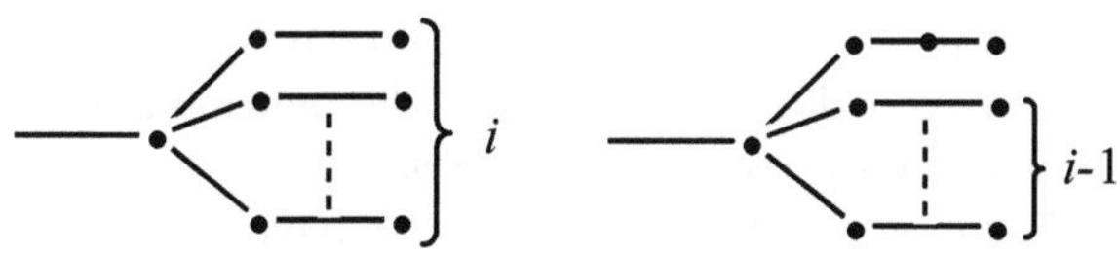

Fig. 27.1 The branches B_i and B_i^*

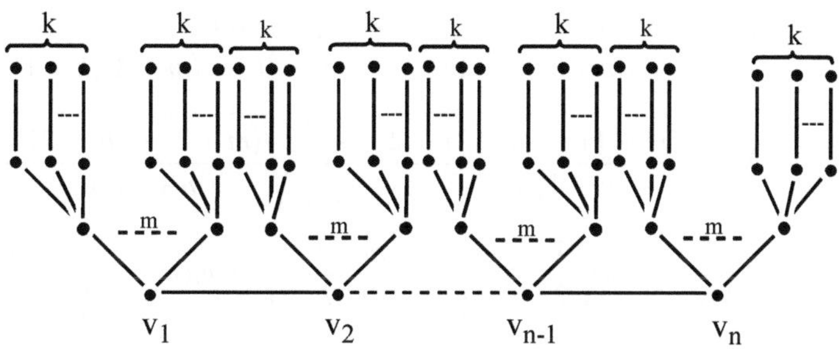

Fig. 27.2 The curtain graph $T(\mathrm{n}, B_{\mathrm{k}}^{m})$

27.3 *ABC* Index of Two Types of Trees

In this section, we provide the general formula for atom bond connectivity index of two types of trees graphs, the first one is associated with the Curtain tree $T(n, B_k^m)$ as follows.

Theorem 3.1 *Let n, k, m be positive integers such that $n \geq 3$, $k \geq 2$ and $m \geq 1$, the atom bond connectivity index of the Curtain graph $T(n, B_k^m)$ is*

$$ABC(G) = kmn\sqrt{2} + 2m\sqrt{\frac{k+m}{(k+1)(m+1)}} + m(n-2)\sqrt{\frac{k+m+1}{(k+1)(m+2)}} + 2\sqrt{\frac{2m+1}{(m+1)(m+2)}} + \frac{n-3}{m+2}\sqrt{2(m+1)}$$

Proof We have mn of branches B_k, first we will compute *ABC* index for the branches of B_k, where B_k contains $2k$ edges, with k of them containing two vertices, the first of degree one and the second of degree two. Another k of the edges contains two vertices each, the first of degree two and the second of degree $k + 1$.

We have mn of the edges linking branches of B_k with vertices of path with $2m$ of them containing two vertices; the first of degree $m+1$ and the second of degree $k + 1$. The remaining of the $(n - 2)m$ edges contain two vertices, the first of degree $k + 1$, and the second of degree $m + 2$.

The path contains $n - 1$ edges. Two of them contain two vertices, the first of degree $m + 1$ and the second of degree $m + 2$. The remaining $n - 3$ edges contain also two vertices which have the same degree $m + 2$.

By using the definition of ABC index of G, we have,

$$
\begin{aligned}
ABC(G) &= mn\sum_{i=1}^{k}\sqrt{\frac{1+2-2}{1\times 2}} + mn\sum_{j=1}^{k}\sqrt{\frac{2+(k+1)-2}{2(k+1)}} \\
&+ \sum_{h=1}^{2m}\sqrt{\frac{(m+1)+(k+1)-2}{(m+1)(k+1)}} + m\sum_{g=1}^{n-2}\sqrt{\frac{(m+2)+(k+1)-2}{(m+2)(k+1)}} \\
&+ \sum_{r=1}^{2}\sqrt{\frac{(m+1)+(m+2)-2}{(m+1)(m+2)}} + \sum_{s=1}^{n-3}\sqrt{\frac{(m+2)+(m+2)-2}{(m+2)(m+2)}} \\
&= mnk\sqrt{\frac{1}{2}} + mnk\sqrt{\frac{1}{2}} + 2m\sqrt{\frac{m+k}{(m+1)(k+1)}} + m(n-2)\sqrt{\frac{m+k+1}{(m+2)(k+1)}} \\
&+ 2\sqrt{\frac{2m+1}{(m+1)(m+2)}} + (n-3)\sqrt{\frac{2m+2}{(m+2)(m+2)}} \\
&= kmn\sqrt{2} + 2m\sqrt{\frac{k+m}{(k+1)(m+1)}} + m(n-2)\sqrt{\frac{k+m+1}{(k+1)(m+2)}} \\
&+ 2\sqrt{\frac{2m+1}{(m+1)(m+2)}} + \frac{n-3}{m+2}\sqrt{2(m+1)}
\end{aligned}
$$

Let $T(n, B_k^{*m})$ be a tree graph obtained by attaching m of a branch to each vertex of path P_n (see Fig. 27.3).

For this type of tree we give its ABC index as in the following theorem.

Theorem 3.2 *Let n, k, m be positive integers such that* $n \geq 3, k \geq 2$ *and* $m \geq 1$, *the atom bond connectivity index of a graph* $T(n, B_k^{*m})$ *is*

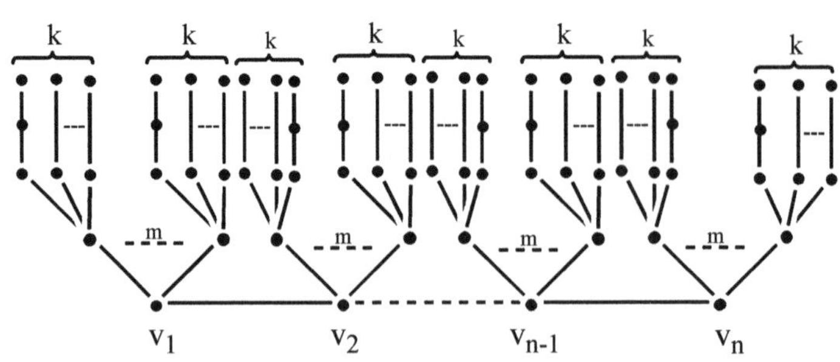

Fig. 27.3 Type of trees $T(n, B_k^{*m})$

$$ABC(G) = (2k+1)mn\frac{1}{\sqrt{2}} + 2m\sqrt{\frac{k+m}{(k+1)(m+1)}} + m(n-2)\sqrt{\frac{k+m+1}{(k+1)(m+2)}} + 2\sqrt{\frac{2m+1}{(m+1)(m+2)}} + \frac{n-3}{m+2}\sqrt{2(m+1)}$$

Proof We have mn branches B_k^*. We compute ABC index for the branches of B_k^*, where B_k^* contains $2k+1$ edges with k of them containing two vertices, the first of degree one and the second of degree two. Each of the other $k+1$ edges contains two vertices, the first of degree two and the second of degree $(k+1)$.

We have mn of the edges linking branches of B_k with vertices of path with $2m$ of them containing two vertices; the first of degree $m+1$ and the second of degree $k+1$, and the remaining of the $(n-2)m$ edges contain two vertices, the first of degree $k+1$, and the second of degree $m+2$.

The path contains $n-1$ edges. Two of them contain two vertices, the first of degree $m+1$ and the second of degree $m+2$. The remaining $n-3$ edges contain two vertices with the same degree $m+2$. By the definition of ABC index of G, we have,

$$\begin{aligned} ABC(G) &= mn\sum_{i=1}^{k}\sqrt{\frac{1+2-2}{1\times 2}} + mn\sum_{j=1}^{k+1}\sqrt{\frac{2+(k+1)-2}{2(k+1)}} \\ &+ \sum_{h=1}^{2m}\sqrt{\frac{(m+1)+(k+1)-2}{(m+1)(k+1)}} + m\sum_{g=1}^{n-2}\sqrt{\frac{(m+2)+(k+1)-2}{(m+2)(k+1)}} \\ &+ \sum_{r=1}^{2}\sqrt{\frac{(m+1)+(m+2)-2}{(m+1)(m+2)}} + \sum_{s=1}^{n-3}\sqrt{\frac{(m+2)+(m+2)-2}{(m+2)(m+2)}} \\ &= mnk\sqrt{\frac{1}{2}} + mn(k+1)\sqrt{\frac{1}{2}} + 2m\sqrt{\frac{m+k}{(m+1)(k+1)}} \\ &+ m(n-2)\sqrt{\frac{m+k+1}{(m+2)(k+1)}} + 2\sqrt{\frac{2m+1}{(m+1)(m+2)}} \\ &+ (n-3)\sqrt{\frac{2m+2}{(m+2)(m+2)}} \\ &= (2k+1)mn\frac{1}{\sqrt{2}} + 2m\sqrt{\frac{k+m}{(k+1)(m+1)}} + m(n-2)\sqrt{\frac{k+m+1}{(k+1)(m+2)}} \\ &+ 2\sqrt{\frac{2m+1}{(m+1)(m+2)}} + \frac{n-3}{m+2}\sqrt{2(m+1)} \end{aligned}$$

27.4 *ABC* Index of Types of Bicyclic Graph

A bicyclic graph is a connected graph in which the number of its edges is one more than the number of its vertices. A Jellyfish graph is obtained by joining a cycle of length n with another cycle of length m by a path of length r followed by putting branches of B_k at each vertex in the two cycles and path. We placed B_k at the terminal vertices in the path. We denote such graph by $B^{B_k}(n, r, m)$ as in Fig. 27.4.

In this section, we provide the general formula for atom bond connectivity index of some bicyclic graphs.

Theorem 4.1 *Let n, m, k, r be positive integers such that* $n,\ m \geq 3,\ k,\ r \geq 2$. *The atom bond connectivity index of a graph* $B^{B_k}(n, r, m)$

$$ABC(G) = (n+m+r-3)\left[2\sqrt{2}k + \sqrt{\frac{k+3}{k+1}}\right] + \frac{\sqrt{6}}{4}(n+m+r-1).$$

Proof In the graph $B^{B_k}(n, r, m)$ we have $2n-1$ branches B_k in the first cycle, $2m-1$ branches in the second and $2(r-2)$ branches in the path. Therefore, we have $2(n+m+r-3)$ branches of B_k. The branch B_k contains $2k$ edges. The first k of them contain two vertices of which the first and the second are both of degree two, and another k of edges containing two vertices of which the first is of degree two and the second of degree $k+1$.

We have $2(n+m+r-3)$ of the edges each of which contains two vertices. The first is of degree four and the second is of degree $k+1$. Likewise, we have $n+m+r-1$ of edges that contain vertices of degree four. The result follows from the definition of *ABC* index of G as shown below.

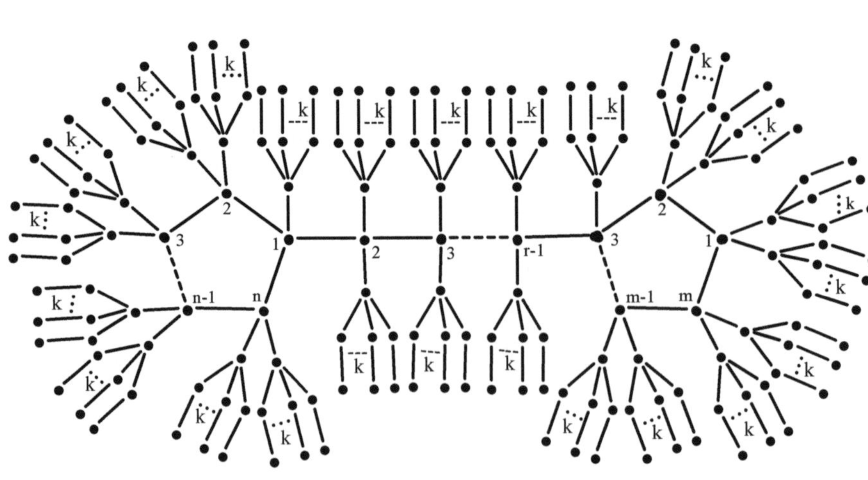

Fig. 27.4 The jellyfish graph $B^{B_k}(n, r, m)$

$$
\begin{aligned}
ABC(G) &= 2k(n+m+r-3)\sqrt{\frac{1+2-2}{1\times 2}} + 2k(n+m+r-3)\sqrt{\frac{2+k+1-2}{2(k+1)}}\\
&\quad + 2(m+n+r-3)\sqrt{\frac{4+k+1-2}{4(k+1)}} + (n+m+r-1)\sqrt{\frac{4+4-2}{4\times 4}}\\
&= 2k(n+m+r-3)\sqrt{\frac{1}{2}} + 2k(n+m+r-3)\sqrt{\frac{1}{2}}\\
&\quad + (m+n+r-3)\sqrt{\frac{k+3}{k+1}} + (n+m+r-1)\frac{\sqrt{6}}{4}\\
&= 2\sqrt{2}k(n+m+r-3) + (n+m+r-3)\sqrt{\frac{k+3}{k+1}} + \frac{\sqrt{6}}{4}(n+m+r-1)\\
&= (n+m+r-3)\left[2\sqrt{2}k + \sqrt{\frac{k+3}{k+1}}\right] + \frac{\sqrt{6}}{4}(n+m+r-1)
\end{aligned}
$$

Let $B^{B_k^*}(n,r,m)$ be a bicyclic graph, which is obtained by joining a cycle of length n with another cycle of length m by a path of length r, and by attaching two branches of B_k^* to each vertex in the two cycles and path except at the terminal vertices in the path where we attach one of B_k^* as in Fig. 27.5.

The following theorem gives the *ABC* index of such a graph $B^{B_k^*}(n,r,m)$

Theorem 4.2 *Let n, m, k, r be positive integers such that* $n,m\geq 3$ *and* $k,r\geq 2$. *The atom bond connectivity index of a graph* $B^{B_k^*}(n,r,m)$

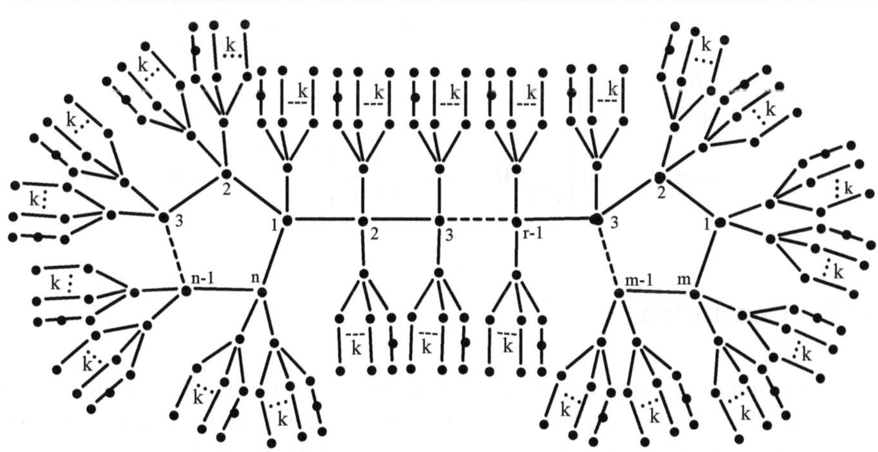

Fig. 27.5 Type of bicyclic graph $B^{B_k^*}(n,r,m)$

$$ABC(G) = (n+m+r-3)\left[\frac{1}{\sqrt{2}}(4k+1)+\sqrt{\frac{k+3}{k+1}}\right]+\frac{\sqrt{6}}{4}(n+m+r-1)$$

Proof In the graph $B^{B_k^*}(n,r,m)$ we have $(2n-1)$ branches B_k^* in the first cycle, $2m-1$ branches in the second cycle and $2(r-2)$ branches in the path. Therefore, we have $2(n+m+r-3)$ branches of B_k. The branch B_k^* contains $2k+1$ edges. One of them is incident with vertices of same degree two; k of them are incident with two vertices of which the first one is of degree one and the second of degree two. The other k of the edges contain two vertices of which the first one is of degree two and the second of degree $k+1$. We have $2(n+m+r-3)$ of edges each of which contains two vertices. The first one is of degree four and the second is of degree $k+1$. Likewise, we have $n+m+r-1$ edges containing vertices of the same degree four.

Now the result follows from the definition of ABC index of G as shown below.

$$\begin{aligned}
ABC(G) &= \sqrt{\frac{2+2-2}{2\times 2}}+2k(n+m+r-3)\sqrt{\frac{1+2-2}{1\times 2}}\\
&\quad +2k(n+m+r-3)\sqrt{\frac{2+k+1-2}{2(k+1)}}+2(n+m+r-3)\sqrt{\frac{4+k+1-2}{4(k+1)}}\\
&\quad +(n+m+r-1)\sqrt{\frac{4+4-2}{4\times 4}}\\
&= \sqrt{\frac{1}{2}}+2k(n+m+r-3)\sqrt{\frac{1}{2}}+2k(n+m+r-3)\sqrt{\frac{1}{2}}\\
&\quad +(n+m+r-3)\sqrt{\frac{k+3}{k+1}}+\frac{\sqrt{6}}{4}(n+m+r-1)\\
&= (4k+1)(n+m+r-3)\frac{1}{\sqrt{2}}+(n+m+r-3)\sqrt{\frac{k+3}{k+1}}\\
&\quad +\frac{\sqrt{6}}{4}(n+m+r-1)\\
&= (n+m+r-3)\left[\frac{1}{\sqrt{2}}(4k+1)+\sqrt{\frac{k+3}{k+1}}\right]+\frac{\sqrt{6}}{4}(n+m+r-1)
\end{aligned}$$

27.5 Conclusion

In conclusion, this study reveals three findings: first, we established the general formulae of ` index for two types of trees graphs. Second, we have shown that this trees graphs are associated with some special chemical trees because no vertex degree greater than fore. Finally, we provided the general formulae to the bicyclic graphs after attaching branches of B_i and B_i^* for all vertices.

References

1. Chen, J., Liu, J.: On atom-bond connectivity index of bicyclic graphs. J. Guangxi Teach. Educ. Univ. **28**, 8–12 (2011)
2. Devillers, J., Balaban, A.T.: Topological indices and related descriptors in QSAR and QSPR. Gordon and Breach, Amsterdam (1999)
3. Estrada, E.: Atom-bond connectivity and the energetic of branched alkanes. Chem. Phys. Lett. **463**, 422–425 (2008)
4. Estrada, E., Torres, L., Rodríguez, L., Gutman, I.: An atom-bond connectivity index: modelling the enthalpy of formation of alkanes. Indian J. Chem. **37A**, 849–855 (1998)
5. Furtula, B., Graovac, A., Vukičević, D.: Atom-bond connectivity index of trees. Discr. Appl. Math. **157**, 2828–2835 (2009)
6. Gutman, I., Tošović, J., Radenković, S., Marković, S.: On atom-bond connectivity index and its chemical applicability. Indian J. Chem. **51A**, 690–694 (2012)
7. Gutman, I., Polansky, O.E.: Mathematical concepts in organic chemistry. Springer, Berlin (1986)
8. Hosseini, S.A., Ahmadi, M.B., Gutman, I.: Kragujevac trees with minimal atom-bond connectivity index. MATCH Commun. Math. Comput. Chem. **71**, 14 (2014)
9. Randić, M.: On characterization of molecular branching. J. Am. Chem. Soc. **97**, 6609–6615 (1975)
10. Todeschini, R., Consonni, V.: Handbook of molecular descriptors. Wiley VCH, Weinheim (2000)

References

Part IV
Application and Education

Chapter 28
A Comparative Study of Problem-Based and Traditional Teaching in Computing Subjects

Sariah Rajuli and Norhayati Baharun

Abstract This study aimed to compare the impact of the problem-based (PBL) with the traditional teaching (TLA) approach on students' academic performance and learning attitudes of computing subjects at the Kolej Profesional MARA Seri Iskandar (KPMSI), Perak, Malaysia. The participants of this study involved 74 students in the second semester of Foundation in Business (FIB 2) and third semester of Higher National Diploma (HND 3). A total of 40 students were allocated in PBL group while 34 students were allocated in TLA group. The participants were selected using a stratified random sampling technique. In this experimental design study, a quantitative method was employed using an adapted pretest and posttest as the research instrument which consists of 30 multiple choice items. Additionally, a structured survey was administered to the students after they are being exposed to a particular teaching method. Findings of the study showed that students instructed with problem-based approach significantly outperformed than those instructed with traditionally designed instruction. This study also suggests that there was a significant relationship between teaching methods and students' academic performance. A significant relationship also found in students' learning attitudes and academic performance where the feedback from the students indicated that items "learn from others" and "peer support" have an impact on their academic performance. No statistically significant relationship was found between teaching methods employed and the students' learning attitudes. The findings from this study will provide educators with an alternative strategy for improving teaching and learning of computing subjects.

Keywords Problem-based approach · Students' academic performance · Students' learning attitudes · Traditional teaching

S. Rajuli (✉)
Kolej Profesional MARA Seri Iskandar, Bota, Malaysia
e-mail: sariah@kpmsi.edu.my

N. Baharun
Universiti Teknologi MARA Cawangan Perak, Perak, Malaysia
e-mail: norha603@perak.uitm.edu.my

A.-R. Ahmad et al. (eds.), *Proceedings of the International Conference on Computing, Mathematics and Statistics (iCMS 2015)*,
DOI 10.1007/978-981-10-2772-7_28

28.1 Introduction

Classrooms today are dictated by standards and assessments. The goal is for students to graduate with a well-rounded education having developed the skills and knowledge to be successful in the next phase in their life. Students need to learn how to read and interpret information. Learning how to think and solve the problems is the point of education. This study is performed to compare the students' academic performance and learning attitudes between two groups of students; problem-based (PBL) and traditional teaching (TLA) approach groups. The study also explores the relationship between students' learning attitudes and their academic performance.

Learning in the classrooms cannot be confined to just the content to be taught for the day, nor the syllabus to be completed in the semester. It will be so unnatural because acquisition of knowledge comes in a package together with the acquisition of other skills. In other words, learners are not just learning but also simultaneously picking up a variety of generic skills. A curriculum requires its future professionals to be fully prepared with a variety of skills such as analytical, creative thinking, teamwork, communication, technical, self-learning, and problem solving. The skills are acquired through various effective practices. Thus, the curriculum should be more dynamic and well-developed in order to acquire the necessary skills rather than teaching information. The need for such skills is vital to prepare proficient future professional in their own field of expertise to face the challenges of increasingly complex information and communication technology (ICT) world. Unfortunately, many students struggle to develop those skills and construct new knowledge especially when the subject materials are inherently intangible, difficult to visualize, and are conceptually different than what students are usually familiar with [1]. Past research has showed an important role of PBL and students' learning attitudes to stimulate learners to learn and find solutions to the problems [2, 3].

28.2 The Context

Kolej Profesional MARA (KPM) is formerly known as the Institut Perdagangan MARA (IPM) was established in May 1977. Since its establishment until the end of July 1982, KPM was located at the Sing Hoe Motor Building, Jalan Ipoh, Kuala Lumpur. Initially, KPM had only 100 students with four lecturers. On June 1, 1998, KPM Kuala Lumpur campus has moved to Beranang, Selangor. The second campus of KPM is located at Bandar Melaka, followed by KPM Indera Mahkota, KPM Seri Iskandar, KPM Bandar Penawar, and KPM Ayer Molek. To date, there are six KPM campuses in Malaysia. A Majlis Amanah Rakyat (MARA) through the Bahagian Pendidikan Tinggi (BPT) has taken many steps to strengthen its education sector in order to support Malaysia into a leading education hub. In 2011, BPT has urged all KPM campuses to implement a PBL teaching method at their colleges (cited in BPT MARA Road Map 2011–2015).

28.3 Problem Statement

Students' academic performance problems are often highlighted in the academic literature and mass media. Therefore, it is important for teachers to be aware and know about the factors related to students' academic performance. Traditionally, the computing subjects at the Kolej Profesional MARA Seri Iskandar (KPMSI) are taught in a lecture-based environment where each computing concept was delivered via theory and practical teaching method. Several teachers used analogy technique to enhance students' understanding. But in a real situation, some computing behaviors and ideas cannot be readily shown [4]. Due to the requirement of considerable amount of resources, it is difficult to expose students to some basic concepts of computing. To overcome the limitations of this traditional teaching method, a problem-based (PBL) approach is broadly used nowadays. The approach generally seems to have a positive effect on computing learning environment. Nonetheless, there was a major issue raised by many researchers; is the approach significantly aided the learning process and led to deeper content understanding? Several studies were conducted in the early 1990s and since the year of 2000, studies had showed a remarkably difference result of PBL effectiveness toward student's academic performance [5]. Furthermore, according to [6], psychological and sociological are the two main factors that contribute to students' academic performance. Psychological factors refer to the internal elements of individual including emotional and cognitive domains, whereas sociological factors refer to the external factors such as socio-environment and peers. Agreed by Mamat and Mazelan [7] on predicting student success, they discovered that student's attitude toward learning was one of the main factors that influenced student's academic performance. In the light of this perspective, this study attempts to investigate the influences of PBL and students' learning attitudes that only focus on six categories namely collaboration with peers, learning from others, peers support, social skills, problem solving and motivation, and interest toward students' academic performance specifically in a computing subject.

28.4 Review of Literature

The PBL model has become both nationally and internationally recognized by universities, researchers, and students as an advanced and efficient learning model. The PBL is based on the information processing model, constructivist learning theories. Learners are given only guidelines as to how to approach the problems [8] where students should metacognitively aware of what information they already know and what information they are need to know in order to solve the PBL problems. In solving the problems, learners will share the information but at the same time they construct their own knowledge [5]. The shift of TLA toward PBL provide a more realistic approach to learning and create an educational methodology which emphasizes real-world challenges, higher order thinking skills, multidisciplinary learning, independent learning, teamwork, and communication skills through a problem-based learning environment. The PBL model consists of three items; (1) the problem (2) the teachers who act as facilitators, and (3) the students who act as the problem solvers [9]. In order to transform the TLA to PBL teaching method, this study adapted the PBL model introduced by Neo and Neo [9].

In addition to teaching method as a factor that may affect the student's academic performance, students' learning attitude is another factor that was identified by numerous researchers which can influence the student's academic performance. Learning attitudes have been shown to impact the academic performance especially between students-centered and teacher-centered learning environment. Research has shown that when students have positive attitude for a certain learning task, forth more effort will be set to accomplish the task [10]. Students will work harder and persist longer with the learning task. As a result they are more likely to be successful than students with negative learning attitude [11]. Several researchers such as [2, 3, 5] illustrated that the more positive one's attitude toward an academic subject, the higher the possibility for the students to perform well academically. From the literature, some researchers developed a theory that could be used to explain the relationship between learning attitudes and academic performance. This preliminary analysis of students' attitudes will determine whether there is a statistically significant association between the teaching methods and students' learning attitudes that may influence their academic performance.

The theoretical framework provides a rationale for predictions about the relationships among variables for this study. Thus, based on the review of literature mentioned previously, the framework of this study can be conceptualized as illustrated in Fig. 28.1.

The study believed that when the PBL method is applied, the problem-driven instruction could motivate students to learn the subject due to the human nature of curiosity and taking on challenge [12]. Through the problem solving process, students are not only acquiring the domain knowledge but at the same time constructing an effective knowledge acquisition and perform better academically. Students' learning attitudes is another one factor that expected to influence the students' academic performance. The study of students' learning attitudes will determine whether it has an association with the teaching method that may influence the students' academic performance.

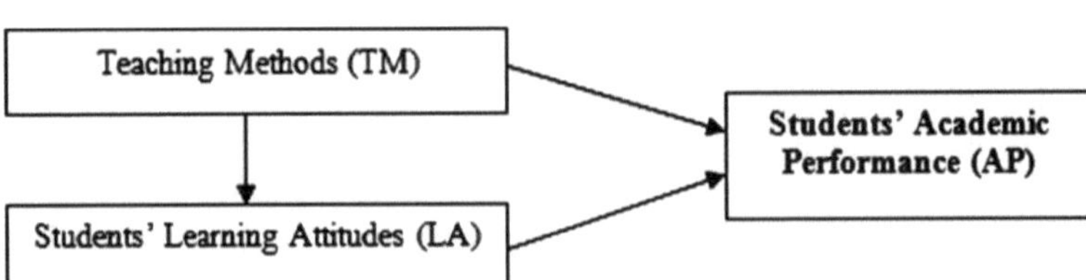

Fig. 28.1 Conceptual framework of the study

28.5 Methodology

28.5.1 Population and Sampling Technique

The population of this study involved the students of the Kolej Profesional MARA Seri Iskandar (KPMSI) enrolled in Foundation in Business (FIB 2) and Higher National Diploma (HND 3) programs. The samples or participants of this study involved 74 students in the second semester of Foundation in Business (FIB 2) and third semester of Higher National Diploma (HND 3). A total of 40 students were allocated in PBL group while 34 students were allocated in TLA group. The participants were selected using a stratified random sampling technique. This type of sampling allows the participants to be grouped according to a variable determined by the researcher. According to [13], the researchers can predetermine the stratified population, and then select the samples that best represent the population under study. The samples or participants selected in each program were randomly picked from the specified sampling frame of this study. The stratified random sampling technique was used to select samples from the selected computing subjects.

28.5.2 Instrumentations

In this experimental design study, a quantitative method was employed via three types of approaches: (1) an adapted pretest and posttest, (2) a structured survey, and (3) students' final exam grades. The following sections were the descriptions of instruments used in this study:

(1) *Pretest and Posttest*
There were two tests, namely, pretest and posttest used to evaluate the academic performance of the students before and after the completion of each teaching method. The tests were based on a 30-multiple choice items adapted [14, 15]. Pretest was administered in classes which was completed by students (on-paper) during the first week of semester before implementing the PBL or TLA teaching method. One topic from each subject's syllabus, that is, Networking Technologies (21470D) and Fundamentals of Database (CSC 1143) has been selected. Though the topics were different, it was tested previously that the level of difficulties in pretest are similar for both the subjects. The purpose of the pretest was to evaluate the academic performance of the students' basic computing skills before the experiment. The total score of test was 30 marks. During tenth week of semester, the students completed the posttest (same items in pretest) in classes. The posttest was used to test for significant learning gains (from pretest to posttest) within each group of PBL and TLA teaching method.

(2) *Structured Survey*
A structured survey was adapted from the recent study conducted by Looi and Seyal [16]. It consists of two sections: Section A provides the information about the respondents' demographic background, and Section B measures the student's attitudes toward their learning on a 5-point Likert scale (maximum score of 5 represents the highest level of agreement, and minimum score of 1 represents the strongest disagreement) of a total 21 items. A Cronbach's alpha coefficient value for each component of students' learning attitudes categorized as collaborative learning (4 items), learning from others (3 items), motivation and interest (3 items), peer support (4 items), social skills (3 items), and problem solving skills (4 items) produced an acceptable reliability level of 0.802, 0.796, 0.843, 0.879, 0.763, and 0.776, respectively.

(3) *Students' Final Exam Grades*
The grading system of computing subjects at KPMSI is based on Qualification and Credit Framework (QCF) by Pearson, the United Kingdom's largest awarding organization. It is also known as outcomes-based grading system. The student's academic performance is evaluated and documented using this grading system. Their final grade is tied to the level of performance documented according to several outcomes. For each outcome, student's performance can be graded as "Redo", "Pass", "Merit", and "Distinction". Both the subjects consist of three types of assessments with four outcomes to be achieved by students. At the end of the semester, student's grade depends on their overall performance of each criteria assessed in all assessments.

28.6 Analysis and Results

In order to perform the statistical analyses required for this study, the software called Statistical Package for Social Sciences (SPSS) Version 20.0 was used. All statistical analyses performed in this study were set to use 5 % level of significance, in other words, it will reject the null hypothesis if the given significance value (*p*-value) is less than 0.05. The analyses consist of four parts of statistical tests according to its purpose as described below:

(i) *Descriptive Analysis of the Participants*

The participants of this study were categorized into two groups of age range; 18–20 years old and 21–23 years old. This was possibly due to the participants from different batch, FIB 2 and HND 3. Out of 74 participants, 58 % of them aged 18–20 years old, and 42 % aged 21–23 years old. Among gender, 74 % were female and 26 % were male. About 61 % of the participants had a current CGPA between 3.00 and 3.50, 37 % were above 3.50, and others below 3.00.

According to the composition of experimental groups; out of 34 students in TLA teaching method, 19 (56 %) students enrolled in FIB 2 subject and 15 (44 %)

students enrolled in HND 3 subject. While out of 40 students in PBL teaching method, 22 (55 %) students enrolled in FIB 2 subject, and 18 (45 %) students enrolled in HND 3 subject.

(ii) *Independent t-test Analysis (Pretest and Posttest) between Groups*

An independent t-test analysis was used to compare the pretest and posttest mean scores obtained by students in PBL and TLA teaching methods. The pretest was conducted to determine whether there is a significant difference in terms of the prerequisite knowledge (basic computing skills) among students in PBL and TLA teaching methods (first week of semester). The result revealed that no significant difference was found in the mean scores ($t_{72} = 0.157$, $p = 0.876$ with equal variances assumed, $F = 0.387, p = 0.536$) for the pretest between the students in PBL (mean score of 11.10 marks) and TLA teaching methods (mean score of 10.97 marks). This indicated that the level of prerequisite knowledge among students in PBL and TLA teaching methods was significantly equivalent.

The posttest was conducted to determine whether there is a significant difference in terms of learning gains among students after implementing the PBL and TLA teaching methods (tenth week of semester). The result revealed that a significant difference was found in the mean scores ($t_{72} = 3.563$, $p = 0.001$ with equal variances assumed, $F = 0.371, p = 0.545$) for the posttest between the students in PBL and TLA teaching methods. This indicated that the performance of students in PBL teaching method (mean score of 19.35 marks) was significantly higher than in TLA teaching method (mean score of 15.82 marks) for the posttest.

(iii) *Dependent t-test Analysis (from Pretest to Posttest) within Groups*

A dependent t-test analysis was used to determine whether there is a significant improvement in the mean scores (from pretest to posttest) within each teaching methods (PBL and TLA). The findings revealed that the students in TLA teaching method demonstrated significant learning gains throughout the semester from a mean score of 10.97 marks on the pretest to a mean score of 15.82 on the posttest ($t_{33} = 8.380$, $p < 0.001$). This indicated that there was an increase in the performance of students in the posttest compared to the pretest at the average increase of 4.85 marks in TLA teaching method.

Similarly in PBL teaching methods, the students demonstrated significant learning gains throughout the semester from a mean score of 11.10 marks on the pretest to a mean score of 19.35 marks on the posttest ($t_{39} = 10.390$, $p < 0.001$). This indicated that there was an increase in the performance of students in the posttest compared to the pretest at the average increase of 8.25 marks. In comparison, the findings revealed that the students in PBL teaching method performed better than the students in TLA teaching method with a mean difference score of 3.4 marks.

(iv) *Chi-square test of Independence Between Variables*

A Chi-Square test of independence is used to examine the relationship between two categorical variables. An analysis of differences between two teaching methods on the students' final grades revealed that the percentage of students in PBL teaching method who obtained a Distinction (24.3 %) was higher than in TLA (8.1 %). The percentage of students who obtained a Merit in PBL teaching method (16.2 %) was also higher than in TLA (9.5 %). While, the percentage of students in PBL teaching method who obtained a Pass grade (13.5 %) was lower than in TLA (28.4 %). The findings revealed that there was a significant impact or association between the teaching methods (TLA and PBL) on students' academic performance (Chi-Square value of 10.804 and p-value of 0.005).

The findings also revealed that most of the learning attitude items have no significant relationship with the students' academic performance except only for the items "understand the difficult material by hearing to classmate discuss it" and "get support and encouragement from classmate to learn." These two items were significantly associated with the students' academic performance with a Chi-Square value of 17.067 with a p-value of 0.009, and a Chi-Square value of 13.461 with a p-value of 0.036, respectively. In other words, the students' attitudes toward learning do not impact their academic performance except for the two items. These findings showed that the student's attitude in learning from others and peer support are effective in helping them to learn and perform better in their academic performance.

28.7 Conclusion

Based on the findings discussed previously, it can be concluded that the PBL teaching method employed significantly impacted the students' academic performance. In comparison with the TLA teaching method, the students in PBL had better academic performance after being exposed and guided in classrooms. This evidence was in line with the studies conducted by Sangestani and Khatiban [17] and Sungur et al. [18] where the students instructed with PBL teaching method significantly outperformed than those instructed with traditionally designed instruction. A group work activity which was employed in PBL teaching method such as "peer support" and "learn from others" also proved a significant impact or relationship with the students' academic performance. Similar to the study conducted by Looi and Seyal [16], these two activities were ranked as the most important activities in the PBL learning environment. At the KPMSI, though the TLA teaching method also focus on group work activity, no significant differences were found in collaborative learning, learn from others, peer supports, and social skill components of learning attitudes. Nevertheless, this study also revealed that different teaching methods employed will not influence the students' learning attitudes. Likewise, a study by McParland et al. [19] reported that no significant differences were found in learning styles or attitudes between students in PBL and

TLA teaching methods. Specifically in this study, both groups were taught by the same instructor to ensure that equivalent motivation was given to the students in learning and solving problems and exercises.

In order to improve the representativeness of future study, it is suggested that the researcher involves the computing students from all KPMs in Malaysia. Therefore, the results generated from the study can be generalized to represent a larger group of population. Further study is recommended to use a variety of pre- and posttest instruments so as to evaluate students' knowledge acquisitions instead of multiple choice questions.

References

1. Zhang, Y., Liang, R., Ma, H.: Teaching Innovation in Computer Network Course for Undergraduate Students with Packet Tracer (2012)
2. Baharom, S., Palaniandy, B.: Problem-Based Learning: A Process for the Acquisition of Learning and Generic Skills. Aalborg University Press, Denmark (2013)
3. Ismail, N.: Defining Vocational Education and Training for Tertiary Level Education: Where does Problem Based Learning Fit in? Aalborg University Press, Denmark (2013)
4. Song, H.G., Gao, X., Qiu, Z.: The computer network curriculum teaching innovation and practice. In: 2nd International Workshop on Education Technology and Computer Science, ETCS 2010, vol 2, pp. 739–741 (2010)
5. Oliveira, A.M.C.A., Dos Santos, S.C., Garcia, V.C.: PBL in teaching computing: an overview of the last 15 years. In: Proceedings—Frontiers in Education Conference, FIE, pp. 267–272 (2013)
6. Ajzen, I.: Attitude structure and behavior. In: Attitude Structure and Function (1989)
7. Mamat, N.J.Z., Mazelan, F.F.: Learning encouragement factors and academic performance. Procedia—Soc. Behav. Sci. **18**, 307–315 (2011)
8. Chang, L.C., Lee, G.C.: A team-teaching model for practicing project-based learning in high school: collaboration between computer and subject teachers. Comput. Educ. **55**, 961–969 (2010)
9. Neo, M., Neo, K.T.K.: Innovative teaching: using multimedia in a problem-based learning environment. Educ. Technol. Soc. (2001)
10. Bandura, A.: Self-efficacy: toward a unifying theory of behavioral change. Psychol. Rev. **84**, 191–215 (1977)
11. Kazemi, F., Shahmohammadi, A., Sharei, M.: The survey on relationship between the attitude and academic achievement of in-service mathematics teachers in introductory probability and statistics. World Appl. Sci. J. **22**(7), 886–891 (2013)
12. Hung, W.: Theory to reality: a few issues in implementing problem-based learning. Educ. Tech. Res. Dev. **59**, 529–552 (2011)
13. Wu, A.C., Chen, J.: Sampling and Experimental Design Fall 2006 Lecture Notes (2006)
14. MS Access Final Exam, http://www.proprofs.com/quiz-school/story.php?title=ms-access-final-exam
15. Lammle, T.: CCNA Cisco Certified Network Associate Study Guide, 4th edn. Sybex Inc., San Francisco (2004)
16. Looi, H.C., Seyal, A.H.: Problem-based Learning: An Analysis of its Application to the Teaching of Programming, pp. 68–76 (2014)
17. Sangestani, G., Khatiban, M.: Comparison of problem-based learning and lecture-based learning in midwifery. Nurse Educ. Today **33**, 791–795 (2013)

18. Sungur, S., Tekkaya, C., Geban, Ö.: Improving achievement through problem-based learning. In: J. Biol. Educ. (2006)
19. McParland, M., Noble, L.M., Livingston, G.: The effectiveness of problem-based learning compared to traditional teaching in undergraduate psychiatry. Med. Educ. **38**(8), 859–867 (2004)

Chapter 29
Network Analysis in Exploring Student Interaction and Relationship in Mathematics Learning Environment

Liew Kee Kor

Abstract Characterizing patterns of interaction in any learning environment is complex. This paper reports a study that examined student interaction patterns in learning mathematics. The study employed Social Network Analysis (SNA) to capture the relational nature of interaction and to identify the patterns and trends of the network dynamics on a group of 17 mathematics students. These students had enrolled in a mathematics course offered in the second semester of 2015. The respondents were each given a name list in which they indicated to whom they had helped and from whom have they sought help throughout the mathematics course. Data were analyzed and interpreted based on the degree of centralities and correlations between degree centralities and achievement. Results show that higher and lower achieving students has the highest out-degree and in-degree, respectively. Also respondents who have lesser number of close friends created a wider network in the course of the study. Future studies may enlarge the population of the study to include teachers and mathematics concepts. The use of SNA software is also highly recommended for more comprehensive data analysis.

Keywords Social analysis network · SNA · Degree centrality · Interaction pattern · Mathematics

29.1 Introduction

The transition from secondary to the university is expected to bring noteworthy changes in the teaching approach and hence, the learning behavior of the students. While school mostly practices teacher-guided learning, university curriculum generally emphasizes more on independence and self-sufficiency learning [1]. In the university environment, students are expected to engage in more independent study

L.K. Kor (✉)
Faculty of Computer and Mathematical Sciences,
Universiti Teknologi MARA, Kedah, Malaysia
e-mail: korlk564@kedah.uitm.edu.my

A.-R. Ahmad et al. (eds.), *Proceedings of the International Conference on Computing, Mathematics and Statistics (iCMS 2015)*,
DOI 10.1007/978-981-10-2772-7_29

time and manage their assignments through a repertoire of learning strategies. In the absence of a supportive group, along with the potential shift from guided to independent learning, some students are finding it difficult to cope with their subject especially Mathematics in the university.

Investigating the mechanism of student interaction or learning patterns in any learning environment is complex. The uses of conventional classroom-based measures of learning such as students' achievement grades or their perception of learning are not able to capture the in-depth relational nature of interaction. Social Network Analysis (SNA) provides a way to visualize the study of social relations among people within community.

This study aims to explore students interacting behaviors exhibited in the mathematics learning environment from the social network analysis perspective. The main objective is to gain insight into the flow of networking elements in the process of independent learning.

The research questions are

(1) Is there any association between achievement and the measures of centrality?
(2) How does the community of practice (gender and the number of close friends) affect the measures of centrality?

29.2 Literature Review

The learning milieu of a classroom is made up of many interacting and complex factors. How a person learns depends largely on the availability of the knowledge and skills, as well as the situation in which the learners practice [2]. This observation is supported by Wenger's [3] social theory of learning which highlighted the importance of communities of practice. In the mathematics learning environment, the community of practice includes all the students in the class, the teachers, the curriculum, text books, and learners with various beliefs about mathematics and mathematics teaching and learning [4].

Social scientists have increasingly recognized the potential of SNA drawing on the basis of graph theory. SNA enriches the explanation of human behavior by giving attentions to the identified "important" actors [5]. In particular the science of groups, SNA is a valuable tool for investigating some of the central mechanisms that underlie intra- and intergroup behavior [6].

Wasserman and Faust [5] stated that SNA envisages an actor's prominence through the show of its greater visibility to the other network actors (an audience). Also, an actor's network location is justified by the direct socio-metric choices made and choices received (out-degrees and in-degrees), as well as the indirect ties with other actors visualized in the network diagram.

29.2.1 Degree Centrality

The degree centrality measures actor out-degree. Accordingly, student with high degree centrality maintains numerous contacts with other students. Wasserman and Faust [5] stated that Group Degree Centrality (GDC) quantifies the dispersion or variation among individual centralities. The value ranges from 0 to 1, reaching the maximum when all others choose only one central actor and the minimum when all actors have identical centralities. The associated formulas are:

$$\text{In-degree centrality (IDC)} \quad \text{IDC}(i) = \sum_{j=1}^{n} x_{ij}(i \neq j) \tag{29.1}$$

$$\text{Out-degree centrality (ODC)} \quad \text{ODC}(i) = \sum_{j=1}^{n} y_{ij}(i \neq j) \tag{29.2}$$

$$\text{Group degree centrality (GDC)} \quad \text{GDC} = \frac{\sum_{i=1}^{n} [\text{DC}(N^*) - \text{DC}(i)]}{(n-1)(n-2)} \tag{29.3}$$

$$\text{Group in-degree centrality (GIDC)} \quad \text{GIDC} = \frac{\sum_{i=1}^{n} [\text{INC}(N^*) - \text{IDC}(i)]}{(n-1)} \tag{29.4}$$

$$\text{Group out-degree centrality (GODC)} \quad \text{GODC} = \frac{\sum_{i=1}^{n} [\text{ODC}(N^*) - \text{ODC}(i)]}{(n-1)^2} \tag{29.5}$$

GDC measures the extent to which the nodes in a network differ from one another in their individual degree centrality in undirected network. The larger the group degree centrality, the more uneven is the degree centrality of the nodes in a network. For a directed network a higher GODC indicates more uneven influence among the nodes in a network, while a higher GIDC indicates greater inequality among the nodes' popularity, as cited in [7]. A prestigious or popular actor receives many ties from others. The measure actor-level degree prestige is the in-degree.

29.3 Methodology

A total of 17 mathematics students (nine boys and eight girls) from the Computer Science Program took part in this study. To gain insight into the network in the learning milieu, the respondents were required to answer two questions: In the course of learning mathematics; (1) Whom did you seek help from when you faced a problem?; (2) Who came to you asking for help when there is a problem? Each respondent was given a class list of names. He/she ticked the relevant names in response to the two questions. This process was repeated three times throughout the semester (early, mid, and end of semester). The achievement scores were taken from the mean of three class tests scheduled in the mathematics course and students' CGPA. Respondents were also required to provide the number of close friend they have in their class.

29.4 Results and Discussions

The social network analysis is a quantitative method that assessed the activities and relationships of network members. The collected data for the social network analysis was analyzed using the In-Degree Centrality (IDC) and Out-Degree Centrality (ODC) for directed network and total Degree Centrality (DC) for undirected network. Group In-Degree Centrality (GIDC), Group Out-Degree Centrality (GODC), and Group Degree Centrality (GDC) were also calculated. Spearman's rho correlation was engaged to check the relationships. Results of the study are presented below.

The respondents were made up of nine males and eight females. Table 29.1 displays the detailed characteristics of each respondent.

From the above table, it is observed that respondent R14 has the greatest out-degree, and might be regarded as the most influential and the one whom most classmates looked up to for help. Respondent R1 has the greatest in-degree and therefore has sought help from a number of classmates in the mathematics learning. This observation is supported by the achievement scores of these two respondents.

Table 29.1 Characteristics of the respondents

Respondents	Gender	CGPA	Test score	Friends	IDC	ODC
R1	F	2.30	60	3	22	3
R2	M	3.35	76	3	13	10
R3	M	2.96	64	4	10	6
R4	M	2.85	68	1	15	12
R5	M	2.90	60	8	5	5
R6	M	2.10	55	4	14	6
R7	M	2.76	68	5	21	7
R8	M	3.02	82	1	14	16
R9	M	2.29	44	4	9	2
R10	F	2.44	58	4	10	1
R11	F	3.81	95	8	10	1
R12	M	3.77	71	4	7	6
R13	F	3.53	79	6	12	10
R14	F	3.60	80	4	13	20
R15	F	2.60	67	4	6	2
R16	F	3.33	69	7	10	8
R17	F	3.26	83	4	9	3

R14 is a higher achiever in mathematics and the overall academic performance (CGPA) than R1.

In analyzing the effect of respondents' number of close friends, an adjacent matrix was built with value 1 indicating a relationship while 0 denoting no relationship. Figure 29.1 below shows that respondents R14 and R17 are the two most popular personalities in the class. The matrix also reveals that respondent R5 is the friendliest person in the class.

The mean scores for the degree of centralities and achievement were analyzed and displayed in Table 29.2. Comparisons between the mean scores were shown in Fig. 29.2.

R	1	2	3	4	5	6	7	8	9	10	11	12	13	14	15	16	17	F
1	0	0	0	0	0	1	0	0	0	1	0	1	0	0	0	0	1	4
2	0	0	0	0	1	1	0	0	1	0	0	0	0	0	0	0	0	3
3	0	1	0	1	0	0	0	1	1	0	0	0	0	0	0	0	0	4
4	0	0	0	0	0	0	0	0	1	0	0	0	0	0	0	0	0	1
5	0	1	1	1	0	1	1	1	1	0	0	1	0	0	0	0	0	8
6	1	0	0	1	1	0	1	0	0	0	0	0	0	0	0	0	0	4
7	1	0	0	0	1	1	0	0	0	1	0	1	0	0	0	0	0	5
8	0	0	1	0	0	0	0	0	0	0	0	0	0	0	0	0	0	1
9	0	1	1	1	0	0	0	1	0	0	0	0	0	0	0	0	0	4
10	1	0	0	0	0	0	0	0	0	0	1	0	1	1	1	1	1	7
11	0	0	0	0	0	0	0	0	0	0	0	0	1	1	0	1	1	4
12	1	0	0	0	1	1	1	0	0	1	0	0	0	0	0	0	0	5
13	0	0	0	0	0	0	0	0	0	0	1	0	0	1	0	1	1	4
14	0	0	0	0	0	0	0	0	0	0	0	1	0	1	0	1	1	4
15	1	0	0	0	0	0	0	0	0	1	1	0	1	1	0	1	1	7
16	0	0	0	0	0	0	0	0	0	0	1	0	1	1	0	0	1	4
17	1	0	0	0	0	0	0	0	0	1	1	0	1	1	1	1	0	7
P	6	3	3	4	4	5	3	3	4	5	5	4	5	7	2	6	7	

R: respondent; P: popular; F: friendly

Fig. 29.1 Adjacent matrix of the close friend network analysis

Table 29.2 Descriptive statistics for the test scores and measures of centralities

Gender	Frequency	Mean				
		ODC	IDC	DC	Math scores	CGPA
Male	9	12.00	7.78	19.78	19.578	2.889
SD		4.822	4.206	7.645	3.369	0.501
Female	8	11.38	6.63	18.00	22.188	3.109
SD		4.779	6.232	8.315	3.761	0.578
Total	17	11.71	7.24	18.94	20.801	2.992
SD		4.661	5.118	7.766	3.697	0.534

SD standard deviation

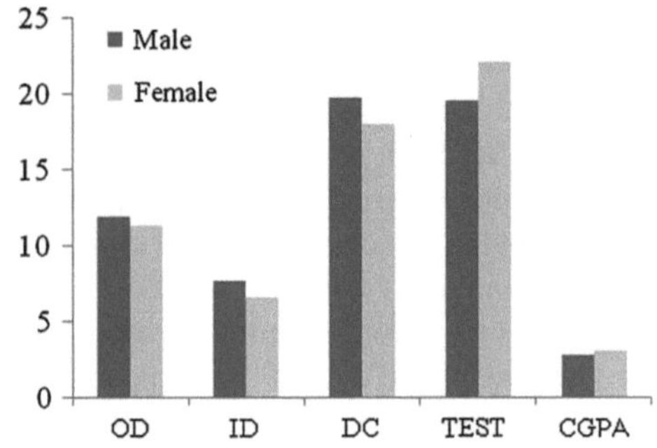

Fig. 29.2 Measures of mean centralities and achievement scores for male and female respondents

Figure 29.2 shows that male respondents exhibited higher mean in the centralities scores than the female respondents indicating that they were more interactive among the community of practice. Comparing the mean DC among the male respondents shows that the male respondents obtained a higher mean of ODC than IDC ($n = 9$; $\text{mean}_{\text{ODC}} = 12$; $\text{mean}_{\text{IDC}} = 7.80$). This means that male respondents tend to seek help from others in the mathematics course than giving help to others. Similar trend is observed among the female respondents ($n = 8$; $\text{mean}_{\text{ODC}} = 11.38$; $\text{mean}_{\text{IDC}} = 6.63$). Another observation is the female respondents scored higher in the achievement than male respondents.

In the correlation analysis shown in Table 29.3, it is observed that CGPA is highly and significantly correlated to the mathematics test scores ($r = 0.857$, $p < 0.01$). Both CGPA and mathematics test scores are only moderately related to IDC ($r_{\text{CGPA}} = 0.532$, $p < 0.05$; $r_{\text{test scores}} = 0.565$, $p < 0.05$). This can be interpreted that respondents who are high achievers in mathematics and CGPA have more classmates seeking help from them in the mathematics course. Thus, this group is popular among the class.

Interestingly, it is also observed that the number of close friends showed a moderate to strong negative relationship with ODC, IDC, and total DC ($r_{\text{ODC}} = -0.610$; $r_{\text{IDC}} = -0.701$; $r_{\text{DC}} = -0.704$, $p < 0.01$). This means that as the degree of centralities increases, the number of close friend decreases. In other words, the respondents in class who have less close friends had made more efforts to interact with the others when solving mathematics problems. This group with

Table 29.3 Correlation coefficients (Spearman's rho) between degree centralities and achievement scores ($n = 17$)

	CGPA	Test scores	Friends
CGPA	1		
Test scores	0.857**	1	
Friends	–	–	1
ODC	−0.243	−0.032	−0.610**
IDC	0.532*	0.565*	−0.701**
Total DC	0.189	0.345	−0.704**

*Correlation is significant at the 0.05 level (2-tailed)
**Correlation is significant at the 0.01 level (2-tailed)

Table 29.4 Correlation coefficients (Spearman's rho) between degree centralities and achievement scores for male respondents (n = 9)

	CGPA	Test scores	Friends
CGPA	1		
Test scores	0.778*	1	
Friends	–	–	1
ODC	−0.351	0.269	−0.560
IDC	0.339	0.817**	−0.719*
Total DC	0.050	0.644	−0.664

*Correlation is significant at the 0.05 level (2-tailed)
**Correlation is significant at the 0.01 level (2-tailed)

Table 29.5 Correlation coefficients (Spearman's rho) between degree centralities and achievement scores for female respondents (n = 8)

	CGPA	Test scores	Friends
CGPA	1		
Test scores	0.833*	1	
Friends	–	–	
ODC	−0.108	−0.265	−0.536
IDC	0.755*	0.563	−0.823*
Total DC	0.310	0.190	−0.825*

*Correlation is significant at the 0.05 level (2-tailed)
**Correlation is significant at the 0.01 level (2-tailed)

less close friends has built a wider network in their interaction in mathematics learning.

Tables 29.4 and 29.5 show that further exploration into the correlation analysis based on genders revealed similar result for male and female respondents. For both groups the number of close friends are highly and negatively ($r_{\text{male-IDC}} = -0.719$; $r_{\text{female-IDC}} = -0.823$, $p < 0.05$) related to IDC indicating that respondents with lesser number of close friends offered more help to the other classmates. In addition, the higher the ODC, the higher the test scores for the male respondents ($r = 0.817$, $p < 0.01$) whereas the higher the CGPA, the higher is the IDC for female respondents ($r = 0.755$, $p < 0.01$).

Lastly, GODC, GIDC, and GDC were similarly measured and correlated to GPA and mathematics test scores. However, the results of the correlation was not significant at $p < 0.05$ level.

29.5 Conclusion and Recommendation

Characterizing patterns of interaction in any learning environment is complex. Literature search has shown that most researchers rely on learner's perception and achievement scores to interpret the degree of student interaction. The setback is that such results may not explain student interaction patterns thoroughly. This study employed SNA to capture the relational nature of interaction. The degree of centralities provided information about the connectedness of each individual in the course of their mathematics learning. Result shows that higher and lower achieving students score the highest out-degree and in-degree measures, respectively, indicating the academically more able students tend to give help to other classmates and the academically weaker ones tend to get help from the others. Male and female student alike shows that the better the academic result the higher is the degree of centrality in the network. Also students who have lesser number of close friends created a wider network in the course of the study. Results obtained in this paper are confined to a case study. Future studies may enlarge the population of the study and to also include teachers and mathematics concepts. The use of SNA software is also highly recommended for a more comprehensive data analysis.

Acknowledgments The study reported in this paper was made possible by the generous support from Exploratory Research Grant Schemes (ERGS) from the Malaysian Ministry of Education.

References

1. Zimmerman, B.J.: Achieving self-regulation: the trial and triumph of adolescence. In: Pajares, F., Urdan, T. (eds.) Academic Motivation of Adolescents, vol. 2, pp. 1–27. Information Age, Greenwich, CT (2002)
2. Putman, R.T., Borko, H.: What do new views of knowledge and thinking have to say about research on teacher learning? Educ. Res. **29**(1), 4–15 (2000)
3. Wenger, E.: Communities of Practice: Learning, Meaning, and Identity. Cambridge University Press, Cambridge (1998)
4. Grootenboer, P., Zevenbergen, R.: Identity as a lens to understand learning mathematics: developing a model. In: Goos, M., Brown, R., Makar, K. (eds.) Proceedings of the 31st Annual Conference of the Mathematics Education Research Group of Australasia (2008)
5. Wasserman, S., Faust, K.: Social Network Analysis: Methods and Applications. Cambridge University Press, Cambridge/New York (1994)
6. Wölfer, R., Faber, N.S., Hewstone, M.: Social network analysis in the science of groups: cross-sectional and longitudinal applications for studying intra- and intergroup behavior. Group Dyn.: Theor. Res. Pract. **19**(1), 45–61 (2015)
7. Jin, H., Wong, K.Y.: A network analysis of concept maps of triangle concepts. In: Sparrow, L., Kissane, B., Hurst, C. (eds.) Shaping the future of mathematics education: proceedings of the 33rd annual conference of the Mathematics Education Research Group of Australasia (2011)

Chapter 30
Perceived Information Quality of Digital Library Among Malaysian Postgraduate Students

Abd Latif Abdul Rahman, Adnan Jamaludin, Zamalia Mahmud and Asmadi Mohammed Ghazali

Abstract Currently, users are already relying on the availability of digital libraries for their information needs to support their daily tasks. One of the most important aspects of the digital library in the university environment is its information quality services to the university community, particularly by students. Therefore, it is crucial to investigate and understand how users perceived information quality in digital library environment. With this knowledge, better library and information services can be offered to the users at large. Samples comprised 534 respondents from four Malaysian intensive research universities and were obtained through questionnaire. This study investigated how user perceived information quality in digital library environment. For analysis of differences, Analysis of Variance (ANOVA) was employed to determine differences between demographics factors and Information Quality variable. The results show that 81.7 % of the respondents said they consider the digital information resources they retrieved from the digital library to be reliable enough to be used for their work. On average, about 81 % agreed that the digital information resources they retrieved are related and correct enough to be used for their work. About 70 % agreed that digital information resources retrieved are accurate, interpretable, comprehensive, dependable, complete, and sufficient for their needs for their study and/or research. The result of Analysis of Variance (ANOVA) shows there is no evidence of a significant difference in mean scores for information quality between age groups ($p = 0.089$) universities ($p = 0.428$) and academic programs, with p-values greater than 0.05.

A.L.A. Rahman (✉) · A.M. Ghazali
Universiti Teknologi MARA Kedah, Kedah, Malaysia
e-mail: ablative@kedah.uitm.edu.my

A.M. Ghazali
e-mail: Asmadi165@kedah.uitm.edu.my

A. Jamaludin · Z. Mahmud
Universiti Teknologi MARA, Shah Alam, Malaysia
e-mail: adnanj@salam.uitm.edu.my

Z. Mahmud
e-mail: Zamal669@salam.uitm.edu.my

A.-R. Ahmad et al. (eds.), *Proceedings of the International Conference on Computing, Mathematics and Statistics (iCMS 2015)*,
DOI 10.1007/978-981-10-2772-7_30

On the other hand, there is evidence of significant differences in mean score for information quality for the race category.

Keywords Information quality · Digital library · Malaysia · Postgraduate students

30.1 Background of the Study

The development of digital libraries in Malaysia has mostly occurred in university libraries. This may be because of the intensity of library use among the university community and also the availability of budgets allocated for the enhancement of library services. Most of university digital libraries consist of a web page or portal that offers all the services that are related to digital types of information resources. The web page is given an address to enable library users to access the services.

For example, the Universiti Kebangsaan Malaysia (UKM) Library provides its digital library services through its library website at the URL address http://www.ukm.my/library/. This website links all information and electronic services such as OPAC searching named GEMILANG, subscribed e-databases, subscribed e-journals, e-books, UKM's internal electronic collections, and inter-library loans. As of 2008, the UKM digital library had a digital collection of 393,366 e-books and e-theses and 30,600 e-journals, and had budgeted about RM 4.15 million on the digital collections. The use of its digital library nearly doubled from 94,000 users in 2006 to 110,000 in 2007 and to 167,040 in 2008.

The digital library services at the Universiti Malaya (UM) Library are offered through its Universiti Malaya Library website (http://www.umlib.um.edu.my) which has links to its Pendeta portal (http://www.pendeta.um.edu.my), and to its Interaktif portal (http://www.diglib.um.edu.my/interaktif). The Interaktif portal offers services such as the online databases, e-journals, e-books, document delivery, inter-library loans, and online registration, while the Pendeta portal offers services such as online catalogue and online renewal. The digital library has more than 40,000 subject-based Web links to resources of more than 100 databases, 12,000 e-journals, 25,000 e-books, and more than 3000 selected web resources.

Using the term 'virtual library', the Universiti Sains Malaysia (USM) Library developed its digital library through its website at URL: http://www.lib.usm.my. This website offers services such as Web OPAC KRISALIS, subscribed databases, subscribed journals, in-house databases, Malaysian Thesis Online, eQUIP, Malaysian Links, References Resources, Virtual Subject Libraries, and Subject Browser. Among the in-house databases are a collection of articles by the Vice Chancellor, past examination question papers, newspaper cuttings, theses, MIDAS Bulletin, Malaysiana, APEX collection, and a special collection of National Laureate Professor Muhammad Hj. Salleh. The USM digital library has spent more than RM13 million on digital collections and as of 2009, it had a collection of

1,343,061 e-books, 511,515 e-journals, and 112 databases. The digital library had recorded 805,034 users and 34,735,196 hits at that time.

Universiti Putra Malaysia (UPM) Library offers its digital library services through the library website at URL: http://www.lib.upm.edu.my which consists of links to OPAC search, subscribed e-databases, subscribed e-journals, and e-books. Nevertheless, all these services are offered within the boundary of the university Intranet, while outside users can access the services through the EZproxy portal offered by the library.

Based on the development of the Malaysian digital library development outlined above, it can be concluded that the digital library of Malaysian universities consists of a web page or a portal developed by each of the university libraries to offer services pertaining to the online digital collection to their library users.

30.1.1 Research on Information Quality

Previous research has shown that Information Quality has a direct positive influence on Information Systems Success [3]. This means that in a situation when the level of information quality is high, it will contribute to high level of Continued Use of the Digital Library. The digital library users expect a high quality of information from a digital library to motivate them to use it in their academic work. High quality is very useful to the digital library users.

A number of studies in information systems (IS) and library science (LS) have devoted attention to the notion of information quality. By understanding the concepts of information quality, digital librarians would facilitate digital library designers and managers to develop more influential digital library systems.

Also, previous researchers on information quality [1–3] have agreed Information Quality can be described by several sub dimensions namely information accuracy, relevancy of information, comprehensiveness of the digital information resources in the digital library, authoritativeness, and comprehensibility of the digital information resources.

30.2 Methodology

30.2.1 Population and Sampling Technique

The population for this research consists of postgraduate students in Master's and Doctoral programmes in four research-intensive universities in Malaysia. The four research universities are

- Universiti Malaya
- Universiti Kebangsaan Malaysia

- Universiti Putra Malaysia
- Universiti Sains Malaysia.

30.2.2 Variables and Measures

30.2.2.1 Information Quality (IQ)

Information quality refers to the usability of the digital information resources retrieved by the users from the digital library. Usability can be examined from the following dimensions and elements [1–3]:

- Accuracy of the digital information resources in the digital library, which relates to correctness, freedom from errors, and accuracy
- Relevance of the digital information resources in the digital library concerns the relatedness of the digital information resources in the digital library to the user's needs
- Comprehensiveness of the digital information resources in the digital library relates to completeness, comprehensiveness, and sufficiency of the resources
- Authoritativeness of the digital information resources in the digital library relates to dependability and reliability
- Comprehensibility of the digital information resources relates to the ability to be understood.

Based on the dimensions and elements above, a total of ten items have been developed for IQ.

30.3 Data Analysis

For the purpose of this research, two stages of analysis were employed, i.e. descriptive analysis techniques in the first stage, and inferential analysis techniques in the second stage.

30.3.1 Characteristics of Respondents

This section describes the respondents' characteristics such as gender, age, race, university, mode of study, academic program, experience in using the digital library and usage of the digital library per week.

Table 30.1 Profile of respondents

Respondents' characteristics	Categories	Number of respondents	Percentage
Gender	Male	288	53.9
	Female	246	46.1
Age (years)	22–30	382	71.5
	31–40	125	23.4
	41–51	27	5.1
Race	Malay	249	46.6
	Indian	36	6.7
	Chinese	80	15.0
	Others	169	31.6
Mode of study	Full time	484	90.6
	Part time	50	9.4

Table 30.1 shows a slightly higher representation of male respondents (53.9 %) compared to female respondents (46.1 %). A majority of the respondents (71.5 %) were between 22–30 years old, while only 5 % were older than 41 year. Based on ethnic composition, 47 % of the respondents were Malay, 15 % Chinese, 7 % Indian, and 32 % comprised other races. About 90.6 % of the postgraduate students were studying on a full-time basis, while 9.4 % were part-time students.

Samples of the 534 respondents who participated in the study, 26.8 % were from UM, followed by 25.8 % (138) from UPM, 25.3 % (135) from UKM and 22.1 % (118) from USM.

30.3.2 Descriptive Analysis of Information Quality

From Table 30.2 it can be seen that 81.7 % of the respondents said they consider the digital information resources they retrieved from the digital library to be reliable enough to be used for their work. On average, about 81 % agreed that the digital information resources they retrieved are related and correct enough to be used for their work. About 70 % agreed that digital information resources retrieved are accurate, interpretable, comprehensive, dependable, complete, and sufficient for their needs for their study and/or research.

30.3.3 Test of Group Differences for Information Quality—Analysis of Variance (ANOVA)

Table 30.3 shows that there is no evidence of a significant difference in mean scores for information quality between age groups ($p = 0.089$) universities ($p = 0.428$) and

Table 30.2 Distribution of information quality

Item No.	Items	Disagree (%)	Neutral (%)	Agree (%)
1	The digital information resources that I retrieved from the digital library are considered ***reliable*** to be used for my work	5.2	13.1	81.7
2	The digital information resources that I retrieved from the digital library are considered ***related*** to be used for my work	5.1	13.7	81.2
3	The digital information resources that I retrieved from the digital library are considered ***correct*** to be used for my work	6.4	12.9	80.7
4	The digital information resources that I retrieved from the digital library are considered ***accurate*** to be used for my work	7.3	13.7	79.0
5	The digital information resources that I retrieved from the digital library are considered ***interpretable*** to be used for my work	5.4	17.4	77.2
6	The digital information resources that I retrieved from the digital library are considered ***comprehensive*** to be used for my work	6.6	16.7	76.7
7	The digital information resources that I retrieved from the digital library are considered ***sufficient for my need*** to be used for my work	8.8	15.4	75.8
8	The digital information resources that I retrieved from the digital library are considered ***dependable*** to be used for my work	6.7	19.7	73.6
9	The digital information resources that I retrieved from the digital library are considered ***complete*** to be used for my work	8.1	19.7	72.2
10	The digital information resources that I retrieved from the digital library are considered as ***free of errors*** to be used for my work	11.6	24.9	63.5

academic programs, with p-values greater than 0.05. On the other hand, there is evidence of significant differences in mean score for information quality for the race category, with a p-value equal 0.000 (<0.005). Since at least one mean is different, a Post Hoc Multiple Comparison test was conducted to determine the pairwise differences between the group means. The post hoc results in Table 30.4 show that there are significant differences in mean score information quality between Malay and Chinese groups (p = 0.000) and between Chinese and Other races groups (p = 0.000).

Table 30.5 shows the summary of findings for differences among dependent and independent variables. It was found that at least one mean is significantly different for information quality between race categories.

Table 30.3 Analysis of variance (ANOVA) test based on information quality, according to categories of user characteristics

Test variable	Group		Mean information quality	(*p*-value)	F-statistics	*p*-value
Information quality	Age	Less than 31	4.6518	0.765	2.434	0.089
		31–40	4.8264			
		More than 40	4.8037			
	Race	Malay	4.7229	0.084	9.043	0.000*
		Indian	4.5972			
		Chinese	4.3188			
		Others	4.8698			
	University	USM	4.8008	0.231	0.925	0.428
		UM	4.6713			
		UKM	4.7052			
		UPM	4.6399			
	Academic program	PhD-By research	4.7781	0.905	2.336	0.051
		PhD-by coursework	4.8286			
		PhD-by mixed mode	5.1000			
		Master by research	4.6280			
		Master by coursework	4.5755			
		Master by mixed mode	4.8174			

*Significant at $\alpha = 0.05$

Table 30.4 Post hoc multiple comparisons of race on information quality

(I) Race	(J) Race	Mean difference (I–J)	Std. error	Sig.	95 % confidence interval	
					Lower bound	Upper bound
Malay	Indian	0.12567	0.14120	0.810	−0.2382	0.4896
	Chinese	0.40414*	0.10177	0.000	0.1419	0.6664
	Others	−0.14693	0.07893	0.246	−0.3503	0.0565
Indian	Malay	−0.12567	0.14120	0.810	−0.4896	0.2382
	Chinese	0.27847	0.15893	0.298	−0.1311	0.6881
	Others	−0.27260	0.14536	0.240	−0.6472	0.1020
Chinese	Malay	−0.40414*	0.10177	0.000	−0.6664	−0.1419
	Indian	−0.27847	0.15893	0.298	−0.6881	0.1311
	Others	−0.55107*	0.10747	0.000	−0.8280	−0.2741
Others	Malay	0.14693	0.07893	0.246	−0.0565	0.3503
	Indian	0.27260	0.14536	0.240	−0.1020	0.6472
	Chinese	0.55107*	0.10747	0.000	0.2741	0.8280

*The mean difference is significant at the 0.05 level

Table 30.5 Summary of results differences in information quality between selected demographic elements

Dependent variables	Independent variables	*p*-value	Conclusion
Information quality		0.000	There are evidence of significant differences in mean score information quality between Malay and Chinese
	Race	0.000	There are evidence of significant differences in mean score information quality between Chinese and Other races

30.4 Discussions and Conclusion

The analyses of data were conducted to thoroughly investigate the Information Quality variable among postgraduate students in Malaysian research universities. The result of descriptive analysis shows majority of respondents considers digital information resources they retrieved from the digital library to be reliable enough to be used for their work. These findings support the argument that students make the greatest use of the digital library in their study and research work.

The result also shows about 70 % agreed that digital information resources retrieved are accurate, interpretable, comprehensive, dependable, complete and sufficient for their needs for their study and/or research. These findings support the argument that students perceived digital library provide a very high quality information quality in term of accuracy, interpretable, comprehensive, dependable, complete, and sufficient to be used for study and research work.

The result of Analysis of Variance (ANOVA) shows there is no evidence of a significant difference in mean scores for information quality between age groups ($p = 0.089$) universities ($p = 0.428$) and academic programs, with p-values greater than 0.05. On the other hand, there is evidence of significant differences in mean score for information quality for the race category. These arguments indicated that local and international students perceived information quality of digital library differently. Therefore, more research should be conducted to understand these differences.

Several limitations are worth noting in the present study, as well as the areas that require further intention in the future research. The first limitation relates to the small sample size, which was derived from Peninsular Malaysian universities only. Specifically, this study was conducted in only four universities—Universiti Malaya, Universiti Kebangsaan Malaysia, Universiti Putra Malaysia and Universiti Sains Malaysia—and only for postgraduate students, and thus, the results obtained may not be applicable and able to be generalized to other samples across different regions or levels of study. Accordingly, the external validity of the findings may be somewhat limited. The use of more diverse samples from different regions of Malaysia (such as North Peninsular Malaysia, South Peninsular Malaysia and East

Malaysia) and other countries such Singapore, Thailand and Indonesia would improve the generalizability of the findings.

In summary, it is hoped that the study findings may offer some insights into the use of Malaysian digital libraries so that librarians and other professionals can better understand the information quality factors that affect students' continued use of these libraries. In this way, appropriate steps and strategies could be taken to improve services and increase students' use of the digital library.

References

1. Al-Hakim, L.: Information quality factors affecting innovation process. Int. J. Inf. Qual. **1**(2), 162–176 (2007)
2. Bailey, J.E., Pearson, S.W.: Development of a tool for measuring user satisfaction. Manage. Sci. **29**, 530–545 (1983)
3. Delone, H.W., Mclean, R.E.: The Delone and Mclean model of information systems success: a ten-year review. J. Manag. Inf. Syst. **19**(4), 9-3 (2003)

Chapter 31
Malaysia's REITs in the Era of Financial Turbulence

Noor Zahirah Mohd Sidek and Fauziah Hanim Tafri

Abstract Real estate investment trust (REIT) in Malaysia is a relatively new investment vehicle. Islamic REIT is defined as REIT that operates within the parameters of the *shari'ah* rules. In this study, we examine whether the Islamic REIT serves as a compliment or a substitute for stock market shares. We rely on the autoregressive distributed lag technique to estimate the impact of stock market proxied by the Kuala Lumpur composite index (KLCI) along with a host of control variables. Results suggest that the Islamic REIT complements the stock market and its movements almost mirror the stock market. Bonds, on the other hand, have no statistically significant impact on the Islamic REIT. The results provide an important insight to the relation between Islamic REIT and the stock market; and carries important policy implications for portfolio management. In this case, investors looking for *shari'ah* compliant investment may rely on the Islamic REIT to diversify their portfolios. However, they should also bear in mind that during financial turbulence, the Islamic REIT would not be able to cushion any downward effect of the stock market due to its complimentary effect. The total clustering errors reduced significantly whereas interdistances between clusters are preserved to be as large as possible for better clusters identification.

Keywords Real estate investment trusts · Complementary effect · Malaysia

N.Z.M. Sidek (✉)
Department of Economics, Faculty of Business Management,
Universiti Teknologi MARA, Kedah, Malaysia
e-mail: nzahirah@kedah.uitm.edu.my

F.H. Tafri
Faculty of Computer & Mathematical Science, Universiti Teknologi MARA,
Shah Alam, Malaysia
e-mail: fauziahanim@salam.uitm.edu.my

A.-R. Ahmad et al. (eds.), *Proceedings of the International Conference on Computing, Mathematics and Statistics (iCMS 2015)*,
DOI 10.1007/978-981-10-2772-7_31

31.1 Introduction

REIT is a relatively new investment vehicle in Malaysia's capital market with the first REIT listed at the end of 2005. Being less than 10 years in existence, the REIT market is relatively underdeveloped with 'thin trading' compared to the more established stock or the derivative markets. Nevertheless, REIT provides an alternative investment for investors looking for avenues to diversify their investment portfolio. With the advent of Islamic finance especially after the repeat financial crisis around the globe, Muslim investors can now look forward towards investment in *shari'ah* compliant products.

Shari'ah compliant products need to be free from certain code of conduct as explicitly prescribed in the Holy Al-Quran. Among others, any *shari'ah* compliant products or services must be free from excessive uncertainty (*gharar*), speculative behaviour which may lead to gambling (*maisir*) and most importantly, the prohibition of *riba* which literally entails any oppressive behaviour that enriches claimants through the payment of or receipt of interest between two counterparties. On the other hand, the Islamic jurisprudence (*fiqh mualamat*) promotes the attainment of the objectives of the Islamic law (*maqasid al-shariah*) namely to protect the life, family, property, sanity, religion, and encourages of distributive justice such as giving special attention to public goods (*maslahah*).

Islamic REIT in Malaysia is designed to be compatible with the Islamic principles and religious norms. The authors are aware of the major four schools—Shafie, Maliki, Hanbali and Hanafi, where applications of such religious norms may vary and differ. Variations and differences in applications are only in non-principle or *muamalah* activities. Such differences or variations may not arise in the basic rulings such as *solah* (five times daily prayers for Muslims), zakat (compulsory payment when certain criteria are fulfilled) and many more. The authors also acknowledge that certain contracts such as *bay' ad-dayn*, *bay' al-'inah* or *al-bai bithaman ajil* (BBA) which are extensively used in Islamic finance products in Malaysia are not readily accepted by the Middle East scholars.

Malaysia is one of the earliest countries to introduce Islamic REIT. On 21 November 2005, the Securities Commission issued the Guidelines for Islamic REIT to propel the development of Islamic REIT in line with the country's vision to become an Islamic financial hub. Among others, this guideline listed at least eight (8) non-permissible rental activities such as financial services based on *riba'*, gambling or gaming, manufacture or sale of non-*halal* products or related products, conventional insurance, entertainment activities that are non-permissible according to the *shariah* law, manufacture or sale of tobacco-based products and their peripherals, stock broking or share trading in non-*shariah* compliant securities, and hotels and resorts. In addition, an Islamic REIT is not permitted to own buildings where the activities conducted in the building are non-*shariah* compliant although the percentages of the non-permissible activities are low. In November 2012, Islamic REIT account for 171 Islamic-based funds or 28 % of the total approved funds with total net asset value of fund worth RM34 billion. This is approximately 11.6 % of the total net asset value worth RM259.981 billion.

Studies on REIT evolved from the relationship between REIT and macroeconomic factors to more industry-based studies on REIT. Earlier literature focuses on the macroeconomic factors affecting the behaviour and performance of REIT. Wing and Payne [1] examine the impact of price, interest rates, output, investment and default risk premiums on REIT. McCue and Kling [2] incorporate prices, nominal short-term interest rates, output and investment as important macroeconomic factors affecting REITs performance in a time series setting. Later, [3] add default risk premiums as another important factor that determines REIT return. On a similar note, [4] examine the impact of monetary shocks on REIT returns. The authors improve previous work by [5–7] by accounting for anticipated and unanticipated effects of interest rate changes. Simpson and Ramachander [8] find that REIT return response in a positive but asymmetric manner to both increase and decrease in inflation. Studies also look at the behaviour of REIT along with the stock and bond returns. Glascock et al. [9] show that REIT mimics the behaviour of stocks rather than bonds especially after the structural changes in the US capital market in the 1990s. In similar veins, [10] investigate the relationship between equity REIT return along with volatilities in commercial banks, savings and loans, and life insurance companies in a bivariate GARCH setting. Results suggest that volatilities in REIT are transmitted to savings and loans, and life insurance company markets.

Studies on Islamic REIT are still scanty since it is a relatively new product with limited data. This study intends to contribute to the vast literature on REIT by examining whether Islamic REIT follows the conventional REIT behaviour. Results show that this contention is plausible where Islamic REIT maps the movement of the stock market in the case of Malaysia in the era of financial instability.

31.2 Empirical Framework

The augmented market model is expressed as follows:

$$\mathrm{IR}_t = \beta_0 + \beta_1 \mathrm{KLCI}_t + \beta_i \sum M_t + \varepsilon_t, \tag{31.1}$$

where IR denotes the Islamic REIT, KLCI captures the stock market index and M represents a host of control variables. A statistically significant β_1 implies that the stock market affect the Islamic REIT market. A positive β_1 suggests that Islamic REIT complements the stock market. If β_1 is negative, this indicates that Islamic REIT is a substitute to the stock market. If Islamic REIT and stock market are complements, Islamic REIT will be subjected to the same cyclical patter of the stock market. A fall in stock market would also indicate that the Islamic REIT will also fall. Therefore, Islamic REIT is not able to mitigate risks despite being providing more alternative for investors to diversify their portfolio. If Islamic REIT is a substitute for the stock market, then Islamic REIT would be able to cushion slumps in the stock market.

31.3 Estimation Method and Data

Unit root and cointegration have extensively been used in the past 20 years in time series econometrics. This is due to the fact that cointegrated variables revert to their mean despite potential outward drift in the short run. These two concepts give rise to a host of estimation strategies ranging from the simplest bivariate two-step test developed by [11], multivariate vector autoregressive technique by [12], the dynamic ordinary least squares proposed by [13], the fully modified ordinary least square method (FM-OLS) by [14, 15] bounds testing procedure.

This study uses the bounds test techniques to capture its advantages. First, bounds test relaxes the strict requirement that all the regressors must be integrated of order 1. With the bounds test, the regressors are allowed to be $I(0)$, $I(1)$ or mutually cointegrated, hence, avoids the pretesting problems. One necessary condition is that the dependent variable must be integrated of order 1 or $I(1)$. The other explanatory variables may be either $I(1)$ or $I(0)$ but not $I(2)$ since this technique is not designed to accommodate $I(2)$ variables [16]. In addition, this technique could not handle seasonal unit roots and explosive roots [14].

Second, this technique is more robust and efficient for small size samples. Third, it is relatively simple compared to Johansen and Juselius multivariate cointegration procedure since it allows the cointegration relationship to be estimated by OLS once the lag order of the model has been identified. Nevertheless the model is still based on VAR.

Based on [14] and in line with the theoretical model discussed in the previous section, the conditional VECM is expressed as

$$\Delta \mathrm{IR}_t = c_0 + \delta_1 \mathrm{KLCI}_{t-1} + \delta_2 M_{t-1} + \delta + \sum_{i=1}^{p} \phi_i \Delta \mathrm{KLCI}_{t-i} + \sum_{j=0}^{q} \varpi_j \Delta M_{t-j} + \varepsilon_t, \tag{31.2}$$

where δ are the long-run multipliers, c_0 is the drift term, p is the optimal number of lag length and ε_t are the white noise error terms. IR, KLCI and M represent Islamic REITs index, Kuala Lumpur stock index and other control variables, respectively. Equation (31.2) can also be interpreted as an ARDL (p, q) model. The lag length for the first differences of IR, KLCI and M can be different, rendering the appropriateness of the use of the ARDL (p, q) model [17]. In this study, the number of lags for p and q are determined based on the Schwarz Bayesian criterion (SBC). This study chooses the SCB criterion instead of Akaike Information Criterion (AIC), Schwarz information criterion (SIC), final prediction error (FPE) or Hannan–Quinn information criterion (HQ) due to parsimony.

The bounds test involves a three-step procedure. First, equation Eq. (31.2) is estimated using the ordinary least squares (OLS) method. This is undertaken to examine whether there exists a long-run relationship between the dependent and independent variables. Under the null hypotheses $H_0 : \delta_1 = \delta_2 = 0$ against the alternative, $H_1 : \delta_1 \neq \delta_2 \neq 0$ the joint significance of the lagged levels of the variables is tested with reference to the F-statistics.

The approximate asymptotic critical values are adapted from [18]. The lower values assume that the variables are purely $I(0)$ and the upper values denote purely $I(1)$ variables. Any F-statistics which is above the upper critical value infer cointegration regardless whether the variables are integrated of order 1 or zero, $I(0)$ or $I(1)$. Conversely, any F-statistics that are below the lower critical values, infer no cointegration which means that there exists no long-run relationship amongst the stipulated variables. If the F-statistic falls between the lower and upper critical values, then no conclusive results can be inferred.

The second step involves the estimation of the conditional long-run ARDL (p, q) model for IR_t as follows:

$$\mathrm{IR}_t = c_0 + \sum_{i=1}^{p} \beta_1 \Delta \mathrm{KLCI}_{t-i} + \sum_{j=0}^{q} \beta_2 \Delta M_{t-j} + \varepsilon_t, \tag{31.3}$$

where all the variables are as previously defined. Finally, the coefficients of the short-run dynamics is derived by estimating the error correction model associated with the long-run estimates, such that,

$$\Delta \mathrm{IR}_t = \mu + \sum_{i=1}^{p} \phi_i \Delta \mathrm{KLCI}_{t-i} + \sum_{j=0}^{q} \varpi_j \Delta M_{t-j} + \eta \mathrm{ect}_{t-1} + \varepsilon_t, \tag{31.4}$$

where ϕ and ϖ are the short-run dynamic coefficients and η is the speed of adjustment.

Data ranges from June 2008 to March 2012 with 46 observations. The Islamic REITs is proxied by the Dow Jones Islamic REITs index (RI) which is only available form June 2008 onwards on monthly basis. Data on the Kuala Lumpur composite index (KLCI) and bond were derived from Bloomberg database. Reserves (RES), 6-month treasury bills (TB), long-term rate (LR) and government spending (GS) were obtained from the Bank Negara Monthly Statistical Bulletin (various issues). The volatility of the exchange rate was estimated using the moving average technique based on the real effective exchange rates from the International Financial Statistics (IMF, June 2012).

This study also employed a host of other control variables such as the risk premia proxied by the difference between the short-term rates and the long-term rates, real effective exchange rates, nominal effective exchange rates, bilateral exchange rates (RM/USD, RM/YEN, RM/Euro and RM/Sing), inflation, changes in inflation, industrial production index (IPI) and changes in IPI. None of these variables appear to have statistically significant effects on the Islamic REITs.

31.4 Results and Discussion

Figures 31.1 and 31.2 map the movement of the Islamic REIT index, Kuala Lumpur composite index and the bond market index. Figure 31.1 shows that the movement of Islamic REIT is in tandem KLCI. Similarly, the Islamic REIT and BOND move concurrently except for a short stint from November 2008 to July 2009 where movements are in the opposite direction. Generally, the Islamic REIT index move in an upward manner especially after July 2009 which shows that Malaysia's capital market had begun to regain momentum following the global financial crisis.

In general, Table 31.1 shows that cointegration is present in all four different specifications. The first regression serves as the baseline regression based on the theoretical model discussed in the previous section. The second, third and fourth specifications serve as a robustness check and to ensure the consistency of the estimated coefficients. In specification 2–4, more control variables were inserted to test whether these variables directly affect the Islamic REIT. The *F*-statistics are greater than the 1 % upper bound critical values for all four specifications. Results are supported by the negative and significant error correction terms as reported in Table 31.4.

Table 31.2 shows the estimated coefficients of the autoregressive distributed lag models. The lagged values of the Islamic REIT have positive and significant impact on RI. Results are consistent in all four models despite the changes in the control variables. On similar note, RES also exhibit positive and significant impact on the Islamic REIT. The estimated coefficients are correspondingly consistent in all four specifications. The short-term interest rate as proxied by TB is significant only in

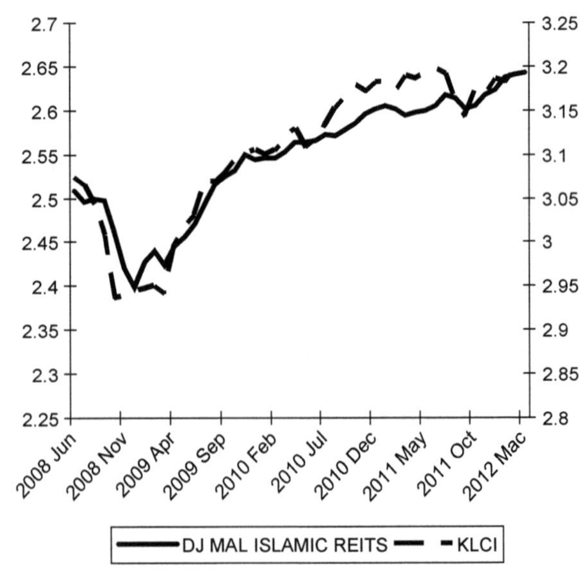

Fig. 31.1 Islamic REITs and KLCI

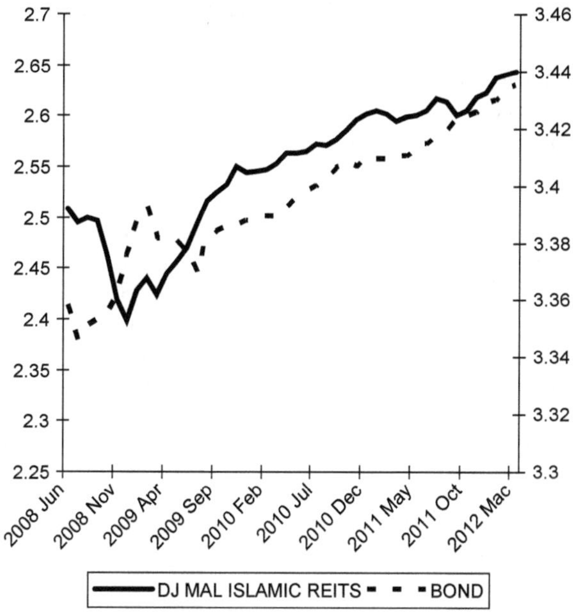

Fig. 31.2 Islamic REIT and BOND

Table 31.1 Bounds tests results

Specification	Variables	F-stats
1	R| KLCI RES TB	325.34
2	R| KLCI RES TB GS	292.59
3	R| KLCI RES TB GS BOND	191.78
4	R| KLCI RES RER TB LR	170.03

the second and third specifications. The coefficients are not consistent with estimations being positive in the third and fourth specifications but negative in the first two specifications. Results imply that interest rates are sensitive towards the changes in control variables, so no conclusive inferences can be made on the effect of interest rates on Islamic REITs.

The long-run estimated coefficients in Table 31.4 show that only KLCI has a positive and significant effect on the Islamic REIT. RES are consistently positive but significant in the first two specifications. Other variables in Table 31.3 are not statistically significant in the long run. This is probably due to the fact that the span of data is only about 4 years which is not adequate to capture the long-run effects. GS, BOND and LR significantly affect the Islamic REIT. GS and LR negative affect the Islamic REIT whilst BOND confers positive effects. These results infer that BOND may also have a complementary impact on the Islamic REIT, at least, in the short run.

The error correction terms (ect_{t-1}) in Table 31.4 are both negative and significant therefore; substantiate the presence of long-run relationship amongst the Islamic REIT, KLCI and the control variables. In addition, [19] suggest that the

Table 31.2 Autoregressive distributed lag estimates: dependent variable—Islamic REITs index (RI)

Specification Variables	1	2	3	4
RI(−1)	0.826*** (7.73)	0.887*** (7.90)	0.817*** (7.46)	0.744*** (6.19)
KLCI	0.364*** (7.83)	0.306*** (5.779)	0.293*** (5.97)	0.349*** (6.13)
TB	−0.003 (−0.84)	−0.024** (−2.35)	0.021** (2.16)	0.010 (0.75)
RES	0.066*** (3.45)	0.060*** (2.95)	0.048 (1.63)	0.058** (2.19)
GS	–	−0.011 (−1.74)	−0.059*** (−4.40)	–
BOND	–	–	0.040*** (3.38)	
LR	–	–	–	−0.33** (−2.06)
RER	–	–	–	0.023 (1.71)
C	−0.196** (−2.38)	−0.071 (−0.69)	−0.515 (−0.44)	0.250 (0.59)

Notes t-statistics are in parentheses. Lags are estimated based on the SBC criteria. *** and ** denote significance at 1 and 5 % levels, respectively

Table 31.3 Long-run estimates: dependent variable—Islamic Reits index (RI)

Specification Variables	1	2	3	4
KLCI	0.747*** (21.9)	0.700*** (17.07)	0.665*** (4.29)	0.585*** (5.07)
TB	−0.007 (−0.79)	−0.013 (−1.32)	−0.016 (−1.05)	0.017 (0.81)
RES	0.135*** (3.505)	0.138*** (3.23)	0.109 (1.35)	0.097 (1.94)
GS		−0.026 (−1.49)	−0.007 (−0.16)	–
BOND	–	–	0.352 (0.39)	–
LR	–	–	–	−0.039 (−1.16)
RER	–	–	–	0.38 (1.84)
C	−0.401** (−2.22)	−0.162 (−0.71)	−1.17 (−0.48)	0.419 (0.62)

Notes t-statistics are in parentheses. Lags are estimated based on the SBC criteria. *** and ** denote significance at 1 and 5 % levels, respectively

Table 31.4 Error correction and diagnostic tests

Specification Variables	1	2	3	4
dRI	0.313*** (3.08)	0.324*** (3.32)	0.257** (2.27)	0.341*** (3.21)
dKLCI	0.364*** (7.83)	0.306*** (5.78)	0.293*** (5.97)	0.349*** (6.134)
dTB	−0.003 (−0.84)	0.019 (1.81)	0.021 (2.16)	0.10 (0.75)
dRES	0.066*** (3.45)	0.060*** (2.96)	0.048 (1.63)	0.058** (2.18)
dGS	–	−0.011 (−1.74)	−0.039*** (−3.38)	–
dBOND	–	–	1.16** (2.33)	–
dLR	–	–	–	0.009 (0.42)
dRER	–	–	–	0.023 (1.71)
dC	−0.196 (−2.38)	−0.69 (0.494)	−0.514 (0.44)	–
ect_{t-1}	−0.488*** (−7.26)	−0.437*** (−6.00)	−0.440*** (−4.66)	–
LM	0.828 [0.623]	1.18 [0.35]	1.96 [0.11]	0.514 [0.88]
RESET	2.57 [0.12]	2.85 [0.26]	2.98 [0.13]	0.278 [0.60]

Notes t-statistics are in parentheses. *p*-values are in square brackets. *** and ** denote significance at 1 and 5 % levels, respectively

error correction terms represent a more efficient indication of cointegration, which support the bounds testing results in Table 31.1. Based on Table 31.4, the lagged values of RI and KLCI are consistently positive and significant in all four specifications. RES is significant in only three out of the four specifications. Other control variables such as TB, LR and RER do not affect the Islamic REIT even in the short run.

31.5 Conclusion

Islamic REIT is a thriving investment vehicle in Malaysia with less than 20 % share of the total net asset value. This study investigates whether the Islamic REIT acts as a complement or substitute to the stock market in Malaysia. The Islamic REIT is proxied by the Dow Jones Islamic REIT Index (RI) whilst Malaysia's stock market is represented by the Kuala Lumpur stock index. Data ranges from June 2008 and spans over to March 2012 with 46 observations. The choice of data span is partly dictated by availability of data. The observations cover the subprime crisis period which epitomizes the start of the extended global financial crisis. Therefore, this study fully captures the behaviour of the Islamic REIT in the era of financial crisis.

This study employs the autoregressive distributed lag technique to examine whether Islamic REIT complements or substitute the stock market. Results infer that Islamic REIT complements the stock market, and its movements almost mirror the stock market. This indicates that any upward or downward movement in the stock market will be reflected in the Islamic REIT market. An important policy implication is that Islamic REIT does provide an alternative vehicle for investors looking for *shariah* compliant products. Although Islamic REIT helps to diversify an investor's portfolio, in the event of downfall in the market, investment in Islamic REIT will not help to cushion the effect of such downfall. Bonds, on the other hand, have no significant relationship with the Islamic REIT. Our results are consistent with the findings of [9] for the National Association of Real Estate Investment Trusts (NAREITs) over the period of 1972–1997. In the future, this study can be extended to compare the behaviour of Islamic REIT during crisis and non-crisis period. However, this is only possible when more data could be collected over longer period of time.

References

1. Ewing, B.T., Payne, J.E.: The response of real estate investment trust returns to macroeconomic shocks. J. Bus. Res. **58**, 293–300 (2005)
2. McCue, T.E., Kling, J.K.: Real estate returns and the macroeconomy: some empirical evidence from real estate investment trust data, 1972-1991. J. Real Estate Res. **9**, 277–287 (1994)
3. Karolyi, G.A., Sander, A.B.: The variation of economic risk premiums in real estate returns. J. Real Estate Financ. Econ. **17**, 245–262 (1998)
4. Bredin, D., O'Reilly, G., Stevenson, S.: Monetary shocks and REIT returns. J. Real Estate Financ. Econ. **35**, 315–331 (2007)
5. Chen, K., Tzang, D.: Interest rate sensitivity of real estate investment trusts. J. Real Estate Res. **3**, 13–22 (1988)
6. Liang, Y., MacIntosh, W., Webb, J.: Intertemporal changes in the riskiness of REITs. J. Real Estate Res. **10**, 427–443 (1995)
7. Devaney, M.: Time-varying risk premia for real estate investment trusts: a GARCH-M model. Q. Rev. Econ. Financ. **41**, 335–346 (2001)
8. Simpson, M.W., Ramchander, S., Webb, J.R.: The asymmetric response of equity REIT returns to inflation. J. Real Estate Financ. Econ. **34**, 513–529 (2007)
9. Glascock, J.L., Lu, C., So, R.W.: Further evidence on the integration of REIT, bond and stock returns. J. Real Estate Financ. Econ. **20**(2), 177–194 (2000)
10. Elyasiani, E., Mansur, I., Wetmore, J.L.: Real estate risk effects on financial institutions' stock return distribution: a bivariate GARCH analysis. J. Real Estate Financ. Econ. **40**, 89–107 (2010)
11. Engle, R.F., Granger, C.W.J.: Cointegration and error correction: representation, estimation and testing. Econometrica **55**(2), 251–276 (1987)

12. Johansen, S.: Likelihood analysis of seasonal cointegration. J. Econom. **88**(2), 301–339 (1999)
13. Stock, J.H., Watson, M.W.: A simple estimation of cointegrating vectors in higher order integrated systems. Econometrica **61**, 783–820 (1993)
14. Phillips, P.C.B., Hansen, B.E.: Statistical inference in instrumental variables with regression with I(1) process. Rev. Econ. Stud. **57**, 99–125 (1990)
15. Pesaran, M.H., Shin, Y., Smith, R.J.: Bounds testing approach in the analysis of level relationships. J. Appl. Econom. **16**, 289–326 (2001)
16. Fosu, O.E., Magnus, F.J.: Bounds testing approach to cointegration: an examination of foreign direct investment, trade and growth relationships. Am. J. Appl. Sci. **3**(11), 2079–2085 (2006)
17. Atkins, F.J., Coe, P.J.: An ARDL bounds test of the long run Fisher effect in the United States and Canada. J. Macroecon. **24**, 255–266 (2002)
18. Narayan, P.: The saving and investment nexus for China: evidence from cointegration tests. Appl. Econ. **37**(17), 1979–1990 (2005)
19. Kremers, J.J.M., Ericsson, N.R., Dolado, J.J.: The power of cointegration tests. Oxf. Bull. Econ. Stat. **5**(3), 325–348 (1992)

Chapter 32
Classifying Bankruptcy of Small and Medium Sized Enterprises with Partial Least Square Discriminant Analysis

Nurazlina Abdul Rashid, Amirah Hazwani Abdul Rahim, Ida-Normaya M. Nasir, Sallehuddin Hussin and Abd-Razak Ahmad

Abstract The impact of firms' insolvency is enormous to all stakeholders involved, especially to the creditors. It increases unemployment rate and harms investors' confidence in the financial market. This paper compares the ability of Partial Least Square Discriminant Analysis (PLS-DA) in classifying bankrupt firms with the benchmark logit model. Dataset are Malaysian Small and Medium Enterprise (SMEs) categorized under the transportation and storage sector. The analysis shows that PLS-DA outperformed the logit model.

Keywords Bankruptcy classification · Small and medium-sized enterprises · Partial least square discriminant analysis · Logit model

---

N.A. Rashid (✉) · A.H.A. Rahim · I.-N.M. Nasir · S. Hussin · A.-R. Ahmad
Faculty of Computer and Mathematical Sciences,
Universiti Teknologi MARA, Kedah, Malaysia
e-mail: azlina150@kedah.uitm.edu.my

A.H.A. Rahim
e-mail: amirah017@kedah.uitm.edu.my

I.-N.M. Nasir
e-mail: normaya@kedah.uitm.edu.my

S. Hussin
e-mail: din-za@hotmail.com

A.-R. Ahmad
e-mail: ara@kedah.uitm.edu.my

A.-R. Ahmad et al. (eds.), *Proceedings of the International Conference on Computing, Mathematics and Statistics (iCMS 2015)*,
DOI 10.1007/978-981-10-2772-7_32

32.1 Introduction

Small and medium-sized enterprises (SMEs) are the backbone of an economy and thus play an important role in the economic development of most countries as well as in Malaysia. SMEs complement the economic cycle or activity of a nation. Most of the existing companies in Malaysia are from the SME sector, which makes the sector an important source of employment [10, 13]. Effective January 1, 2014, SMEs in Malaysia are defined as companies with annual turnover less than 50 million ringgits and hiring not more than 200 employees for the manufacturing sector. For the services sector, companies that are classified as SMEs are those hiring not more than 75 workers and with annual sales not exceeding 20 million ringgits. With this new definition of SMEs, effectively places 98.5 % of all business entities in Malaysia as SMEs which contribute more than 56 % of the total workforce [14, 15].

Due to lack of financial information on SMEs, most studies on bankruptcy are on large or listed firms. Large listed firms are required to submit full accounts by the regulators. SMEs on the other hand are not under the same requirement. However, SMEs are prone to failure more often than its larger counterparts [7]. The main objective of this paper is to develop a classification model to classify bankrupt Malaysian SMEs based on a set of financial ratios by using partial least squares discriminant analysis and compared it to the logit model. Being able to determine potential firm failure will help bankers manage their credit portfolios.

The next section reviews the literature on bankruptcy. Section 32.3 describes the methodologies, while Sect. 32.4 discusses the empirical result. Section 32.5 presents the conclusion and discussion.

32.2 A Review on Bankruptcy Studies

Corporate failure and insolvency are alternative terms to bankruptcy. According to Bryan [5], bankruptcy occurs when a firm is unable to meet its loans obligations. The firm will normally apply either for a period of relief to reorganize its debts or to liquidate the assets. The impact of insolvency is enormous to all stakeholders involved—owners of the firms, its shareholders, employees, lenders, suppliers, clients, the society, and the local government. It increases the unemployment rate and harms investors' confidence in the financial market.

Studies on bankruptcy started with the works of Beaver [4] and Altman [3]. Their works inspired almost all bankruptcy literatures thereafter. Beaver conducts a univariate analysis to determine financial ratios that can identify potential firms' failures. He uses an equal number of matched firms which consist of 79 bankrupt and 79 nonbankrupt firms and analyzed 14 financial ratios individually to classify bankrupt firms.

While Beaver analyzes the covariates individually, Altman introduces Multiple Discriminant Analysis (MDA) in predicting bankruptcy and eventually built the ever popular Z-score model. He employs 33 bankrupt firms and match pairs to 33 nonbankrupt firms. Five out of the twenty-two financial ratios used as covariates in the study are found to have predictive abilities in predicting bankruptcy. MDA, however, has several pitfalls, one of which is that MDA assumes that the independent variables are normally distributed. Violation of this assumption will cause the significance tests to be bias. To overcome the shortcoming, Ohlson [12] introduces logistic regression in bankruptcy studies. He uses 2058 nonbankrupt and 105 bankrupt industrial firms as samples and the information were collected from 1970 to 1976. The logit model is more reliable as its response variable engages with binary outcome. Besides the statistical models, models based on artificial intelligence such as neural network, support vendor machine, evolution algorithm, and case-based reasoning were used in bankruptcy studies.

SMEs are not required by the regulators to submit full financial disclosures. As such there is limited financial data on SMEs that are available compared to public firms, making most study on bankruptcy as studies on bankruptcy of large public firms. Edmister [8] is one of the few studies that used MDA to develop a bankrupt model for SMEs. The author uses 19 financial ratios collected from year 1954 through 1969. Altman et al. [2] also use ratios of SMEs—categorized into profitability, leverage, liquidity, asset, and efficiency—but built a logit model instead of MDA.

Nasaruddin and Ahmad [11] did a bankruptcy study of British small publishing and printing firms. The data consists of 5850 firms where 2595 (44.4 %) of the firms are bankrupt while the other 3255 (55.6 %) are nonbankrupt firms. The authors used financial ratios as well as nonfinancial information consisting of account qualification, the information whether the account is lodged late and the age of the firms. Three methods were used—the classification and regression tree (CART) model, the artificial neural network model, and the logit model. The CART model is the best classifier with a 60.4 % overall accuracy rate.

32.3 The Methodology

32.3.1 Partial Least Squares Discriminant Analysis (PLS-DA)

Partial least squares (PLS) regression provides a dimension reduction strategy to regroup and eventually reduce the number of predictive variables X, associated with the response variable Y by looking for the orthogonal X-components $t_h = Xw_h^*$ and the Y-components $u_h = Yc_h$. The objective is to maximize the covariance between t_h and u_h. PLS was developed as an alternative approach to ordinary least squares (OLS) regression.

Problems with multicollinearity arise in linear regression which makes the estimates of the model parameters become unstable [6, 9]. PLS is popularly used with high dimensional data (the number of variable p is larger than the number of observation n), nevertheless it also suitable for data with larger observation as long as high multicollinearity occurs. PLS maintains most of the useful information available from a dataset while reducing the number of predictive variables.

To classify data that is normally distributed, linear discriminant analysis is the most appropriate technique. The weakness of this technique, however, is that it cannot perform well when the dataset consists of large number of variables relative to the amount of measurements taken. A better alternative is partial least squares discriminant analysis (PLS-DA). However, to avoid over fitting in PLS-DA, the high number of dimensions is reduced through the dimension reduction strategy. PLS-DA is a PLS regression where Y is a set of binary variables describing the categories of a categorical variable on X. For each response variable, y_k, a regression model on the X-components is written as

$$\sum_{h=1}^{m} (Xw_h^*)c_h + e = XW * c + e$$

where w_h* is a p dimension vector containing the weights given to each original variable in the k-th component, and c_h is the regression coefficient of y_k on h-th X-component variable.

The first step in PLS-DA is a dimension reduction process using partial least squares (PLS). PLS is comparable to principal component analysis (PCA), a commonly used dimension reduction technique. The difference between the two is that PCA explains the observed variability in the predictor variables, without considering the response variable. In contrast with PCA, PLS components are chosen such that the sample covariance between the response and a linear combination of the predictors is maximized. The second step in the PLS-DA technique involves a classification process using linear discriminant analysis (LDA). LDA is a classification technique based on gross variability 'within groups' and 'among groups'. The combination of PLS and LDA will result in a dimension reduction and classification outcome.

This research uses sensitivity, specificity, and area under a receiver operating characteristic curve (AUC) as indicators to evaluate the model performance. Sensitivity and specificity are the percentages of correct prediction of the minority and majority classes, respectively. AUC measures the model ability between sensitivity and specificity. Sensitivity, also known as true positive rate (TPR), is defined as a percentage of the minority class classified as true minority class. Sensitivity is written as

$$\text{Sensitivity} = \frac{\text{TN}}{\text{FP} + \text{TN}}$$

Table 32.1 Confusion matrix for binary classification

		Predicted class	
		Positive (minority)	Negative (majority)
Actual class	Positive (minority)	True positive (TP)	False negative (FN)
	Negative (majority)	False positive (FP)	True negative (TN)

Specificity, also known as true negative rate (TNR), is defined as a percentage of the majority class classified as true majority class. Specificity is given as

$$\text{Specificity} = \frac{\text{TP}}{\text{TP} + \text{FN}}$$

The simplified explanation on the concept of sensitivity and specificity can be referred to a confusion matrix as shown in Table 32.1.

32.3.2 *Data and the Analysis Strategy*

The dataset is obtained from the database of Companies Commission of Malaysia (SSM). It uses financial information of companies categorized under the transportation and storage classification only. There are 44 failed and 2451 nonfailed SMEs under this classification. The financial information collected is from year 1999 to 2013. All bankrupt firms failed in year 2013. This paper implements match-pair approach like most previous bankruptcy studies. Since there are 44 bankrupt firms, we selected 44 nonbankrupt firms and match-paired with the bankrupt firms based on total assets. Then, the dataset was split into estimation and validation by 70 and 30 %, respectively.

Financial ratios are used in this study as the independent variables. There are a large number of financial ratios used in the literature to predict bankruptcy. Previous studies showed that a firm is more likely to go into bankruptcy if it is unprofitable, has high leverage, and suffers from liquidity difficulties. This paper considered 14 financial ratios that are available in the database. The financial ratios are categorized into five categories according to prior research, which are measures of profitability, liquidity, leverage, efficiency, and asset. Table 32.2 listed the financial ratios. There is one ratio under liquidity and efficiency, three ratios under leverage and asset, and six ratios under profit.

32.4 Empirical Results

Table 32.3 shows the summary of statistics, mean, and test of significance of each of the financial variables. The results show that bankrupt firms are less liquid, less efficient and less profitable, have higher leverage and inconsistent assets. It also

Table 32.2 Financial ratio and its categories

Label	Financial ratio	Category
liq3	Sales to current asset	Liquidity
lev1	Total liabilities to total asset	Leverage
lev4	Earnings before tax to interest expenditure	Leverage
lev6	Current liabilities to total asset	Leverage
prof1	Net income to sale	Profit
prof2	(Net income + interest expenses)/sales	Profit
prof4	Net income to total asset	Profit
prof5	Earnings before tax to total asset	Profit
prof7	Return of equity to return of asset	Profit
prof8	Earnings before tax to sale	Profit
ef1	Sale to total asset	Efficiency
as1	Sale to working capital	Asset
as2	Sale to asset	Asset
as3	Interest expenses to earnings before tax	Asset

Table 32.3 Variable mean and test of significance

Financial ratio	Bankrupt (mean)	Nonbankrupt (mean)	*t*-stat	*p*-value
liq3	3.9688	5.5404	3.724**	0.000
lev1	1.3422	0.7750	4.891**	0.000
lev4	1.6160	1.2599	4.172**	0.000
lev6	0.7584	0.6804	11.010**	0.000
prof1	−0.0062	0.1876	1.062	0.292
prof2	−0.0099	0.1671	0.942	0.350
prof4	−0.5358	0.1031	−1.082	0.284
prof5	−0.2207	0.0469	−0.764	0.448
prof7	−0.4764	1.0623	0.130	0.897
prof8	−0.0331	0.0300	−0.188	0.851
ef1	2.0176	2.9183	3.667**	0.001
as1	−0.8351	−4.2786	−0.365	0.717
as2	−3.9475	1.4184	−0.489	0.627
as3	−0.1285	0.1567	0.024	0.981

*, ** significant at 10 and 5 %, respectively

shows that five financial ratios are statistically significant at least at 10 % level in predicting bankruptcy. The ratios are sales to current asset, total liabilities to total asset, earnings before tax to interest expenditure, current liabilities to total asset, and sale to total asset.

Model 1 of Table 32.4 presents the results of the stepwise logistic regression model while model 2 shows the results of the partial least square discriminant analysis model. For the logic model, there are three financial ratios that are

Table 32.4 Results

Model 1 Logit = 1/(1 + exp(−(0.894 − 3.719 * prof 4 − 0.328 * ef1 + 0.0247 * as2))) Goodness of fit −2 Log likelihood	55.301
Cox & Snell R Square	0.351
Nagelkerke R Square	0.468
Hosmer and Lemeshow Test (d.f, *p*-value)	5.023 (8, 0.755)
Model 2 *F*(0) = 0.606 + 0.0058*liq3 − 0.220 * Lev1 + 0.731 * prof1 + 0.0008 * prof2 + 0.230 * prof4 + 0.1596 * prof5 + 0.001449 * prof7 + 0.01748 * ef1 − 0.0007 * as1 + 0.0026 * as2 + 0.113 * as3 + 0.0244 * prof8 *F*(1) = 0.393 − 0.0057 * liq3 + 0.220 * Lev1 − 0.007 * prof1 − 0.0008 * prof2 − 0.230 * prof4 − 0.15 * prof5 − 0.0015 * prof7 − 0.017 * ef1 + 0.0007 * as1 − 0.002 * as2 − 0.11 * as3 − 0.024 * prof8	

statistically significant in predicting bankruptcy—earnings before tax to interest expenditure, sale to total asset, and sale to asset. This means that, lower profitability and less efficiency will lead to a higher probability of a firm becoming bankrupt. The Hosmer and Lemeshow test shows that the model is adequate and effective in predicting the dichotomous variable. The Cox & Snell R Square and Nagelkerke R Square of Model 1 are 0.351 and 0.468, respectively.

The results of Model 2 shows that bankruptcy of a firm occurs when the firm faces less liquidity, less profitability, less efficiency, high leverage, and inconsistent assets. These results are consistent to that of Abdullah [1].

Table 32.5 summarizes the results of the classification accuracy of the two models. Model 2 has a higher overall accuracy rate (OAR) for both the estimation and validation samples compared to model 1. The difference of the accuracy rate for the estimation sample is 4 % while for the validation sample is 7 %. Model 2 also has the lowest type II error in estimation and validation samples. The results show that the AUC of model 1 is 84 % and of model 2 is 83 %. There is a slight difference of 1 % in the AUC.

Table 32.5 Summary of classification accuracy

	Measurements	Model 1 (Logit)	Model 2 (PLS-DA)
Estimation	Overall accuracy rate (OAR)	0.72	0.76
	Type I error	0.28	0.31
	Type II error	0.28	0.17
Validation	Overall accuracy rate (OAR)	0.53	0.60
	Type I error	0.47	0.53
	Type II error	0.47	0.17
	Area under curve (AUC)	0.84	0.83

32.5 Conclusion and Discussion

This paper investigates the predictive ability of PLS-DA in using financial information in classifying bankrupt Malaysian SMEs from healthy firms. The PLS-DA model is compared to the benchmark logit model. Due to a small number of bankrupt SMEs, a match-pair approach was employed in this study. There are five out of fourteen financial ratios that are significantly predictive of bankruptcy. Bankrupt firms have lower liquidity and profitability, have higher leverage and inconsistent asset, and are less efficient. The overall accuracy rate of classification for the PLS-DA is higher than the logit model at 76 % for the estimation sample and 60 % for the validation sample.

There are some limitations in this study. First, financial data were representative of transportation and storage sectors of SMEs only. Second, this study uses PLS-DA and compared that to logistic regression as a benchmark model. Future work might include nonstatistical models as well. Finally, under-sampling technique was used in this study. In future, this study will be extended for oversampling technique and the result will be compared to under-sampling technique.

The implication of this research is two-fold. The management of small firms may use the information to monitor the performance of their firms to avoid bankruptcy while bankers may use the information to determine credit credibility of potential borrowers.

References

1. Abdullah, N.A. H., Zainudin, N., Ahmad, A.H., Md Rus, R.: Predictors of financially distressed small and medium-sized enterprises: a case of Malaysia. Int. Proc. Econ. Dev. Res. **76**(18), 108–112 (2014)
2. Altman, E., Sabato, G., Wilson, N.: The value of non-financial information in small and medium-sized enterprise risk management. J. Credit Risk **6**(2), 1–33 (2010)
3. Altman, E.I.: Financial ratios, discriminant analysis and the prediction of corporate bankruptcy. J. Finance **23**, 589–609 (1968)
4. Beaver, W.H.: Financial ratios as predictors of failure. In: *Empirical Research in Accounting: Selected Studies*, Supplement to Vol. 5. J. Account. Res. (pp. 71–111) (1966)
5. Bryan, D.: Bankruptcy risk, productivity and firm strategy. Rev. Account. Finance **12**(4), 309–326 (2013)
6. De Jong, S.: SIMPLS: an alternative approach to partial least squares regression. Chemometr. Intell. Lab. Syst. **18**(3), 251–263 (1993)
7. Dietsch, M., Petey, J.: Should SME exposures be treated as retail or corporate exposures? A comparative analysis of default probabilities and asset correlations in French and German SMEs. J. Bank. Finance **28**, 773–788 (2004)
8. Edmister, R.O.: An empirical test of financial ratio analysis for small business failure prediction. J. Financ. Quant. Anal. **7**(2), 1477–1493 (1972) (Supplement: Outlook for the Securities Industry)
9. Geladi, P., Kowalski, B.R.: Partial least-squares regression: a tutorial. Anal. Chim. Acta **185**, 1–17 (1986)

10. Idar, R., Mahmood, R.: Entrepreneurial and marketing orientation relationship to performance: the SME perspective. Interdiscip. Rev. Econ. Manage. **2**(1), 1–8 (2011)
11. Nasaruddin, N., Ahmad, A.: Using data mining to classify bankrupt small publishing and printing UK Firms. In: Proceeding International Conference on Computing, Mathematics and Statistics 2013, Penang (2013)
12. Ohlson, J.A.: Financial ratios and the probabilistic prediction of bankruptcy. J. Acc. Res. **18** (1), 109–131 (1980)
13. Radam, A., Abu, M.L., Abdullah, A.M.: Technical efficiency of small and medium enterprise in Malaysia: a stochastic frontier production model. Int. J. Econ. Manage. **2**(2), 395–408 (2008)
14. SME: SME Corporation Malaysia. Available online at www.smecorp.gov.my/vn2/. Accessed on 7 Dec 2015
15. Sundaily: Najib Announces New Definition of SME. Available online at www.thesundaily.my/news/768182. Accessed on 7 Dec 2015